Jakob von Uexküll and Philosophy

Dismissed by some as the last of the anti-Darwinians, his fame as a rigorous biologist even tainted by an alleged link to National Socialist ideology, it is undeniable that Jakob von Uexküll (1864–1944) was eagerly read by many philosophers across the spectrum of philosophical schools, from Scheler to Merleau-Ponty and Deleuze and from Heidegger to Blumenberg and Agamben. What has then allowed his name to survive the misery of history as well as the usually fatal gap between science and humanities?

This collection of essays attempts for the first time to do justice to Uexküll's theoretical impact on Western culture. By highlighting his importance for philosophy, the book aims to contribute to the general interpretation of the relationship between biology and philosophy in the last century and explore the often-neglected connection between continental philosophy and the sciences of life. Thanks to the exploration of Uexküll's conceptual legacy, the origins of cybernetics, the overcoming of metaphysical dualisms, and a refined understanding of organisms appear variedly interconnected.

Uexküll's background and his relevance in current debates are thoroughly examined as to appeal to undergraduate and postgraduate students, as well as postdoctoral researchers in fields such as history of the life sciences, philosophy of biology, critical animal studies, philosophical anthropology, biosemiotics and biopolitics.

Francesca Michelini is Senior Research Fellow at the University of Kassel (Germany). Her main fields of research are the antireductionist theories of life and the bridging of continental philosophy and science. She is author of many publications on the topic of philosophical anthropology, philosophy of the life sciences, teleological explanations in nature, and autonomy in biology (among others, *The Living and the Deficiency. Essays on Teleology* 2011, in Italian).

Kristian Köchy is a biologist and Professor of Theoretical Philosophy at the University of Kassel (Germany). His research focuses on the areas of philosophy of science and the history of the life sciences, natural philosophy and the philosophy of animal–human relations. He is author of an introduction on *Biophilosophy* (2008, in German) and coeditor of a three-volume collection on the philosophy of animal research (*Philosophie der Tierforschung*, 2016–2018).

History and Philosophy of Biology

Series Editor: Rasmus Grønfeldt Winther is Associate Professor of Philosophy at the University of California, Santa Cruz (UCSC).

This series explores significant developments in the life sciences from historical and philosophical perspectives. Historical episodes include Aristotelian biology, Greek and Islamic biology and medicine, Renaissance biology, natural history, Darwinian evolution, Nineteenth-century physiology and cell theory, Twentieth-century genetics, ecology, and systematics, and the biological theories and practices of non-Western perspectives. Philosophical topics include individuality, reductionism and holism, fitness, levels of selection, mechanism and teleology, and the nature-nurture debates, as well as explanation, confirmation, inference, experiment, scientific practice, and models and theories vis-à-vis the biological sciences.

Authors are also invited to inquire into the "and" of this series. How has, does, and will the history of biology impact philosophical understandings of life? How can philosophy help us analyze the historical contingency of, and structural constraints on, scientific knowledge about biological processes and systems? In probing the interweaving of history and philosophy of biology, scholarly investigation could usefully turn to values, power, and potential future uses and abuses of biological knowledge.

The scientific scope of the series includes evolutionary theory, environmental sciences, genomics, molecular biology, systems biology, biotechnology, biomedicine, race and ethnicity, and sex and gender. These areas of the biological sciences are not silos, and tracking their impact on other sciences such as psychology, economics, and sociology, and the behavioral and human sciences more generally, is also within the purview of this series.

Evolutionary Moral Realism
John Collier and Michael Stingl

Jakob von Uexküll and Philosophy
Life, Environments, Anthropology
Edited by Francesca Michelini and Kristian Köchy

For more information about this series, please visit: www.routledge.com/History-and-Philosophy-of-Biology/book-series/HAPB

Jakob von Uexküll and Philosophy
Life, Environments, Anthropology

Edited by Francesca Michelini and Kristian Köchy

Routledge
Taylor & Francis Group

LONDON AND NEW YORK

First published 2020
by Routledge
2 Park Square, Milton Park, Abingdon, Oxon OX14 4RN

and by Routledge
52 Vanderbilt Avenue, New York, NY 10017

Routledge is an imprint of the Taylor & Francis Group, an informa business

First issued in paperback 2021

British Library Cataloguing-in-Publication Data
A catalogue record for this book is available from the British Library

Library of Congress Cataloging-in-Publication Data
A catalog record for this book has been requested

ISBN: 978-0-367-23273-3 (hbk)
ISBN: 978-1-03-208215-8 (pbk)
ISBN: 978-0-429-27909-6 (ebk)

Typeset in Times New Roman
by Apex CoVantage, LLC

Contents

Contributors

Ralf Becker is Professor of Philosophy at the University of Koblenz-Landau (Germany). His research focuses on the areas of philosophical anthropology, philosophy of culture, and philosophy of science. He is the author of *Der menschliche Standpunkt* (2011) and coeditor of *Zeitschrift für Kulturphilosophie* (since 2020).

Cornelius Borck is a historian, a philosopher of science and medicine, and Director of the Institute for History of Medicine and Science Studies at the University of Lübeck (Germany). His research focuses on the epistemology of experimentation in science, medicine, and the arts. He is the author of an introduction to *Philosophy of Medicine* (2016, in German) and the editor of a volume on Hans Blumenberg (*Hans Blumenberg beobachtet: Wissenschaft, Technik und Philosophie*, 2013).

Carlo Brentari is Assistant Professor of Philosophical Anthropology and Ethics at the University of Trento (Italy). His research focuses on contemporary human–animal studies, German philosophical anthropology of 20th century, animal and environmental ethics, and classical ethology. He is currently working on Nicolai Hartmann's philosophy of nature. He is author of the monograph *Jakob von Uexküll. The Discovery of the Umwelt between Biosemiotics and Theoretical Biology* (2015).

Brett Buchanan is Director of the School of the Environment and Professor of Philosophy and Environmental Studies at Laurentian University (Canada). His research focuses on contemporary continental philosophy, environmental humanities, and philosophies of nature. Recent publications include coedited books, translations, and articles on philosophical ethology, extinction studies, and field philosophy. He is the author of *Onto-Ethologies: The Animal Environments of Uexküll, Heidegger, Merleau-Ponty, and Deleuze* (2008).

Jui-Pi Chien is a semiotician and Professor of Comparative Literature, Comparative Aesthetics and the Arts at National Taiwan University (Taiwan). Her research focuses on the integration of linguistic, aesthetic, and evolutionary strands of thinking in light of current neuropsychology. By way of bridging pillars of (bio)semiotics, Saussure, Peirce, and Uexküll, she has been developing renewed methodologies that serve to boost our appreciation of chances, oddities, and ambiguities found in various fields of study.

Felice Cimatti is Full Professor of Philosophy of Language at the University of Calabria (Italy). His research mainly focuses on the areas of philosophy of language, philosophical anthropology and philosophy of animality. His most recent books include *A Bio-semiotic Ontology: The Philosophy of Giorgio Prodi* (2018) and *Philosophy of Animality. Unbecoming Human* (2020).

Ezequiel A. Di Paolo is Research Professor at Ikerbasque, the Basque Foundation for Science (Spain). He has authored more than 150 publications on embodied cognitive science, philosophy of mind, psychology, and robotics. His interests include enactive approaches to mind and life, embodied intersubjectivity, and language. He is coauthor of two recent books on enactive cognition, *Sensorimotor Life: An Enactive Proposal* (2017) and *Linguistic Bodies: The Continuity between Life and Language* (2018).

Maurizio Esposito is Assistant Professor of Epistemology at the Center of Natural and Human Sciences of the Federal University of ABC, São Paulo (Brazil). He is mainly interested in the history and philosophy of the life sciences. He has published in 2013 a monograph on the history of organismal biology, *Romantic Biology 1890–1954*, and various articles exploring the relation between biology and society as well as the connections between humanities and natural sciences.

Juan Manuel Heredia is Assistant Professor of Ethics at the University of Buenos Aires and a postdoctoral researcher for the CONICET (Argentina), currently working at the Centro de Historia Intelectual (University of Quilmes). He has published articles on Jakob von Uexküll, Gilbert Simondon, and Gilles Deleuze, among others, focusing in particular on the archeological approach in contemporary philosophy.

Kristian Köchy is a biologist and Professor of Theoretical Philosophy at the University of Kassel (Germany). His research focuses on the areas of philosophy of science and the history of the life sciences, natural philosophy, and the philosophy of animal–human relations. He works in particular on the interrelations between biological research approaches and philosophical framework concepts. He is author of an introduction on *Biophilosophy* (2008, in German), and coeditor of a three-volume collection on the philosophy of animal research (*Philosophie der Tierforschung*, 2016–2018).

Hans-Peter Krüger is Professor of Political Philosophy and Philosophical Anthropology at the Philosophy Department of the University of Potsdam (Germany). He is a coeditor of the *Deutsche Zeitschrift für Philosophie*, of the *International Yearbook for Philosophical Anthropology*, and the book series *Philosophische Anthropologie*. His main areas of research include philosophical anthropologies, classical pragmatisms and neo-pragmatisms, and political and social philosophies of public communication.

Kalevi Kull is Professor of Biosemiotics of the Department of Semiotics, University of Tartu (Estonia). His research deals with semiotic approaches in biology, mechanisms of diversity, theory of general semiotics, history of biosemiotics, and ecosemiotics. He is a coeditor of the journal *Sign Systems Studies* and of a semiotic book series (*Biosemiotics*; *Semiotics, Communication and Cognition*; *Tartu Semiotics Library*).

Marco Mazzeo is Professor of Philosophy of Language at the University of Calabria (Italy). His research mainly focuses on the areas of philosophy of language, philosophical anthropology and philosophy of perception. He has also translated Uexküll's book *Streifzüge durch die Umwelten von Tieren und Menschen* into Italian (2010).

Francesca Michelini is Senior Research Fellow at the University of Kassel (Germany). Her main fields of research are the antireductionist theories of life and the bridging of continental philosophy and science. She is author of many publications on the topic of philosophical anthropology, philosophy of the life sciences, teleological explanations in nature, and autonomy in biology (among others: *The Living and the Deficiency. Essays on Teleology* 2011, in Italian).

Tristan Moyle is Senior Lecturer in Philosophy at Anglia Ruskin University (UK). His research focuses on 19th- and 20th-century European philosophy, ethics, and the philosophy of nature. As part of a project to develop a nonreductive form of naturalism, he has published articles on Uexküll, Heidegger, and Merleau-Ponty, among others.

Agustín Ostachuk is Researcher at Universidad Nacional de San Martín, working at the CEJB and at the LICH-CONICET (Argentina). His research focuses on evolution, from both a philosophical and a biological point of view, and more widely, on the interrelation and integration between biology and philosophy. His research interests include the history and philosophy of biology, biophilosophy, theoretical biology, evolutionary biology, evolutionary developmental biology, and complexity.

Foreword

Philosophizing with animals

Brett Buchanan

In setting out to write this Foreword to *Jakob von Uexküll and Philosophy*, I cannot help but turn to Uexküll's own Foreword that he wrote for his popular book *A Foray into the Worlds of Animals and Humans*. The opening sentence is an acknowledgment that his book "does not claim to serve as the introduction to a new science" (Uexküll 2010, 41), which can be read as both an apology and a dare, both an admission and a tease. Given that he goes on to characterize this book as a "travelog," it is perhaps unsurprising that he chooses to downplay its scientific significance, which is not the same as dismissing its worth or merit, or that of travel writing for that matter. With the rise of ethology over the course of the 20th century, "the ethological revolution," as Dominique Lestel has framed it (Lestel 2001; Chrulew 2014), casts Uexküll in another light. Even if we were to accept his word that he does not introduce a new science within this book, or any of his others for that matter, we could claim that for many he introduces something else entirely: a new philosophy and way of doing philosophy.

Throughout all his writings, Uexküll poses, responds to, and transforms fundamental philosophical problems. Ontologically, he is interested in the question, "What is an animal?" Similarly, he asks the epistemological questions, "What can we know about animals, and how do we know?" But in addition to being a biologist doing philosophy, what cannot be overemphasized is how he challenges these traditional branches of philosophy by thinking these questions through the perspectives of animals themselves, such that the ontological becomes translated into how animals are subjects and agents themselves, and not mere objects, through their world-forming relationships with their environments, and the epistemological becomes a question of what the animals themselves know, not just what we know about them. He offers, in other words, an animal philosophy, and he does so not as a scientist dispassionately breaking apart the world around him but by donning the guise of an exploratory guide leading us into the otherworldly realms of nonhuman life. His appreciation of animal life surpasses any sense of neutral detachment toward other beings, and instead of eliminating the possibility that animals are capable of leading meaningful lives, as the mechanistic views of his day would have it, Uexküll begins his study from one of the original philosophical positions: one of absolute wonder at the natural world, a wonderment that he shares with his readers.

Like his philosophical influences before him, Uexküll marvels at the unexplored and unknown worlds around him, including all the "flitting," "buzzing," and "dancing" animals that populate the meadows, woods, bush, soils, seas, farms, and zoos he describes and that awakens in him a curiosity to explore and understand them better. But unlike his philosophical predecessors who remain predominantly within anthropocentric worldviews, ironically trapped within soap bubbles of their own making, Uexküll opens up his wonder to all animal life and, in so doing, argues that there is much more to be studied and known than just a single, objective world. Rather, "millions of environments" exist out there, each one revealing a "new, infinitely rich field of research," and each one with an animal within it. Even if many are now familiar with Uexküll's ethological writings, it is his contributions to ethology as philosophy that remain provocative and generative, and it starts, I would suggest, with believing that animals can be agents and thinkers in their own right as opposed to the dominant trend of assuming that they are animal-machines.

His thought is refreshing because he gives animals and human investigators the benefit of the doubt: rather than assuming that they have no inner life or environment of their own or that they are ultimately unknowable (which is often conflated with not existing), Uexküll opens a speculative venture into the lives of nonhuman animals, suggesting not only that animals can and do have agency and unique worlds of their own but that they can also be knowable. Knowable not in some absolute, universal sense wherein the lives of animals become transparent to human observation and thus potentially dismissed as insignificant or controlled for ulterior motives but knowable through modes of speculative ethology and field philosophy:[1] by observing the connections and interactions that animals create with their environments (Buchanan, Chrulew, and Bastian 2018); by noticing how individual animals perceive and interact with their surroundings differently, often in ways that differentiate them as singular beings, unique from species-defining characteristics; by paying attention to how they engage with and respond to their *Umwelten* in meaningful and significant ways; or by methodically leading both novices and experts, academics and nonspecialists, to new revelations about animals through his situated storytelling. And he does so by giving animals the chance to present themselves as interesting and in thinking along with them rather than just about them.[2]

One can certainly say, then, that Uexküll's ethology is already a hybrid form of philosophical speculation that combines multiple questions and methods in epistemology and ontology. His ethology is philosophical just as much as he creates a new philosophy out of his field observations. Similarly, his writing is infused with philosophical references, from Plato to Kant, and there is no question of his impact on philosophical thought since the time of his writing. His thinking has permeated semiotics, phenomenology, neo-Kantianism, philosophy of biology, and continental philosophy, as well as human–animal studies, anthropology, environmental humanities, science and technology studies, architectural studies, performative dance, and more. For someone who so convincingly portrayed the symbiotic and sympoietic relationships between living beings, and for someone who emphasized how animals make meaning within their environments, it should be no surprise that his writings have reached out and connected with so many diverse philosophers and modes of thought. He is all about relations and connections.

Part of this, I believe, has to do with his generosity as a thinker, and the epigraph to the first edition of *The Theory of Meaning* is a case in point. Quoting a passage from Plato's *Sophist* (246a–b), Uexküll chooses a point in the dialog in which the Eleatic visitor, in conversation with Theaetetus, recalls a battle between gods and giants over the ontological question of what is, with the former advocating a materialist position of bodily existence and the latter holding a more idealist position.[3] The epigraph points to many themes that will emerge in Uexküll's book, but what I wish to highlight is where Uexküll ends his quote. He pauses on the inability of some to listen to new ideas, to ideas that challenge or threaten one's worldview. The explicit reading is that Uexküll's critics to whom he addresses this book ought to listen to what he is saying first, instead of dismissing his thinking outright (which is not to say that his thought is beyond criticism). But it is also more implicitly and metaphorically about listening to the animals he recounts: rather than hurling boulders at one another and ignoring the animals in front of us, we might do better by opening ourselves up to one another, to listen to what the animals have to say and, perhaps, even to allow ourselves to be changed by them. Through his writings, Uexküll exhibits a willingness to reach out beyond divisions, to challenge our shortcomings, and to forge relationships between disciplines, theories, and animals. His speculative ethology takes us into the environments of animals and shows us how to listen and learn from them. And I cannot think of a better philosophical exercise than this.

Notes

1 For more on the possibilities of "field philosophy" and "speculative ethology," see Buchanan, Chrulew, and Bastian (2018). I draw the term *speculative ethology* from Matthew Chrulew.
2 This theme may be found in Despret's early writings, such as *Naissance d'une théorie éthologique* (1996), through to *Penser comme un rat* (2009) and *What Would Animals Say if We Asked the Right Questions?* (2016).
3 The epigraph reads: "The materialists pull everything down from the sky and out of the invisible world onto the earth as if they wanted to clench rocks and oak trees in their fists. They grasp them, and stubbornly maintain that the only objects that exist are those that are tangible and comprehensible. They believe that the physical existence of an object is existence itself, and look down smugly on other people – those who acknowledge another area of existence separate from the physical. But they are totally unwilling to listen to another point of view." This is a fairly liberal English translation (that I suspect Uexküll has worded himself), that I have quoted from the English translation of Uexküll's *The Theory of Meaning* (1982, 25).

References

Buchanan, Brett, Matthew Chrulew, and Bastian, Michelle (eds.) (2018) 'Introduction: Field philosophy and other experiments'. *Parallax* 24 (4), 383–391.
Chrulew, Matthew (2014) 'The philosophical ethology of Dominique Lestel'. *Angelaki* 19 (3), 61–73.

Despret, Vinciane (1996) *Naissance d'une théorie éthologique. La Danse du cratérope écaillé*. Paris: Les Empêcheurs de penser en rond/Le Seuil.

Despret, Vinciane (2009) *Penser comme un rat*. Versailles: Quæ.

Despret, Vinciane (2016) *What Would Animals Say If We Asked the Right Questions?* Minneapolis, MN: University of Minnesota Press.

Lestel, Dominique (2001) *Les origines animales de la culture*. Paris: Flammarion.

Uexküll, Jakob von (1982) 'The theory of meaning'. *Semiotica* 42 (1), 25–82.

Uexküll, Jakob von (2010) [1934, 1940] *A Foray into the Worlds of Animals and Humans with a Theory of Meaning*. Translated by Joseph D. O'Neil. Minneapolis/London: University of Minneapolis Press.

Introduction

A foray into Uexküll's heritage

Francesca Michelini

By some dismissed as the last of the anti-Darwinians, his fame as a rigorous biologist even tainted by an alleged link to National Socialist ideology, it is undeniable that Jakob von Uexküll was eagerly read by many philosophers across the spectrum of philosophical schools, from Scheler to Merleau-Ponty and Deleuze, from Heidegger to Blumenberg and Agamben. What has then allowed Uexküll's name to survive the misery of history as well as the usually fatal gap between science and humanities? This collection of essays attempts for the first time to do justice to Uexküll's theoretical impact on Western culture.

1 Uexküll's *Umwelt* between tradition and topicality

One hundred years have gone by since our eccentric zoologist, Estonian born from an aristocratic family of German descent, would publish one of his most impactful books: *Theoretical Biology* [*Theoretische Biologie*] (1920). In this work, he notably expands his account on the notion of environment – *Umwelt*. Already tackled in the first decade of the 20th century, more precisely in the other great theoretical contribution of his, Umwelt *and Inner-World of Animals* [*Umwelt und Innenwelt der Tiere*] (1909), this notion will keep him busy to the end of his life. This is also the one biological concept that was bound to have considerable influence on zoology, philosophy, psychology, and medicine in the 20th century.

Besides being a pioneer of ethology – as openly stated by Konrad Lorenz – in virtue of his notion of *Umwelt*, Jakob Johann Baron von Uexküll (1864–1944) is today acknowledged as one of the founders of modern ecology. He was the first, in fact, to systematically and consistently apply the concept in biology. Previous references to *Umwelt* could be found mainly in the realm of history and culture or in sociology, and no frequent use was made of it in everyday language.[1] One exception in this respect, among Uexküll's contemporaries, is the biologist and geographer Friedrich Ratzel (1844–1904). In all likelihood, we owe to him the first replacement, in the life sciences, of the back then widely spread notion of *milieu*[2] with that of *Umwelt*, as well as its conscious differentiation from the usage made of this notion by philosophers and sociologists. Although some evidence suggests he might have had an influence on Uexküll (see Mildenberger 2014, 2f.), it is unclear whether or to what extent this was so. At any rate, one might remark that the notion of *Umwelt* is never included in biology textbooks before World War I, with the only great exception, notably, of Umwelt *and Inner-World of Animals*, where for the first time it is applied by Uexküll to the animal world.

The Baltic biologist's great lesson is indeed that living organisms are the center of their own *Umwelt*. This latter stands for the world of experiences and relations the animal creates, of which it is the sensorial and operational fulcrum. Everything an organism perceives makes up, according to Uexküll, for its "perception world" [*Merkwelt*] and everything an organism does is its "effect world" [*Wirkwelt*]: "These two worlds, of perception and production of effects – so Uexküll sums up – form one closed unit, the environment [*Umwelt*]" (Uexküll 2010, 42). Between the two worlds subsists a complex form of circular action, which Uexküll calls "functional circle" [*Funktionskreis*]. From the simplest organisms – such as the *Paramecium*, the rhizostome, and more in general those which are labeled by Uexküll as "reflex animals" – with only one functional circle to the most complex organisms with many of them, one should also remark that, although endowed with individualized traits, an *Umwelt* is species-specific.

With only three functional circles, the tick is the most renown example Uexküll provides. There is probably not another biological example that has had as much fortune among philosophers and beyond (see, for instance, the different readings of this example given by Merleau-Ponty, Canguilhem, Deleuze and Guattari, and Agamben – Chapters 8, 9, 10 and 12 in this volume). Uexküll tells us that a tick can stay inactive on a tree branch for a very long time – up to 18 years, as calculated by an experiment carried out at the Zoological Institute in Rostock (see Uexküll 2010, 52) – and there wait for a warm-blooded animal. When one passes by, a smell-based stimulus – the butyric acid secreted by the sebaceous glands of the mammals – pushes the tick to let itself fall on the body of the animal; thanks to a temperature-sensitive organ, the tick knows whether it has fallen onto a warm-blooded object, and if this is the case, by using its sense of touch it starts looking for a hairless spot on the animal's skin and digs its head in it as to suck the animal's blood. After laying eggs and having so achieved its mission, the tick dies (Uexküll 2010, 45).

What for Uexküll is most interesting and defines the work of the biologist as something more than the simple physiological study of animal life is to understand why, among hundreds of effects produced by a mammal body, only three, that is the butyric acid, the warm-related stimulus, and the touch-related stimulus, are for the tick perception-mark carriers [*Merkmalträger*] (Uexküll 2010, 50), in other words, meaning-carriers. To these correspond three effect marks [*Wirkmale*], that is, let itself fall, walk on the animal body, pierce through it. The articulation of the tick's three functional circles is so that, once the process has started, the operative mark produces a new perception mark that extinguishes the previous perception mark. For example, "[t]he falling tick imparts to the mammal's hairs, on which it lands, the effect mark 'collision,' which then activates a tactile feature which, in its turn, extinguishes the olfactory feature 'butyric acid'" (Uexküll 2010, 50).

Venturing into the field of biology means then, according to Uexküll, venturing into animal worlds which are invisible and unknown to us (Pollmann 2013), rather than simply physiologically and "objectively" grasp their functioning. Uexküll explains that, in order to see things from the point of view of a relatively simple animal like the tick, one has to imagine that the whole rich surrounding world – our world or what we perceive as such – shrinks to the size of an elementary structure, precisely, *its Umwelt* (Uexküll 2010, 51). The world surrounding the tick has nothing to do with its *Umwelt*. It is just an *Umgebung* – translated in English as *surrounding* – this word designating what happens hereabouts a living being, in other words, what the human being as observer perceives as the generic environmental surrounding of the living being. But this is not its own world.

Unlike the *Umgebung*, the *Umwelt* dimension is entirely subjective: "each *Umwelt* forms a closed unit in itself, which is governed, in all its parts, by the meaning it has for the subject" (Uexküll 1982, 30). If Uexküll is today renown in the animal-human studies mainly for this – for having proclaimed the animal the subject of its *Umwelt* – less known is possibly the fact that this "discovery" ensues from his transformation of Kantian philosophy – the influence of which has been aptly emphasized in recent years (for instance by Brentari 2015).

Uexküll belongs indeed to the tradition of retrieval of Kant's philosophy in physiological terms. Among his illustrious predecessors one can count Karl Ernst von Baer (1792–1876), supporter of the subjective nature of time; Johannes Müller (1801–1858), famous for his research on specific sense energy; and, last but not least, Hermann von Helmholtz (1821–1894), who had a direct influence on Uexküll, notably concerning the doctrine of sense-qualities.

The main goal of Uexküll's *Umwelt* research is – as he puts it – to provide Kantian philosophy with new foundations in biological terms. With one fundamental difference: unlike Kant, who placed the transcendental subject in the center of his Copernican revolution, Uexküll gives center stage to living beings, in their corporeity, with their sense organs and their nervous systems, if available (Uexküll 1926, XV). To said program of naturalization of Kant's philosophy (Gens 2014, 17) or physiologization of the transcendental Uexküll remained always true: from the programmatic statements of the introduction to *Theoretical Biology* up to *A Foray into the Worlds of Animals and Humans*, where he claims that without a living subject there is no time nor space (Uexküll 2010, 52). Nor it is the case that this important reference is eclipsed by remarks concerning the influence on Uexküll's theory also of other philosophical elements, notably some echoes of the philosophy of nature (deemed not so prevalent, pace Mildenberg 2007, 10; see also Langthaler 1992). This is why in this volume, among Uexküll's several philosophical sources, priority has been given to Kant. Esposito's essay (Chapter 2) on Uexküll and Kant provides a full account of how the relation between the transcendental subject and the world (whether human or animal) is transformed by Uexküll in a semiotic or interpretative one.

By looking at animals as subjects of their individual *Umwelten*, Uexküll is able to question a series of ingrained preconceptions (Mazzeo 2010, 9). For instance, the idea that all existing species have the same sensorimotor space and time. According to Uexküll, there is no single and undifferentiated dimension shared by all living beings but, rather, as many *Umwelten* exist as living species. In this respect, Uexküll's theory is an antidote to any hierarchical account on the living, both, obviously, in their ancient and already-obsolete form of the *scala naturae*, and, more importantly, in some of their naif configurations within the Darwinism of Uexküll's time (on Uexküll in "anti-hierarchical" strategies, see in this volume, Plessner's, Heidegger's and Deleuze and Guattari's examples, in the essays, respectively, by Krüger, Chapter 5; Michelini, Chapter 7; and Cimatti, Chapter 10). Furthermore, Uexküll's theory goes against the anthropomorphic preconception according to which human modes of sense and action can be used as reference parameters for the life of any other animal. Uexküll teaches us for instance that there is no such thing as the slowness of snails in itself. A snail is slow only in our world (Uexküll 2010, 72). Finally, Uexküll's *Umwelt* theory leads to considering living beings never in isolation from their environment and always as an integral part of this latter. Uexküll refers exclusively to animal–environment systems. In Helmuth Plessner's effective wording, in the *circle of life*, "one half is formed by the organism," the other by the environment (Plessner 2019, 178).

For all these reasons, it comes as no surprise that in recent times Uexküll has been the object of something close to a revival. Granted that his true rediscovery is connected to the biosemiotic reading of his theories already at least since the end of the 1970s, since the end of the 1990s, some hints to Uexküll's notion of *Umwelt* can be found also in the specific field of animal behavioral research (see Saidel 2002; Timberlake 2002). Admittedly Uexküll's theories have influenced the development of the theory of cognition (see Maturana 1996, 221; Stanley N. Salthe takes Uexküll more generally as an antecedent to the theories of Maturana and Varela; see Salthe 1993), and they have been employed even recently in neuro-biology and cognitive sciences (e.g., Roepstorff 2001). In particular for his notion of functional circle, he is counted among the pioneers of cybernetics (Emmeche 2001; Lagerspetz 2001; Ziemke and Sharkey 2001; on Uexküll and cybernetics, see Köchy, Chapter 3, in this volume). In the last decade, a new English transla-tion has been published of two among his most popular books, *A Foray into the Worlds of Animals and Humans* and *The Theory of Meaning* (Uexküll 2010), with an introduction underlying the implications of these texts for current debates in biology (Sagan 2010). Furthermore, the first monographic texts entirely devoted to Uexküll's life and thought have recently seen the light (Mildenberg 2007; Buchanan 2008; Gens 2014; Brentari 2015). And the new edition of *Key Thinkers on the Environment* has finally made an amendment to the previous inexcusable absence of the Estonian biologist and has included him in their anthology (Palmer Cooper and Cooper 2018). Undoubtedly, the general dissatisfaction with mecha-nistic theories of the organism or theories that are exclusively based on genetics has played a role in the revival of more organism-centered ideas on the living, and among these is also Uexküll's understanding of the living being as perceiving and acting subject. In this regard, Kalevi Kull argues that the Uexküllian approach is particularly well suited to the 21st-century changes in evolutionary biology (see Kull, Chapter 13, in this volume).

Under many respects, Uexküll's ideas display striking similarities with the most recent niche-construction theories (Odling-Smee, Laland, and Feldman 2003) and proposals have been made endorsing the integration of Uexküll's *Umwelt* theory with contemporary Darwinism, the latter being nowadays way less monolithic than in Uexküll's time (see, e.g., Gutmann 2014; Brentari 2015, 238f.). Concerning Uexküll's contribution to ecological thinking, it has been rightly emphasized that his ideas foster the basic intuition that preserving biodiversity does not necessarily mean only preserving "the animals" but, rather, protecting their environments, in other words, the semiotic and operational worlds in which life develops (Brentari 2015, 241).

2 Jakob von Uexküll's contradictions

Although Uexküll's approach bears rich developments in contemporary terms, it also entails some aspects that are hard to adjust to the current understanding of science. These difficulties have also been counted among the intrinsic contradictions of his thought (Harrington 1996, 63), and they would stem not necessarily from Uexküll's arguments but from their being rooted "in certain cultural assumptions and historical reference points of the time," actually shared by many scientists and intellectuals of the Weimar Republic (Harrington 1996, 34). Nevertheless, it is undeniable that some difficulties also stem from the fact that the notion of *Umwelt* is not systematically nor univocally applied by Uexküll, as the essays in this collection try to prove (for an overview of the several meanings of *Umwelt*, see Tønnessen, Magnus, and Brentari 2016).

One might even say that each of the multiple hints and inputs offered by Uexküll's theory to the contemporary debate have, so to speak, a flip side, which not only cannot be incorporated but also cannot be easily discarded. For instance, his understanding of the plurality and diversity of animal and human worlds, which plays today a key role when it comes to discussing biodiversity, is actually burdened by the thorny issue of reciprocal communication between the several environments. This question is never suitably dealt with by Uexküll, who ultimately dwells in a monadological framework concerning animal and human subjects, leading to a strong form of environmental solipsism (see Heredia's contribution, Chapter 1, in this volume). Although seductive for the current understanding of ecological niches, the theory concerning the sensorial and operational correspondence between an animal species and its environment appears to be tainted by the metaphysical assumption that a perfect correspondence, or even better the conformity to a plan [*Planmäßigkeit*], subsists among them (criticism of this aspect has been expressed by Di Paolo in the Afterword to this volume). Furthermore, when insisting on how a living organism functions and on its building-plan [*Bauplan*], to the detriment of its genesis, Uexküll seems to make appeal to the current evolutionary developmental biology (see Sagan 2010); however, the whole evolutionary topic of the origin and extinction of living beings is by him entirely neglected. By picking up on Buchanan's witty words, one could well say that Uexküll ultimately "opens many more questions than he is able to answer" (Buchanan 2008, 3).

With these and other fundamental thorny issues deal the essays in this volume. In our introductory pages we would like, nevertheless, to linger on some of the nitty-gritty of two among the greatest "scandals," according to general understanding, in Uexküll's theories, that is to say, the aspects that are least adjustable to current scientific and philosophical approaches: Uexküll's anti-Darwinism and his political biology. Although one can easily agree on the baffling nature of the previously mentioned problems and even on the intrinsic limitations of Uexküll's thinking (Brentari 2015, 239), it is nevertheless useful to free the way from some misunderstandings and unilateral interpretations.

Concerning the relation to Darwinism, Uexküll has been labeled as the last relevant supporter of an anti-Darwinian biology and philosophy of life (Mildenberger 2007, 4). His loathing of Darwinism surfaces already in his writings from 1903/1904, and it is likely to be the result of his taking position in the disputes between Darwinians and anti-Darwinians (vitalists) in his university in Dorpat. Overall, it is undeniable that his work suffers from an often anti-evolutionary vision of animal life, a lack of sensibility for the issues of transformation, origin, and extinction of species and organisms (Mazzeo 2010, 11). However, as soon as one ponders over how was it possible that a leading biologist at the time could be anti-Darwinian, other fundamental factors emerge that should be given due consideration. For instance, it should be made clear that Uexküll's radical criticism of Darwinism does not question selection as a key factor of evolution but, rather, its sufficiency as an explanatory tool, notably due to its causal focus mainly on the mechanistic level (Potthast 2014, 210f.). That Uexküll has often been labeled as "vitalist" has led to the belief that he was also antimechanistic. This is not the case though. He rather believed that physiology – as the science of physical and chemical causes – should be necessarily complemented by the assessment of the living being as perceiving and acting subject. This is what according to Uexküll makes up for biology's specificity and autonomy.

One should also take into account that Uexküll's criticism invests, in particular, a given form of adaptationism that he credits to the Darwinism of his time, the idea, that is to say, that it is possible to take a living being in isolation and assess its "passive" adaptation to an equally isolable environment. Generally employed – unlike Uexküll's intentions – to trace a distinction between the environment of animals and the world of human beings (see, for instance, in this volume the essays on Scheler, Chapter 4; Plessner, Chapter 5; Cassirer, Chapter 6; Heidegger, Chapter 7; and Blumenberg, Chapter 11), Uexküll's criticism of Darwinism has found the most favor among 20th-century philosophers. It should also be remarked, incidentally, that Darwin has never advocated such a naif form of adaptationism, as made clear by his writings, among which, for instance, *The Formation of Vegetable Mould through the Action of Worms* (1881), where emphasis is laid on how the animals contribute to creating the environment they belong to.

In general terms, one could claim that Uexküll's criticism is aimed at specific aspects or distinctive consequences – also from a political viewpoint – of Darwinism, more than against Darwinism as such. This is why it should not be surprising that, unlike some of his contemporaries, Uexküll "never developed a systematic critique of Darwin's theory" (Harrington 1996, 38). In this respect, as previously anticipated, it is hard to trace a clear-cut opposition between Uexküll's theories and evolutionism, especially in its most recent form. One may even claim, on the contrary, that Uexküll's idea concerning the coproduction of organism and *Umwelt*, with due transformations, is already part of the current evolutionary paradigm (Potthast 2014, 212). When it comes to Uexküll's theory, an evolutionary intake might lead, as it has already been remarked, to the removal of some major difficulties, such as, for instance, avoiding the solipsistic isolation of the transcendental subject by means of the inclusion of phylogenetic and diachronic analyses (see Brentari 2015, 240).

It is, however, mainly at the political level that Uexküll's contradictions emerge with vehemence (on the relationship between ideology and science in Uexküll, see Stella and Kleisner 2010). Besides the first edition of the *Theoretical Biology*, exactly one hundred years ago Uexküll also published the first edition of his controversial political account, *State Biology. Anatomy, Physiology and Pathology of the State [Staatsbiologie: Anatomie-Physiologie-Pathologie des Staates]* (2nd ed. 1933). Together with other treatises published in those years, such as Karl Binding's *Zum Werden und Leben der Staaten* (1920), Eberhard Dennert's *Der Staat als lebendiger Organismus* (1920), and Edward Hahn's *Der Staat, ein Lebewesen* (1926), Uexküll's text can be taken as one of the most developed biopolitical accounts on the state understood as organism (Esposito 2008, 17–19). According to Uexküll, the state – the German state – is sick and should be purged from infection, contamination, illness, just like a living organism when one wishes to return it to a healthy condition. The biologist identifies as pathologies both Western democracies, in particular, the Weimar Republic and the Bolshevik revolution, as well as, in general, all revolution. Further "weakening" and "parasitic" elements are seen as deriving from other races. According to Uexküll, one should retrieve the "German traits" that he describes, in other texts such as *Volk und Staat* (1915), in idealistic terms as conveying a deep sense of unity and responsibility. His aristocratic aversion for the democratization and the new parliamentarism of German society, in addition to fears of communism, brings him as far as welcoming Hitler's rise to power. As other intellectuals of his time, he wishes to see the end of the pathological decadence of Germany (Harrington 1996, 62).

Nevertheless, Uexküll never joins the Nazi scientific community (Mazzeo 2010, 16), and his affiliation to National Socialism remains a matter of controversy, unlike what suggested by Harrington's sometimes unilateral account (on Uexküll's ambivalent relationship with Nazism, see Heredia, Chapter 1, in this volume). Whereas his notion of *Umwelt* is often adopted by scientists close to Nazi ideology, it is also seen with great suspicion due to its social matrix leaning potentially toward Marxist positions. Its application to the animal kingdom, furthermore, is seen by party members such as Joseph Goebbels, future minister of propaganda, as "stupid and absurd," inasmuch as it distracts from the "imperative of the here and now" and the task of steering German people toward their "true duties" (G. von Uexküll 1964, 169).

Even his ideas on race are way more contradictory than what is usually assumed. On the one hand, in his correspondence with the English writer Houston Stewart Chamberlain (1855–1927) – whose theories on race will be notoriously assimilated by Nazi ideologists – one can read several statements on the superiority of the German race. Uexküll and Chamberlain were great friends, and Uexküll wrote the foreword to one of Chamberlain's books. On the other hand, at least since 1933, Uexküll makes increasing display of his aversion for the policies and racist ideology of the Nazi Party. In a letter to Chamberlain's widow, to which no answer is recorded, he decries the racial policies and the removal of Jewish scholars from German universities, among them also Ernst Cassirer, as "barbarity of the worst kind" (G. von Uexküll 1964, 172). Said barbarity, on a side note, he makes all effort to keep separate from the remembrance of his friend, whose theories, according to him, would have had nothing to do with the current degeneration, as they were rather inspired to the "purest idealism" (G. von Uexküll 1964, 171). A little earlier, while writing to his former assistant, Lothar Gottlieb Tirala, now director of the Institut für Rassenhygiene in Munich, he had vehemently deprecated the doctrine of the race as "miserable materialism." One could also add that in the same letter, Uexküll hints to the fact that his theory of the building-plan of the organism would avoid entirely such a degeneration (see G. von Uexküll 1964, 169). Finally, in 1934 Uexküll makes his disagreement with the National Socialist ideology official. In a conference hosted in Nietzsche's house in Weimar, Uexküll talks about the freedom of German universities and displeases the attending Nazi officials, who interrupt him and prompt him to leave. From that moment on he was kept under close surveillance (G. von Uexküll 1964, 174f.).

In conclusion, it is not the notion of *Umwelt* – misunderstood by those who link it both to Marxism and the National Socialist theory of "vital space" (*Lebensraum*) – to be responsible for Uexküll's political contradictions, which are instead rooted in his aristocratic background. He is a conservative seeking political stability, an opponent of democracy and crowds, an individualist who is against egalitarianism – this latter notion is, for him, abstract and in contradiction with biological reality – but he is also a loud advocate of freedom of thought, of moral ideals of justice and meritocracy, and even of a fierce critic of "heartless" capitalism as well as of its market-driven ethics. After all, he is "in public opposition to Darwinism" because it "is perceived by him as a type of biological liberalism" (see Heredia, Chapter 1, in this volume).

3 The design of this volume

Despite all the moot cases in his theories, Uexküll's charm is still felt in our time. Whereas we tend toward specialization, Uexküll combines the passion of the observing zoologist with the rigorous approach of an experimental physiologist, the mind-set of an empirical biologist with the theoretical breathing space of a philosopher. Furthermore, whereas most contributions of our time fall under several types of dogmatism, he always comes across as an outsider, someone who does not belong to one scientific movement or one philosophical school. Finally, in opposition to biological reductionism of any kind, he offers such a comprehensive vision of living beings, that even the "holistic" label – often superficially attached to him – is not enough to make justice of the richness of his approach.

Uexküll has been long taken as a forebear or founder of biosemiotics, the field of study that examines how sign systems are produced and interpreted within nature. However, although his impact on the development of this field is now indisputable (see Kull, Chapter 13, in this volume), Uexküll's overarching influence on philosophy in the last hundred years has not yet been studied in a satisfactory manner or has been limited to isolated aspects and the scattered investigation of his relation to individual philosophers. This book aims precisely to fill this gap, presenting for the first time in the scientific literature a comprehensive picture of the philosophical reception of Uexküll's ideas. By collecting essays on the most relevant philosophers who have dealt with Uexküll and by providing a framework for their interpretation, the book aims to generate a global picture of his previously unaccounted and yet enormous impact on philosophy.

Uexküll's footprints and/or those of his notion of *Umwelt* can be found, in full honesty, also in other philosophers who have not been explicitly included in this collection, in most cases due to the limited number of direct references – sometimes even just a hint – in their texts. This is the case for instance of Edmund Husserl, Hermann Keyserling, Nicolai Hartmann, Hans-Georg Gadamer, Arnold Gehlen, Jan Patocka, José Ortega y Gasset, Susanne Langer, and, more recently, Arne Næss, Peter Sloterdijk, and Daniel Dennet. Due to intrinsic limitations, our focus has been on those philosophers for which Uexküll's influence has produced the most distinctive results. In addition to the number of direct references, an important criterion in our choices has taken into account the role played by Uexküll's biology in each theory, at times even in their genesis, in most cases as the preferred validating reference or as a trigger for further development, sometimes even in opposite directions compared to the biologist, nevertheless in close confrontation with him.

Accordingly, while Uexküll – as remarked by Becker, Chapter 4 – suitably served Scheler's cultural-philosophical, ethical, and political interests, other philosophers, such Cassirer, Heidegger, and Agamben, although from different viewpoints, pushed his theory to its extreme consequences, in order to bend it to their own philosophical agendas. Others, such as Plessner, Blumenberg, and Deleuze and Guattari have radically transformed the sense of Uexküll's theory application (see, for instance, Borck, Chapter 11, in this volume on the transformation of Uexküll's bioepistemology into phenomenology by Blumenberg). Very significantly, then, thinkers such as Canguilhem (compared in this volume to the physician and philosopher, Kurt Goldstein, by Ostachuk, Chapter 9) and Merleau-Ponty, have taken the *Umweltlehre* as a fundamental critical reference for the development of their own philosophies of life.

In our project, we have aimed to reach a threefold goal. First of all, by high-lighting the importance of Uexküll for philosophy, we wanted to contribute to the general interpretation of the relationship between biology and philosophy in the last century, thus addressing the issue of the topicality of Uexküll's scientific work. In particular, the emphasis is here placed on a generally neglected field of research, which connects continental philosophy and the sciences of life, to the aim of providing suitable theoretical tools to nonreductionist theories of life. As Brett Buchanan rightly puts it,

> [w]hereas continental philosophy is better known for its engagement with the history of philosophy, the arts, ethics, politics, and its critiques of meta-physics, it is only recently that a more concerted emphasis has been placed again on its diverse relations with the sciences. As an example, we can observe the proliferation in recent years of studies bearing on biophilosophy, zoontology, geophenomenology, geophilosophy, ecophenomenology, animal others, and many others, to say nothing of the growing field of animal studies itself.
>
> (Buchanan, 2008, 4)

Second, the exploration of the relation established by philosophers from several schools of thought with Uexküll's biology is meant to allow also the overcoming of traditional boundaries between philosophical schools, such as philosophical anthropology, phenomenology, existentialism, and philosophy of life, encour-aging, vice versa, the investigation of the fruitful intersections and connections between different lines and styles of thought. "Uexküll's case" reveals, in our opinion, the existence of a "transversal" constellation of thinkers, who encounter the zoologist while working on the same fundamental issues: What is life? And how can we define the living? What is the relation between the animal *Umwelten* and the human world(s)? Does such a differentiation still make sense or is it totally obsolete? Does something like a "human nature" even exist? And if yes, how can we define it? How can we understand animals without falling into naif anthropomorphisms? Despite the variety of answers, the formulated questions suggest a shared ground of theoretical issues, which still have center stage, with all due transformations, in contemporary debates.

Within this perspective, we finally deliberately avoided tracing a clear-cut dis-tinction between biosemiotics and philosophy, which, in our opinion, cannot be considered as totally separate disciplines. Along these lines, the closing essay by Jui-Pi Chien (Chapter 14) argues for the richness and inevitability of a joint account on Uexküll's potential contribution to current theories as well as to the challenges connected to 21st-century evolutionary biology (see also Kull, Chap-ter 13, in this volume).

Preceding the investigation of Uexküll's influence on philosophy, a section composed of three essays has been devoted to his historical and scientific back-ground. This section has been devised, in particular, for the readers who access here for the first time the *"Umwelt"* of Jakob von Uexküll. An intellectual biography is here provided, including his retrieval and (at times ambiguous) transformation of Kantian philosophy, his position within the scientific milieu of his time, and, in particular, his impact on comparative physiology, ethology, cybernetics, and environmental ethics.

Now at the end of our endeavor, we wish to thank the Hessen State Ministry of Higher Education, Research and Art for having funded the early stages of this project within the context of the LOEWE Cluster Project, "Animal-Human Being-Society," at the University of Kassel. We thank Rasmus Grønfeldt Winther for supporting our volume, sending it out for blind review, and accepting it for publication in the book series *History and Philosophy of Biology*. Our gratitude goes also to Carlo Brentari and Kalevi Kull for their enthusiasm, their kind support, and their precious advice throughout the crafting of this volume. We would also like to thank Lenny Moss for the help he offered at the very beginning of this project. To Tessa Marzotto Caotorta goes our warmest thanks for her tireless and competent help extending well beyond simple linguistic counseling. Finally, we would like to close our introduction in heartfelt remembrance of Jean Gayon (1949–2018), who supported the birth of this project with great interest and attention but unfortunately could not see its end.

Notes

1 Tønnessen, Magnus, and Brentari (2016, 138) mention that the German term *Umwelt* is a "comparatively young word. It appears for the first time in the year 1800, in the ode *Napoleon* by the Danish poet J.-I. Baggesen (1764–1826) [...] *Umwelt* is the (prevailingly hostile) natural context that surrounds the poet."

2 As is well known, it was Claude Bernard, in biology, who between 1854 and 1857 defined the notion of *milieu intérieur* to refer to the main internal liquids essential to animal life, a concept that will be later identified with homeostasis.

References

Brentari, Carlo (2015) *Jakob von Uexküll: The Discovery of the Umwelt between Biosemiotics and Theoretical Biology*. Dordrecht/Heidelberg/New York/London: Springer.

Buchanan, Brett (2008) *Onto-Ethologies: The Animal Environments of Uexküll, Heidegger, Merleau-Ponty, and Deleuze*. Albany/New York: SUNY Press.

Emmeche, Claus (2001) 'Does a robot have an Umwelt? Reflections on the qualitative biosemiotics of Jakob von Uexküll'. *Semiotica* 134 (1/4), 653–693.

Esposito, Roberto (2008) [2004] *Bios: Biopolitics and Philosophy*. Translated by Timothy C. Campbell. Minneapolis: University of Minnesota Press.

Gens, Hadrien (2014) *Jakob von Uexküll, explorateur des milieux vivants: Logique de la signification*. Paris: Hermann.

Gutmann, Mathias (2014) 'Uexküll and contemporary biology: Some methodological reconsiderations'. *Sign System Studies* 32, 169–186.

Harrington, Anne (1996) *Reenchanted Science: Holism in German Culture from Wilhelm II to Hitler*. Princeton: Princeton University Press.

Lagerspetz, Kari Y. H. (2001) 'Jakob von Uexküll and the origins of cybernetics'. *Semiotica* 134 (1/4), 643–651.

Langthaler, Rudolf (1992) *Organismus und Umwelt. Die biologische Umweltlehre im Spiegel traditioneller Naturphilosophie.* Hildesheim/Zürich: Olms.

Maturana, Humberto (1996) *Was ist Erkennen?* München/Zürich: Piper.

Mazzeo, Marco (2010) 'Il biologo degli ambienti: Uexküll, il cane guida e la crisi dello Stato'. In: Jakob von Uexküll *Ambienti animali e ambienti umani. Una passeggiata in mondi sconosciuti e invisibili.* Translated by Marco Mazzeo. Macerata: Quodlibet, 7–33.

Mildenberger, Florian (2007) *Umwelt als Vision. Leben und Werk Jakob von Uexkülls (1866–1944).* Stuttgart: Steiner.

Mildenberger, Florian (2014) 'Zur ersten Orientierung'. In: Florian Mildenberger and Bernd Herrmann (eds.) *Jakob von Uexküll: Umwelt und Innenwelt der Tiere.* Berlin/Heidelberg: Springer, 1–12.

Odling-Smee, John, Laland, Kevin N., and Feldman, Marcus W. (2003) *Niche Construction: The Neglected Process in Evolution.* Princeton: Princeton University Press.

Palmer Cooper, Joy A. and Cooper, David E. (2018) *Key Thinkers on the Environment.* Abingdon/New York: Routledge.

Plessner, Helmuth (2019) [1928] *The Levels of Organic Life and the Human: An Introduction to Philosophical Anthropology.* Translated by Millay Hyatt. New York: Fordham University Press.

Pollmann, Inga (2013) 'Invisible worlds, visible: Uexküll's Umwelt, film, and film theory'. *Critical Inquiry* 39 (4), 777–816.

Potthast, Thomas (2014) 'Lebensführung (in) der Dialektik von Innenwelt und Umwelt. Jakob von Uexküll, seine philosophische Rezeption und die Transformation des Begriffs "Funktionskreis" in der Ökologie'. In: Nicole C. Karafyllis (ed.) *Das Leben führen?: Lebensführung zwischen Technikphilosophie und Lebensphilosophie. Für Günter Ropohl zum 75. Geburtstag.* Berlin: Nomos Verlag, 197–218.

Roepstorff, Andreas (2001) 'Brains in scanners: An Umwelt of cognitive neuroscience'. *Semiotica* 134, 747–765.

Sagan, Dorion (2010) 'Introduction. Umwelt after Uexküll'. In: Jakob von Uexküll (2010) [1934, 1940] *A Foray into the Worlds of Animals and Humans with a Theory of Meaning.* Translated by Joseph D. O'Neil. Minneapolis/London: University of Minneapolis Press, 1–34.

Saidel, Eric (2002) 'Animal minds, human minds'. In: Marc Bekoff, Colin Allen and Gordon M. Burghardt (eds.) *The Cognitive Animal: Empirical and Theoretical Perspectives on Animal Cognition.* Cambridge, MA: MIT Press, 53–58.

Salthe, Stanley N. (1993) *Development and Evolution: Complexity and Change in Biology.* Cambridge, MA: MIT Press.

Stella, Marco and Kleisner, Karel (2010) 'Uexküllian Umwelt as science and as ideology: The light and the dark side of a concept'. *Theory in Biosciences* 129 (1), 39–51.

Timberlake, William (2002) 'Constructing animal cognition'. In: Marc Bekoff, Colin Allen, and Gordon M. Burghardt (eds.) *The Cognitive Animal: Empirical and Theoretical Perspectives on Animal Cognition.* Cambridge, MA: MIT Press, 105–114.

Tønnessen, Morten, Magnus, Rijn, and Brentari, Carlo (2016) 'The Biosemiotic Glossary Project: Umwelt'. *Biosemiotics* 9 (1), 129–149.

Uexküll, Gudrun von (1964) *Jakob von Uexküll. Seine Welt und seine Umwelt. Eine Biographie.* Hamburg: Christian Wegner Verlag.

Uexküll, Jakob von (1915) 'Volk und Staat'. *Die neue Rundschau* 26 (1), 53–66.

Uexküll, Jakob von (1920) *Staatsbiologie: Anatomie-Physiologie-Pathologie des Staates*. Hamburg: Hanseatische Verlagsanstalt.

Uexküll, Jakob von (1926) [1920] *Theoretical Biology*. Translated by Doris L. Mackinnon. London: K. Paul, Trench, Trubner & Co.; New York: Harcourt, Brace & Company.

Uexküll, Jakob von (1982) [1940] 'Theory of meaning'. *Semiotica* 42 (1), 25–82.

Uexküll, Jakob von (2010) [1934, 1940] *A Foray into the Worlds of Animals and Humans with a Theory of Meaning*. Translated by Joseph D. O'Neil. Minneapolis/London: University of Minneapolis Press.

Ziemke, Tom and Sharkey, Noel E. (2001) 'A stroll through the worlds of robots and animals: Applying Jakob von Uexküll's theory of meaning to adaptive robots and artificial life'. *Semiotica* 134, 701–746.

Part I

Jakob von Uexküll and his historical background

1 Jakob von Uexküll, an intellectual history

Juan Manuel Heredia

1 Introduction

In order to understand the historical and epistemological significance of Uexküll's theoretical and methodological approach, it is necessary to elucidate the epochal system of thought within which it arises and develops, as well as the specific conceptual problem which it sought to respond to. In principle, we should be aware that the archaeological ground within which Uexküll's work unfolds is radically different from that of the 19th century. This difference ensues from the premises on which 19th-century theories were based being radically questioned and from the correlative growth of synchronic formal approaches to different scientific fields. While restructuring Foucault's archaeological approach, the philosopher and historian Elías Palti (2004, 2017) has conceptualized this epistemological mutation distinguishing an order of knowledge typical of the 19th century ("Age of History") and a system of thought established at the beginning of the 20th century ("Age of Forms"). Unlike the classical *episteme* of the 17th and 18th centuries (exemplified at biological level by Linnaeus's taxonomy, the preformationist theory and the fixist assumptions), in the 19th century, one observes the dominance of diachronic and genetic schemas, which no longer explain beings and nature according to an order of representation but, rather, in terms of a historicity that constitutes them and explains their development in the continuity of time following an order of succession endowed with an immanent logic. As far as biology is concerned, as Canguilhem (2000, 106) has pointed out, von Baer's ontogenesis and Darwin's phylogenetic evolutionism share one underlying assumption: the predominance of time and history. In the same vein, Cassirer points out that in the 19th century:

> The historical, barely tolerated previously, was not actually to *supplant* the rational, for there is no rational explanation of the organic world save that which shows its origins. The laws of *real* nature are historical laws.
>
> (Cassirer 1950, 173)

This order of knowledge begins to show clear signs of decomposition at the end of the 19th century with a series of convergent developments in the natural sciences and already in the first decades of the 20th century, one observes the establishment of a new epistemological configuration (Palti 2004, 66–72, 2017, 125–140). This displacement installs an interscientific setting, in which synchronic totalities (e.g., structures, forms, systems) become the main object of research. Correlatively, a process is triggered based on which scientific categories lose substantiality throughout. This is clear in theoretical physics with Maxwell's and Faraday's concept of electromagnetic field, and then with Einstein's theory of general relativity (which makes time the fourth dimension of a spatial system), and in the biological realm, it brings us to the 1890–1930 period, what Julian Huxley has called "the eclipse of Darwinism"; in the psychological field, it is expressed by the emergence of *Gestaltpsychologie* and, in linguistics, by the theory of F. de Saussure, and so on. This general displacement is accompanied and reinforced by the emergence of the phenomenon of discontinuity in physics and biology, where the ideas of quantum leap, mutation, and sudden change become decisive elements to dismiss the historicist-evolutionist 19th-century premises and produce a dissociation between the notions of totality and finality. These verifiable but not explainable discontinuity phenomena directly posed the problem of how to rationally conceive the passage from one form to another. Compared to causal sequences and teleologically ordered phases that, in the 19th century, explained the articulation between order and change on the basis of a historical temporality, in the 20th century, new emphasis is laid instead on ruptures.

In theoretical biology, this vast process of conceptual rearrangement can be illustrated very succinctly in reference to three displacements. First, from 1892 to 1905, while working on inheritance, August Weismann presents his theory of germinal plasm, and, by distinguishing germinal cells and somatic cells, he refutes the evolutionary theory of the inheritance of acquired traits. Second, the increasing awareness of the inadequacy of mechanical models of physico-chemical causality when dealing with the organism's regulation and regeneration processes opens the door to a rehabilitation of vitalist perspectives. Mention should be here made of Gustav Wolff's and Hans Driesch's ontogenetic studies that, in detriment of phylogenetic approaches, reactivate the embryological perspective opened by von Baer at the beginning of the 19th century and thus reopen the problem of teleology. Third, further progress in the studies on inheritance leads to further departing from the premises that organized the 19th-century historicist-evolutionist model of explanation. This shift finds its central reference in the figure of Hugo de Vries, who through botanical experiments shows that

> [s]pecies are not gradually transformed, but remain unaltered through successive generations. They suddenly produce new forms that are distinctly different from their parents and that are subsequently as perfect, constant, well-defined and pure as may be expected in any given species.
>
> (cited in Jacob 1973, 221)

This finding mobilizes a deep critique of genealogical approaches and will lead to a paradoxical situation in which the genesis of species can be thought according to discontinuous mutations of a contingent character, leaps which, as Cassirer (1961, 179f.) points out, can be established and verified but cannot be explained in causal, evolutionary or teleological terms. And it is precisely this paradoxical situation in relation to genetic explanations that will lead to a rebirth of metaphysics at the beginning of the 20th century and will allow us to understand Uexküll's anti-Darwinism, as well as his project of structural biology.

2 Genesis of a structural vitalism (1892–1905)

Jakob von Uexküll was born on September 8, 1864, into a family belonging to the German Baltic nobility in the state of Keblaste (now Mikhli) in Estonia, at the time under the control of the Russian Empire. Raised in a rural environment and in contact with nature, he moves with his family in 1875 to the German city of Coburg, where he attends the local *Gymnasium* for about three years and has his first readings of Kantian philosophy (Brentari 2015, 22). At the age of twenty, after returning to Estonia and finishing high school in Reval (now Tallinn), he begins to study at the Faculty of Natural Sciences of the University of Dorpat (now Tartu), where Karl E. von Baer, the vitalist biologist and founder of modern embryology, had taught some decades before. In this period, Uexküll adopts an atheistic, materialistic, and deterministic approach (G. von Uexküll 1964, 24), even refusing to participate in the religious ceremony held for his mother's funeral (Harrington 1996, 38). At the beginning of his university education, Uexküll is seduced by the Darwinian theory, but with the development of his studies and dismayed by the simplistic, speculative, and Haeckelian vision of Darwinism that the zoologist Julius von Kennel was teaching at Dorpat, he turns to the study of physiology to the detriment of zoology and favors empirical research over the formulation of general theories (Brentari 2015, 25).

After being awarded the title of *Kandidat der Zoologie* in Dorpat and specializing in the field of marine fauna, in 1888, Uexküll begins to study physiology at Heidelberg University and works at the *Physiologisches Institut* as Wilhelm Kühne's assistant, with whom he learns experimental research methods applied to the study of frogs. He graduates as a physiologist in 1890 with a dissertation on the parietal organ ("third eye") of frogs (Le Bot 2016, 195) and, from 1892 to 1907, regularly publishes in the journal, *Zeitschrift für Biologie*, the results of his research in neuromuscular physiology (Kull 2001, 16–20; see also Köchy's contribution, Chapter 3, in this volume). In this period, Uexküll alternates his work in Heidelberg with research stays at the Zoological Station in Naples. And it is in Naples, in 1891, that, while applying the experimental research methods developed by Kühne to the study of octopuses and sea urchins, Uexküll discovers Driesch's neovitalism. After that, for some time, his theoretical efforts are aimed at reconciling the methodological demands of mechanism and empirical research with von Baer's and Driesch's vitalist positions (Mildenberger 2007, 53f.).

In 1899, Uexküll travels to Paris to study the chronophotographic method developed by physiologist Etienne Marey for the study of animal body motion and then implements this technique in the analysis of the body movements of butterflies and fish (Brentari 2015, 27). In the same year, he publishes along with the physiologists Albrecht Bethe and Theodor Beer an article in which the anthropomorphic and psychological terminology used in sensory physiology is questioned, and a new nomenclature to objectively designate the reception of stimuli is proposed (Rüting 2004, 40). The article made waves and influenced the development of North American behaviorism, as well as Pavlov's and Bekhterev's reflex concept (Harrington 1996, 42).[1] With Kühne's death in 1900, Uexküll's situation in Heidelberg becomes precarious and his research starts to lose institutional support: in 1902, he is denied access to the laboratory of the Physiologisches Institut, and in 1903, the Zoological Station of Naples rejects his request for funding, leaving him with no other option but to self-finance his research stays in Beaulieu, Roscoff, Berck-sur-Mer and Biarritz (Brentari 2015, 28).

In 1902, Uexküll meets the woman who will become his wife and lifelong companion, the German aristocrat Gudrun von Schwerin, with whom he will have three children. In the same year, he publishes "In battle over animal psyche" [*Im Kampf um die Tierseele*, Uexküll 1902], where he sets out his first biological interpretation of Kantian transcendental philosophy (Kull 2001, 18). This text clearly marks a step further toward a deeper understanding of sensory physiology. The development of these theoretical investigations, his increasingly critical attitude toward mechanism and Darwinism, and the correlative defense of vitalism are regarded with suspicion by the institutionalized world of biology in Germany. His work begins to be perceived as excessively speculative, and it becomes the subject of the same antivitalist criticism which was at that time aimed at Driesch. In this hostile environment, Uexküll will become aware of the impossibility of reconciling mechanism with neovitalism and will adopt a vitalist position (Mildenberger 2007, 71).

In 1905, Uexküll publishes his first book, *Guidelines for the Study of the Experimental Biology of Aquatic Animals* [*Leitfaden in das Studium der experimentellen Biologie der Wassertiere*], a text in which – as Brentari (2015, 57–59) has shown – he opposes any kind of metaphysical speculation, defends the legitimacy of theoretical biology as producer of reflective judgments, and proposes a division of tasks between physiology and biology. In this latter respect, he maintains that, while the object of physiology is the organism's material and energetical functioning – analyzing it as a machine and subsuming it to causal schemes – biology deals with the Gestalt of organisms and thematizes "the relationship that holds together the performance of all organs, from the impact of stimuli by the receptors to the effectors' resulting answer" (Uexküll 1905, 9, my transl.). In his 1905 work, Uexküll introduces the difference between receptors [*Rezeptoren*] and effectors [*Effektoren*], and presents a fundamental notion, that of building-plan [*Bauplan*]. Although shrouded in teleological assumptions, related to some extent to the concept of reflex and thereby liable of being perceived as an anachronistic "fixism," this notion has a great epistemological significance. Particular emphasis should be given to the fact that it does not intend to uncritically restore the old final causality, nor is it justified by some kind of intelligent design. On the contrary, as Cassirer (1950, 199–205) has shown, Uexküll's notion of *Bauplan* seeks to methodologically ground the possibility of a structural biology, that is, it supports a synchronic and formal approach to the functional unity of organisms. In this sense, far from being merely an anti-Darwinism and antimaterialism metaphysical reaction, it is the expression of a wider theoretical shift connected to contemporary epistemological transformations. Years later, while relying on the distinction between "conformity to a plan" [*Planmäßigkeit*] and "goal directedness" [*Zielstrebigkeit*], Uexküll will present this approach in terms of "static teleology."

3 Sensory physiology, subjective biology and perception worlds (1907–1920)

Deprived of institutional financial support to his research and ambivalently regarded by his colleagues, in 1905 Uexküll settles down with his wife in Heidelberg and starts to outline his theoretical biology. In 1907, Heidelberg University awards him with an honorary doctorate in Medicine for his research on muscular physiology and neuromotor regulation, studies that led to what would later be known as "Uexküll's law," which stipulates that nerve excitement flows more easily in relaxed and stretched muscles than in contracted ones (Lagerspetz 2001, 646). In these years, Uexküll adopts a more philosophical perspective and, broadening his knowledge of sensory physiology, builds a subjective biology inspired by Kant. In this regard, in 1907 he publishes the article "Sketches of a Coming Conception of the World" [*Die Umrisse einer kommenden Weltanschauung*, Uexküll 1907], where he introduces the notions of "subjective anatomy," "subjective physiology," and "subjective biology" and challenges Darwinism and the physico-chemical explanations of biological phenomena (Uexküll 1913, 123–154). Furthermore, in 1908, he publishes another article in which he makes a diagnosis of the state of affairs in the life sciences and identifies the new problems and challenges that theoretical biology must address in the context of the eclipse of Darwinism. In fact, in "New Questions in Experimental Biology" [*Die neuen Fragen in der experimentellen Biologie*, Uexküll 1908a], he flatly rejects the Darwinian theory of accidental variations in favor of Hugo de Vries's theory of mutations (Uexküll 1913, 17–19). Likewise, he questions the ideas of adaptation [*Anpassung*] and struggle for existence, and states that living beings are perfectly fitted into [*Einpassung*] the environment because they are already pre-adapted to the perceptual marks that correspond to them in it (Uexküll 1913, 20f.).

In the 1908 article, the notion of conformity to a plan is presented as the true object of biological research, opening a reflexive space that exceeds physiological analysis. Against evolutionist and phylogenetic perspectives, it poses a synchronic approach which thematizes living beings as functional units, "by conformity to a plan of an organism" – says Uexküll (1913, 26, my transl.) – "should be understood nothing but a certain disposition of the different parts of an object that make it a unit" and adds that such unit "is always 'functional'" (that is to say, it responds to an overall plan and cannot be reduced to a set of physico-chemical mechanisms). On the other hand, Uexküll (1913, 33) explicitly states that the problem of conformity to a plan should not be confused with the idea of goal-directed tendency: while the first is the product of experimental observation of animals' behavior in space and according to simultaneity orders, the second includes time as additional factor and implies larger ontological commitments. Criticizing Darwinism but also keeping a safe distance from the theoretical reappropriations of the ancient Aristotelian entelechy, Uexküll eloquently points out that

> [w]e can only understand those machines whose wheels are placed one next to another in space; machines whose wheels lie part in the future and part in the past are totally incomprehensible to us.

> (Uexküll 1913, 33, my transl.)

Uexküll will return to this distinction in a 1910 article ("The New Objectives of Biology," in German: *Die neuen Ziele der Biologie*, Uexküll 1910a) where, after stating that the crisis of Darwinism "has shown as insoluble" the problem of the origin of species and has displaced the center of interest from phylogenesis toward ontogenesis (Uexküll 1913, 35f.), he states that two lines of research in experimental biology are opened: one outlined by the "dynamic finality" – whose reference is von Baer, and Driesch's neovitalism – and the other one by the "static finality" (Uexküll 1913, 45–51). Uexküll highlights that his concept of static finality is neither causal nor mechanical but purely structural. As a matter of fact, it seeks to thematize the form of organisms and reflexively reconstructs their functional units in terms of conformity to a plan. Thus, spatializing teleology[2] and separating the concept of totality from that of finality, Uexküll proposes a synchronic biological approach to the study of species and makes it independent from phylogenetic problems. Likewise, with respect to the ontogenetic and neovitalist approaches that mobilize a dynamic finality, although Uexküll does not openly reject them and uses them to come to the idea of protoplasm (Brentari 2015, 65–69), he clarifies that his structural perspective is methodological and not ontological (Uexküll 1909, 12f.).

Uexküll's approach is developed within the framework of a return to Kant's transcendental philosophy, here revived in the light of observations concerning the activity of animals. This approach promotes a shift from physiology toward theoretical biology, which entails taking animals not as objects or machines but, rather, as subjects endowed with specific forms of sensitivity and *a priori* conditions of perception, which determine their insertion into the surroundings [*Umgebung*] notably through the constitution of an *Umwelt*. Depending on its building-plan, Uexküll maintains that each species has a certain perception world [*Merkwelt*] and that each animal is sensitive only to a limited set of perception marks [*Merkmale*] that give rise to certain vital activities, while the rest of the environment, its physical properties and its multiplicity of stimuli, remains totally indifferent and inaccessible.[3] A second order of subjectivity is therefore recognized behind the unique individuality of each animal species; in other words, it is assumed that specific transcendental conditions prescribe what an animal can and cannot perceive.[4] Uexküll presents this theory in 1909 as he publishes his second book, Umwelt *and Inner-world of Animals* [*Umwelt und Innenwelt der Tiere*], where he recapitulates his empirical research on marine animals' sensory physiology; analyzes the perception worlds of jellyfish, sea urchins, and octopuses; and introduces the concepts of *Umwelt*, inner world [*Innenwelt*], and counter-world [*Gegenwelt*].[5] The idea according to which living beings inhabit a specific *Umwelt* and not a universal and homogeneous physical space will have an enormous impact on the intellectual milieu of his time and will imply a radical shift from assessing living beings as objects or statistical populations to taking them as complex and active systems of relationships with the world. Put it otherwise, the environment was no longer necessarily an objective physical-chemical space or place for the struggle for existence and started to be considered in terms of its meaning for a specific biological subject (Canguilhem 1971, 129–154). Eventually, this radical shift will imply recognizing the existence in nature of a countless number of specific perception worlds that are beyond human perception and experience (Uexküll 1909, 5f.).

In 1910, Uexküll publishes two articles in *Die neue Rundschau* magazine. In "Die Umwelt" (Uexküll 1910b), he delves into the philosophical implications of this notion and states that from a methodological point of view, the research on *Umwelten* must, through experimental observation, identify which qualities of an object act on an animal and then make progress in the determination of its perception world and the vital sense assumed by the identified perception marks (Brentari 2015, 80). In the second article, "Mendelism" [*Mendelism*, Uexküll 1910c], Uexküll develops a vitalist interpretation of Gregor Mendel's laws of inheritance as opposed to Darwinian evolutionism, concluding that these laws confirm that the inheritance process is governed not by causality but by a natural super-mechanical factor that operates according to fixed formulas and quantities, thus proving conformity to a plan in living Nature (Uexküll 1913, 281–283).

Once acknowledged by the scientific community for his biological theories, in 1913, Uexküll competes for the management of the Kaiser-Wilhelm-Institut für Biologie, but the jury and the authorities choose to appoint the cell biologist Theodor Boveri, a less controversial figure in the biological field (Harrington 1996, 34). To a similar outcome, in the same year, Uexküll submits to the Kaiser-Wilhelm-Gesellschaft, a funding request for about 200,000 marks in order to develop his research on the *Umwelten* and to establish an aquarium (Brentari 2015, 29; Rüting 2004, 40). The desire to build an aquarium is justified not only by his affinity with marine fauna but also by methodological reasons. Uexküll argues that it is not possible to adequately research the *Umwelten* of terrestrial and aerial animals in artificial spaces such as zoos, where "expatriated animals within their cages, in a foreign air and soil, seem more like ghosts than living beings" (Uexküll 1913, 110, my transl.) and states that, on the contrary, aquariums offer excellent conditions to be able to advance in this study because in them it is possible to reconstruct the *Umwelten* of aquatic animals "without loss of their vital conditions" and "without it becoming a prison" (Uexküll 1913, 110, my transl.). The Kaiser-Wilhelm-Gesellschaft refuses to finance Uexküll's ambitious project but grants him funds for 10,000 marks to develop his research for the next three years, a time during which he will make research stays in Beaulieu, Rapallo, and Biarritz (Brentari 2015, 29f.). Uexküll receives bitterly both institutional decisions and perceives them as a defeat of German biology in the race against the North American one (G. von Uexküll 1964, 96).

In 1913 Uexküll also publishes his third book, *Ideas for a Biological Conception of the World* [*Bausteine zu einer biologischen Weltanschauung*], which gathers a series of articles published between 1907 and 1913 and in which he brings the analysis of the notion of *Umwelt* to a deeper level. In a chapter titled "The Perception Worlds of Animals" [*Die Merkwelten der Tiere*, Uexküll 1913], Uexküll refers to the misunderstandings arisen from such notion and attempts to deal with them emphasizing its eminently perceptive character (Uexküll 1913, 71–73). In this period of Uexküll's intellectual production, the notion of *Umwelt* overlaps with the one of a perception world and, in fact, in many articles included in his 1913 book, Uexküll mostly uses the term *Merkwelt* where he would later use *Umwelt* (Kull 2001, 22).

The beginning of the Great War and the "Spirit of 1914" propagation awakes in Uexküll a deep nationalist passion and, when England enters the war, the conflict becomes for him a spiritual and ideological dispute in which the value of German culture needs to be defended against the threat of capitalist democracies and their matching biological theories: Darwinian mechanism and materialism. In this context, Uexküll intensifies his interventions in the public debate and, through the publication of numerous essays in popular newspapers and magazines, extends the application field of his biological theory to the political and ethical world. In 1915 he publishes the article "People and State" [*Volk und Staat*, Uexküll 1915] where, in contrast to state's artificiality, he regards the natural existence of people as the vital source of collective existence and warns that, as a result of bad policies, people can decompose and turn into irrational masses (Harrington 1996, 56f.). It also emphasizes that the basis of each people are not individuals but families. In 1917, "Darwin and the English moral" [*Darwin und die englische Moral*, Uexküll 1917] is published, an article in which Uexküll combines his criticisms of Darwinism with his rejection of English politics and morals. Following the same path, after remembering the atrocities and crimes committed by England in Ireland and India (Rüting 2004, 43), Uexküll (1917, 229) compares the ethical value of German idealism to the crass Anglo-Saxon materialism.

In general terms, in these years Uexküll develops a biopolitical perspective based on which he articulates the conservative criticism of parliamentary democracies (perceived by him as "the rule of the crowd"), the romantic rejection of heartless capitalism and its market ethics, and the biological questioning of the Darwinian "economy of nature." Uexküll presents this perspective systematically in 1920 when he publishes *State Biology* [*Staatsbiologie*, Uexküll 1933] and, prior to that, in the same year with the publication of *Biological Letters to a Lady* [*Biologische Briefe an eine Dame*, Uexküll 1920]. In the tenth letter of this book, Uexküll sums up the main ideas of his biological theory of the state in clear opposition to Darwinism, which is perceived by him as a form of biological liberalism (Heredia 2011). It is worth recapitulating here some of these biopolitical ideas.

First, against the economy of nature that derives from Darwin's notion of competition and struggle for existence "without a plan," and against the assumption that society is self-instituted by economic exchange, Uexküll claims that

> [t]he State – which must create the physical conditions for existence, housing, clothing and food – cannot be a mere symphony of free sounds, but must form a true structure consisting of very different cells adjusted one to the other, and obeying a common rule of functioning.
>
> (Uexküll 1920, 105, my transl.)

He adds that, like a "working community," the human state must synergistically organize the different "professional worlds" that constitute the economy of society.

Second, this organization and planning task has to be based on "biological-technical laws" of association between working worlds and not on moral ideals or on circumstantial parliamentary majorities criteria. Uexküll acknowledges the ideal of *liberté, égalité, fraternité*, but he clarifies that it constitutes an individual moral imperative incapable of founding a lasting collective order. After ironically pointing out that only a small Ceylon tribe lives according to the Rousseauian ideal of the French Revolution, he adds that

> [t]he rest of the peoples, who live in less favorable natural conditions, are compelled to procure clothing, housing and food through common work before being able to dedicate themselves to the fulfillment of moral demands.
>
> (Uexküll 1920, 98f., my transl.)

Third, Uexküll emphasizes that we must neither confuse people with the state nor dilute politics in society and points out that this indistinction entails fatal consequences for community organization (anomie, social disintegration, and social degeneration being the most extreme). At this point, as a condition for the possibility of implementing socioeconomic planning for the work world and the controlled introduction of new techniques and tools, Uexküll claims that the State has a properly executive dimension, a political instance of sovereign leadership and decision (the "head of the monarchy"), which – beyond parliamentarism and its contingent majorities – must be capable of ensuring "the *Umwelten* a constant common direction" and "a common rule of functioning" (Uexküll 1920, 104, 105, my transl.).

4 The life of worlds and the question of behavior (1920–1936)

With the publication of *Theoretical Biology* in 1920, a new phase of Uexküll's intellectual history begins. The book was a success: it was reprinted several times, was quickly translated into English (published in 1926), and gave rise to a second revised edition (published in 1928). In this work, he summarizes his empirical research and his theoretical reflections (G. von Uexküll 1964, 133), refines the theory of *Umwelten*, formulates a series of new concepts, and succeeds in systematically presenting his philosophical-biological theories. It should be noted that Uexküll begins and ends the book by making reference to Kant. In the introduction (see also Esposito, Chapter 2, in this volume), he states,

> The task of biology consists in expanding in two directions the results of Kant's investigations: (1) by considering the part played by our body, and especially by our sense-organs and central nervous system, and (2) by studying the relations of other subjects (animals) to objects.
>
> (Uexküll 1926a, xv)

And, in order to do that, "we must observe the subject while, as its activity dictates, it is in process of receiving impressions and making use of them" (Uexküll 1926a, xvi). In this way and only through experimental observation, the conformity to a plan of the animal subject can be reflexively reconstructed without leading to psychological speculations. On the other hand, in the last pages of the book, Uexküll writes,

> Why did Kant write no *Critique of Will-Power*? Because we know nothing about will-power [...] We have only the vague sensation that impulses of the will are in play; but we do not know them.
>
> (Uexküll 1926a, 360)

He then adds that, unlike physics and chemistry, biology, "proceeding as it does from the only stable basis, the sense-qualities" (Uexküll 1926a, 361), allows us through experimental observation to know the specific conditions of animal perception, their functional circles, their *Umwelten* and, in this way, brings us closer to the enigma of vitality. But, although it must postulate its existence, it cannot give us a conclusive answer regarding the nature of impulses that nest in the will (Uexküll 1926a, 361).

This double reference to Kant allows us to get a glimpse of the productive tension within which Uexküll's inquiries develop. This tension between an epistemological level and a metaphysical one, between the knowable and the conceivable, between the structural approach and the genetic approach, is clearly shown in *Theoretical Biology*, where the author not only systematizes his concept of conformity to a plan but also presents a full theory on impulse. Although retrieving some classic motifs of (neo-)vitalism, Uexküll's metaphysics is mainly developed against the backdrop of a new interest in the external life of animals and the problem of action, giving rise to original concepts such as "functional circle" [*Funktionskreis*] and "sequence of impulses" [*Impulsfolge*]. Both layers of Uexküll's inquiries are shown in his approach to species. On the one hand, these appear as immutable realities (Uexküll 1920, 80), more perfect than each individual (Uexküll 1926a, 242) and structured according to a plan; on the other hand, their genesis is unknowable, and this has become an "insoluble" problem for science (Uexküll 1913, 36) opening the door to metaphysics. Likewise, at an ontogenetic and individual levels, Uexküll claims that animals appear as independent subjects, structured according to a plan – that is, endowed with certain rules and *a priori* forms that are expressed in their building-plans – and, correlatively, that their genesis and the direction of their behavior can only be grasped in reference to a super-mechanical factor (i.e., the sequence of impulses), which performs the rules of formation, operates as a categorical imperative at a behavioral level, and has "the power to convert an extra-spatial and extra-temporal plan into a physical phenomenon" (Uexküll 1926a, 216).[6] This twofold theoretical horizon (structural method and vitalist demand, conformity to a plan and *Naturfaktor*, being and becoming, etc.) frames Uexküll's theoretical biology as a form of structural vital-ism. Compared to (neo-)vitalism, it clearly shows originality as it investigates the performance of rules and the super-mechanical factors not only at the level of embryogenesis, ontogenesis, and regeneration processes but also and fundamen-tally at the level of behavior. What is at stake for Uexküll is to be able to think the vitality of the adult animal, and in so doing, he produces a novel theory of actions[7] and indicates explicitly that his intention is to observe how sequences of impulses operate not only in the genesis but also in the practical functioning of the fully formed animal, that is, in direction (Uexküll 1926a, 271, 273f.).

Said theory of actions is supported by the important concept of "functional cir-cle," with which Uexküll thematizes the exterior life of animals and the way in which the effects of the actions performed (or in the process of being performed) impact on (self-)perception, direction, and behavior control. He thereby conceptu-alizes the process through which the animal builds its *Umwelt* and presents, against the physical-chemical causality model and the notion of conditioned reflexes, a model of retroactive causality which dynamically articulates two previously antag-onistic elements: the fixity of specific building-plans and the vital constructivism of animals when territorializing and constituting their *Umwelten*.[8] The notion of a functional circle thus enables us to think – at a behavioral level – a circular causality and a practical interaction between the empirical and the transcendental, between the living individual and the specific conditions, between the effect world and the perception world. This model also enables the conceptualization in higher animals of plastic actions, actions based on experience and controlled actions that, on the basis of an impulsive becoming, are capable of producing respectively new schemes of perception, new rules of action and differential guidelines for the self-regulation of behavior in the density of the present time.[9]

From a historical and intellectual point of view, the notion of a functional circle has attracted great interest inasmuch as both the retroactive causality and the oper-ative scheme it entails can be said isomorphic to the cybernetic notion of negative feedback (Lagerspetz 2001; Emmeche 2001; Rüting 2004; see Köchy, Chapter 3, in this volume). Likewise, such a notion suggests a displacement in Uexküll's problematic areas from perception toward behavior. An analogous shift has been developed, incidentally, also by *Gestaltpsychologie*. As a consequence, the notion of counter-world is abandoned and the actions of higher animals and their results can be seen gaining increasingly more ground over the *a priori* perception world and taking over the direction and control of behavior while enriching the under-standing of the perception world with schemes and rules.

Thanks to arrangements by the physiologist Otto Cohnheim, Uexküll is invited to the prestigious International Congress of Physiology held in Edinburgh in

1923, and between 1924 and 1925, he is appointed as "scientific assistant collabo-rator" at the University of Hamburg and becomes responsible for the reconstruc-tion of the Aquarium of the Hamburg Zoological Garden. With tenacity and great organizational capacity, Uexküll manages to equip and revitalize the aquarium: he enriches it with diverse marine animals, obtains financing from private spon-sors, and manages to build a research lab where there once was a kiosk (Brentari 2015, 34). As a result of this intense work, in 1926 Uexküll founds the Institut für Umweltforschung (Institute of Environmental Research), which is active until 1934 under his direction and becomes a prolific research center.

During this period, Uexküll regularly publishes different essays, reviews, and articles that are not confined to the exposition of his theory of *Umwelten*, nor are they circumscribed to philosophical-biological topics, as they also cover cultural, moral, and political issues. Some of these texts are presented in artistic forms such as theatrical dialogs. An example is provided by the play *God or Gorilla* [*Gott oder Gorilla*, Uexküll 1926b], in which Uexküll reevaluates the moral experi-ence withheld by Christianity and defends a sort of romantic sacredness against the threats of a demoniac "gorilla-machine," a false idol, which, in the context of the Weimar Republic, symbolizes all the existential evils produced by modernity, inasmuch as they dissolve spiritual values and replace them with the sanctifi-cation of materialistic science, technology, and market mechanisms (Harrington 1996, 63–65).

In 1934 Uexküll publishes *A Foray into the Worlds of Animals and Humans*, in which he presents his theory of *Umwelten* to the public, updating it with new notions and incorporating the results of a series of experimental investigations. In the book's prologue, Uexküll already makes it clear that his concept of *Umwelt* is not reduced to the phenomenon of perception; it also includes the field of animal action through the notion of a functional circle:

> [E]verything a subject perceives belongs to its *perception world*, and every-thing it produces, to its *effect world*. These two worlds, of perception and production of effects, form one closed unit, the environment [*Umwelt*].
>
> (Uexküll 2010, 42)

Uexküll's enhanced interest for the problem of behavior finds in this book new developments through the notions of effect image [*Wirkbild*] and effect space [*Wirkraum*], as well as with the ideas of familiar path [*Der bekannte Weg*], home [*Heim*], and territory [*Heimat*]. In short, unlike what was the case in previous stages of his inquiries, from the beginning of the 1920s, the preponderance of the perception world in the notion of *Umwelt* is progressively attenuated and leads to the thematization of the effect world and the exterior life of animals. This also explains the progressive transition in Uexküll's texts from physiology to theoreti-cal biology. Likewise, a shift can be detected in Uexküll's increasing interest for superior animals. In this respect, one should mention the fact that in the early 1930s, he gets acquainted with Konrad Lorenz's experimental studies on jack-daws, grey geese, and starlings (Uexküll 2010, 43, 111–113, 176f.).

In his book from 1934, moreover, Uexküll includes an account of the several affective dimensions one can attach to animal behavior, in order to understand different subjective dispositions. He thereby refers to the color [*Färbung*] of perceptive marks with a certain tone [*Ton*], as to explain how different actions are possible with regard to the same stimulus. According to Uexküll, the same perception mark can therefore assume different meanings depending on the mood [*Stimmung*] of the living being, and this differential motivation prevents us from thinking behavior as a chain of tropisms. Here Uexküll openly criticizes Jacques Loeb's mechanistic perspective (Loeb 1906), which assumes that behavior is based on blind reactions and reflexes produced by physico-chemical stimuli devoid of biological meaning.

The publication of *A Foray into the Worlds of Animals and Humans* makes waves, and this text becomes in time Uexküll's most-read book in Germany and his most-translated work. One should also add that the book is dedicated to Otto Cohnheim, who "lost his appointment as a university professor because of racial politics" (G. von Uexküll 1964, 187). This allows us to approach the topic of Uexküll's ambivalent relationship with Nazism. In principle, due to his conservative worldview, his criticisms of the Weimar Republic, his aversion to Bolshevik communism, and his resentment with Western democracies after the German defeat in the Great War, among other reasons, Uexküll regarded Hitler's ascension to power with moderate optimism, and in the second revised edition of *State Biology* (1933), he hopes the new chancellor can save Germany from the international greed of capitalist forces (Uexküll 1933, 78; Rüting 2004, 42). This positive expectation can be linked to the deep friendship Uexküll maintained with Houston Stewart Chamberlain,[10] as well as to the Nazi regime's receptivity to the biological interpretations of sociopolitical phenomena (Brentari 2015, 38).

However, these affinities will soon erode, and a series of disagreements will keep Uexküll away from National Socialist politics and ideology. Already in May 1933, the biologist sends a letter to Chamberlain's widow in which he expresses his dismay over the use Nazism had made of her late husband's ideas to justify the persecution of Jews (G. von Uexküll 1964, 173; Schmidt 1975, 127). He also objects to the unfair dismissal of prestigious Jewish intellectuals and scientists, strongly rejects the race theory held by the regime – which he describes as "the worst form of barbarism" – and asks her to persuade Hitler to stop this policy of segregation (G. von Uexküll 1964, 173; Mildenberger 2007, 158f.). His criticism of racist ideology also appears in a letter sent to Lothar G. Tirala – a former student and assistant who becomes a fervent supporter of the regime and is appointed professor at the Institute of Racial Hygiene in Munich – in which Uexküll questions the simplicity of Nazism's racial doctrine and its relapse into a form of "miserable materialism" which, by exclusively considering the hereditary material, ignores the building-plan of organisms (G. von Uexküll 1964, 169; Brentari 2015, 39f.). Both letters show that the criticism of the regime's racist theory is not only ethical-political but also epistemological. According to Uexküll, the biological ground to this theory is weak, it contradicts Mendel's discoveries and is based on a wrong diagnosis. The core problem would not rest on the purity of blood and its hereditary transmission but the rootlessness of individuals and families, the disorganized mixture of races that have particular *Umwelten*, and the consequent degeneration of the people in an undifferentiated mass. His defense of the theory of *Umwelten*, on the other hand, is accompanied by a growing concern regarding the budget cuts that could affect the Institut für Umweltforschung and any biological research that did not fit in the ideology of Nazism.[11]

Although the figure of Uexküll is subjected to reappropriation and vindication by some Nazi sectors (Harrington 1996, 68f.), there are significant episodes that reveal his tensions with the regime. In 1933, Joseph Goebbels publishes in the official Nazi newspaper, *Völkischer Beobachter*, a review in which he strongly questions Uexküll for his conference, "The Dog's Olfactory Field" [*Das Duftfeld des Hundes*]. The title of the article signed by Hitler's minister of propaganda is eloquent: "Excremental excesses of a German professor" [*Kötereien eines deutschen Professors*]; in its content, Goebbels criticizes the fact that the professors of German universities are dealing with silly and insignificant subjects instead of assuming "the imperative of the now" and contributing to the political-moral development of the German people (Brentari 2015, 39). In 1934, Uexküll is invited by the Academy of German Law – managed by Hans Frank, Hitler's lawyer – to a cultural event at Nietzsche's house in Weimar. With Elisabeth Nietzsche as host and to the presence of renowned Nazi intellectuals, Uexküll gives a lecture in which he criticizes the Nazi restrictions on the university, reaffirms the value of academic freedom, and states,

> Nowadays, as a criterion of vitality and skill we are expected to return the blow we receive. This criterion, however, as biology teaches us, only applies to effector organs. The eye that is punched can only become blind, but cannot punch back. And the task of universities is precisely to be the eyes of the state.
>
> (G. von Uexküll 1964, 175, my transl.)

According to Gudrun von Uexküll (1964, 175), after this statement, the conference was interrupted by the audience and Uexküll left. He was afterward subjected to constant vigilance by the party. In 1936, in fact,

> [h]is book of personal recollections *Niegeschaute Welten* [*Worlds Never Seen*], which contained words of appreciation for the Russian Jews and the Baroness Rothschild [...], was officially banned from being displayed in the windows of bookshops.
>
> (Brentari 2015, 42)

5 The problem of solipsism and the idea of nature (1938–1944)

After retiring in April 1936 and affected by health problems, Uexküll and his family move to the island of Capri, where the biologist will spend his last years. In this period, two themes stand out in his publications. Both questions were already outlined in *A Foray into the Worlds of Animals and Humans* (1934): (a) the problem of solipsism ensuing from his theory, in other words, the idea that each animal subject lives in a "bubble" that represents its *Umwelt* and contains all the signs accessible to the subject, and (b) the metaphysical idea of Nature as an instance prior to *Umwelten*. In his book from 1934, making reference to Lorenz's research on birds, Uexküll thematizes some apriorities in relation to the social relationships of animals and the development of companionship relations between subjects of heterogeneous species, explaining them in terms of a confusion of perception images (Uexküll 2010, 108–113). Finally, in the conclusions to the same book, Uexküll postulates the existence of a third order of subjectivity, Nature, which can be distinguished from the second order of subjectivity (i.e., the *Umwelten* of each species) and operates as an instance of coordination and harmonic articulation of worlds (Uexküll 2010, 135). Thus, there would be three interrelated levels of subjectivity (individuals, species, and Nature).

The problem of solipsism entailed by the Uexküllian concept of *Umwelt* was pointed out not only by Lorenz (Brentari 2015, 217–224; see also Köchy, Chapter 3, in this volume) but also by many of the philosophical interpretations which have emerged since the end of the 1920s: Max Scheler emphasizes the closed nature of animals' *Umwelten* and contrasts them with the human ability to "open themselves to the world" (Scheler 1991, 40; see also Becker, Chapter 4, in this volume), Martin Heidegger points out that the animal is "poor in world" and lives in a "captivation" state while man is "world-forming" (Heidegger 1995, 177, 258f., see Michelini, Chapter 7, in this volume), Kurt Goldstein questions the fixed and static quality of *Umwelten* and claims that what should be investigated further is the genesis of *Umwelten* through a set of not-predetermined activities and performances (Goldstein 1995, 85, see Ostachuck, Chapter 9, in this volume).

As Brentari (2015, 164–168) shows, Uexküll explicitly addresses the problem of solipsism and that of the articulation between heterogeneous *Umwelten* in two philosophical dialogs from the late 1930s (*The Immortal Spirit of Nature* [*Der unsterbliche Geist in der Natur*, Uexküll 1938, 1946, 1947] and *The Omnipotent Life* [*Das allmächtige Leben*, Uexküll 1950]) and provides an answer to the question by applying Leibniz's monadological scheme. In the first work, Uexküll compares the multiplicity of *Umwelten* that populate Nature to the drops of dew that bathe a field in the morning and points out that "each of these myriads of drops mirrors all the world with the sun, the mountains, the forests and the shrubs, a magical world within itself." He then adds that if you imagine that

[e]ach one of these innumerable drops does not only shine in the diversity of the shimmering colors, but also possesses its own subjective tone, the one that distinguishes all living beings, then you will understand that the theory of the environment has nothing to do with the silly solipsism.

(Uexküll 1938, 47f., my transl.)

Furthermore, in order to provoke a vivid intuition of the preestablished harmony that Nature mobilizes, Uexküll completes his argument by presenting an analogy to a great theatrical play in which the roles of each character are predetermined, as the functional circles of each species, but the actors that assume and personify these roles change from generation to generation. In other words, there are always new living subjects executing and interpreting the play, and by doing so, they add their singular tone.

The thematization of the interactions between living worlds and the order of Nature as a third order of subjectivity is systematically developed by Uexküll's last theoretical book. With the publication in 1940 of *Theory of Meaning* [*Bedeutungslehre*, Uexküll 1982], Uexküll goes beyond his theory of *Umwelten* and focuses on the analysis of the relational dynamics of Nature through the concept of meaning. In this respect, he calls upon one of his favorite analogies, that of musical composition, and points out that the task of biology would be to "write the score of nature" (Uexküll 1982, 62). Along the same lines, while providing an account of natural articulations, he presents the relationships between "meaning receptors" and "meaning-carriers" in terms of point and counterpoint. This new approach aims to achieve a "theory of natural composition" (or "technique of nature"), in which the relations of meaning that link the vital forms to one another are more important than the quality of the forms themselves or the peculiarity of their *Umwelten*. Although the latter do not stop playing a role in the theory, the adopted point of view is that of Nature and the intention is to be able to unravel its interconnections and networks (Uexküll 1982, 43). In so doing, Uexküll's moves the main focus from the perception marks or the rules of action which define a living world to the "rules of common meaning" which link two or more specific worlds. In other words, he finally prompts the reader to investigate the common link – the code – that enables tick and mammal, or bat and butterfly, or fly and spider, to be elements of a meaningful process that encompasses them. Correlatively, within the framework of the signifier score of Nature, the notion of meaning becomes the "natural factor" that articulates heterogeneous living worlds and, from this new perspective, the functional circles become "meaning-circles, whose task lies in the utilization of the meaning-carriers" (Uexküll 1982, 36), and perception and action become "the life tasks of the animal subject acting as a meaning-receiver" (Uexküll 1982, 74).

After a long intellectual adventure, Uexküll dies on the island of Capri on July 24, 1944.

Notes

1 A few years later, Uexküll will disavow the mechanistic remnants present in the article and the fact that sense organs were being considered as mechanical devices (G. von Uexküll 1964, 163f.).

2 In *Die neuen Ziele der Biologie* (1910a) Uexküll expresses this point of view through a philosophical-biological dialog: "If today three naturalists walked outdoors in the open air, it could happen that one of them was an Aristotelian, the second a Platonist, and the third a Kantian. 'To live is to bring an end in itself' the Aristotelian will say. The disciple of Plato will let his gaze slide serenely through the peaks of the distant mountains and will answer: 'Yes, a non-temporal end'. And the disciple of Kant will nod silently" (Uexküll 1913, 51, my transl.). On the other hand, with regard to the idea

of "spatialisation of teleology," it is important to point out that forms are not necessarily spatial but can also be taken in their temporal deployment, that is, as regular behavioral patterns or as melodies. These forms, however, order a sequence of movements according to a "nontemporal end" and, in this sense, suppose a subordination of the order of succession to the order of simultaneity. As the famous example of Christian von Ehrenfels shows, a melody is not a mere chronological succession of sounds but a structural organization of concatenated notes, and this implies that the process depends on the structure (and not the other way around). In fact, Uexküll presents the idea of melody as a sequence ordered according to a plan and not according to a temporal end.

3 Uexküll's argument on this issue is analogous to the one developed by *Gestaltpsychologie*, where perception is not explained according to the cumulative reception of stimuli, or "atoms of sensation," but as the immediate capture of forms and significant structures which are imposed on attention.

4 From an archaeological point of view, one can find a connection between this biological interpretation of the transcendental and the phenomenological interpretations that – based on the concept of world – will unfold in German philosophy (i.e., there is a process in which the specific and invariable transcendental becomes intersubjective and temporalized).

5 In the inner world, Uexküll distinguishes a perception organ (connected to the receptors), an effect organ (connected to effectors) and a central organ (or conductor) that processes and circulates waves of excitation between the perception and effect organs. The inner world translates into a language of nervous signs the stimuli that affect the receptors and thus constitutes a network of nervous connections that is not confused with the events and objective entities of the outside world (Uexküll 1909, 59). While lower animals have a simple nerve network, higher animals have a greater capacity to process and articulate the waves of excitations and, differentiating different kinds of receptor networks, constitute schemes that allow them to perceive forms (i.e., collections or series of excitations). Schematization gives rise to what Uexküll calls "counter-world," where "there is only a selection of those forms that are important for the animal's life" (Uexküll 1913, 231, my transl.). Thus, the greater an animal's counter-world, the more schemes will populate its perception organ, and the more forms its perception world will have (Uexküll 1913, 231f.).

6 Uexküll distinguishes three "sequences of impulses" (or, as he also calls them, melodies) that are irreducible to causal analyses and that manifest the execution of certain rules according to a plan. There are melodies of formation (related to the action of genes), functional melodies (related to the reflex and instinctive actions), and, finally, sequences of impulses that, through retroactive processes, produce differential rules of action and govern the direction of behavior.

7 In *Theoretical Biology*, Uexküll presents a theory of actions by differentiating seven types (reflex action, form action, instinctive action, plastic action, action based on experience, controlled action, receptor action), and distinguishing them based on the differential relationship that the impulsive rule of direction maintains with the receptor organ and/or effector organ (Uexküll 1926a, 271–280). The common motive underlying this plurality of actions is emphasizing how insufficient the mechanistic physiological explanations which analyze living beings as objects or machines are, thus denying their vitality and subjective activity (this one, according to Uexküll, is expressed through autonomous rules differing from the "mechanical structure" of the body they are part of, noncontingent rules operating as categorical imperatives and that cannot be confused with psychological ends). Regarding Uexküll's theory of actions, see also Brentari (2015, 115–121).

8 Uexküll points out that the exterior life of animals is organized according to four major functional circles: (a) the circle of medium (which refers, fundamentally, to the signals an animal receives, for example, when passing from an aquatic environment to an

aerial one, and which urge it to return to its natural environment), (b) that of food, (c) that of enemies, and (d) that of sexuality and reproduction (Uexküll 1926a, 127–129).

9 It should be noted that, according to Uexküll, the variation potentiality in these types of actions, and the possibility of generating differential forms of behavior, is a purely individual phenomenon and does not modify the transcendental conditions of the species (Uexküll 1926a, 284).

10 The correspondence between Uexküll and Chamberlain has been analyzed by Schmidt (1975) and occupies a prominent place in Harrington's (1996) interpretation of the Uexküllian worldview.

11 In a letter sent to Driesch in 1934, Uexküll points out that "[t]here is danger that we are going to fall victim to the new race research" (cited in Harrington 1996, 70).

References

Brentari, Carlo (2015) *Jakob von Uexküll: The Discovery of the Umwelt between Biosemiotics and Theoretical Biology*. Dordrecht/Heidelberg/New York/London: Springer.

Canguilhem, Georges (1971) [1952] *La connaissance de la vie*. Paris: Vrin.

Canguilhem, Georges (2000) [1977] *Idéologie et rationalité dans l'histoire des sciences de la vie. Nouvelles études d'histoire et de philosophie des sciences*. Paris: Vrin.

Cassirer, Ernst (1950) *The Problem of Knowledge: Philosophy, Science, and History since Hegel*. Translated by William H. Woglom and Charles W. Hendel. New Haven: Yale University Press.

Cassirer, Ernst (1961) [1942] *The Logic of the Humanities*. Translated by Clarence Smith Howe. New Haven: Yale University Press.

Emmeche, Claus (2001) 'Does a robot have an Umwelt? Reflections on the qualitative biosemiotics of Jakob von Uexküll'. *Semiotica* 134 (1/4), 653–693.

Goldstein, Kurt (1995) [1934] *The Organism: A Holistic Approach to Biology Derived from Pathological Data in Man*. New York: Zone Books.

Harrington, Anne (1996) *Reenchanted Science: Holism in German Culture from Wilhelm II to Hitler*. Princeton: Princeton University Press.

Heidegger, Martin (1995) [1929–1930] *The Fundamental Concepts of Metaphysics: World, Finitude, Solitude*. Translated by William McNeill and Nicholas Walker. Bloomington and Indianapolis: Indiana University Press.

Heredia, Juan Manuel (2011) 'Etología animal, ontología y biopolítica en Jakob von Uexküll'. *Filosofia e História da Biologia* 6 (1), 69–86.

Jacob, François (1973) [1970] *The Logic of Life: A History of Heredity*. Translated by Betti E. Spillmann. New York: Pantheon Books.

Kull, Kalevi (2001) 'Jakob von Uexküll: An introduction'. *Semiotica* 134 (1/4), 1–59.

Lagerspetz, Kari Y. H. (2001) 'Jakob von Uexküll and the origins of cybernetics'. *Semiotica* 134 (1/4), 643–651.

Loeb, Jacques (1906) *The Dynamics of Living Matter*. New York/London: Columbia University Press.

Le Bot, Jean-Michel (2016) 'Renouveler le regard sur les mondes animaux. De Jakob von Uexküll a Jean Gagnepain'. *Tétralogiques* 21, 195–218.

Mildenberger, Florian (2007) *Umwelt als Vision: Leben und Werk Jakob von Uexkülls (1866–1944)*. Stuttgart: Steiner.

Palti, Elías (2004) 'The "return of the subject" as a historico-intellectual problem'. *History and Theory* 43 (1), 57–82.

Palti, Elías (2017) *An Archaeology of the Political: Regimes of Power from the Seventeenth Century to the Present*. Columbia: Columbia University Press.

Rüting, Torsten (2004) 'History and significance of Jakob von Uexküll and of his institute in Hamburg'. *Sign Systems Studies* 32 (1/2), 35–72.

Scheler, Max (1991) [1928] *Die Stellung des Menschen im Kosmos*. Bonn: Bouvier.

Schmidt, Jutta (1975) 'Jakob von Uexküll und Houston Stewart Chamberlain: Ein Briefwechsel in Auszügen'. *Medizinhistorisches Journal* 10, 121–129.

Uexküll, Gudrun von (1964) *Jakob von Uexküll. Seine Welt und seine Umwelt*. Hamburg: Wegner.

Uexküll, Jakob von (1902) 'Im Kampfe um die Tierseele'. *Ergebnisse der Physiologie* 1 (2), 24.

Uexküll, Jakob von (1905) *Leitfaden in das Studium der experimentellen Biologie der Wassertiere*. Wiesbaden: Bergmann.

Uexküll, Jakob von (1907) 'Die Umrisse einer kommenden Weltanschauung'. *Die neue Rundschau* 18, 641–661.

Uexküll, Jakob von (1908a). 'Die neuen Fragen in der experimentellen Biologie'. *Rivista di Scienza 'Scientia'* 7 (4), 17.

Uexküll, Jakob von (1909) *Umwelt und Innenwelt der Tiere* (1st ed.). Berlin: Springer.

Uexküll, Jakob von (1910a) 'Die neuen Ziele der Biologie'. *Baltische Monatsschrift* 69 (4), 225–239.

Uexküll, Jakob von (1910b) 'Die Umwelt'. *Die neue Rundschau* 21, 638–649.

Uexküll, Jakob von (1910c) 'Mendelismus'. *Die neue Rundschau* 21, 1589–1596.

Uexküll, Jakob von (1913) *Bausteine zu einer biologischen Weltanschauung. Gesammelte Aufsätze*. München: Bruckmann.

Uexküll, Jakob von (1915) 'Volk und Staat'. *Die neue Rundschau* 26 (1), 53–66.

Uexküll, Jakob von (1917) 'Darwin und die englische Moral'. *Deutsche Rundschau* 173, 215–242.

Uexküll, Jakob von (1920) *Biologische Briefe an eine Dame*. Berlin: Verlag von Gebrüder Paetel.

Uexküll, Jakob von (1926a) [1920] *Theoretical Biology*. Translated by Doris L. Mackinnon. London: K. Paul, Trench, Trubner & Co.; New York: Harcourt, Brace & Company.

Uexküll, Jakob von (1926b) 'Gott oder Gorilla'. *Deutsche Rundschau* 208, 232–242.

Uexküll, Jakob von (1933) *Staatsbiologie: Anatomie-Physiologie-Pathologie des Staates*. Hamburg: Hanseatische Verlagsanstalt.

Uexküll, Jakob von (1936) *Niegeschaute Welten. Die Umwelten meiner Freunde. Ein Erinnerungsbuch*. Berlin: Fischer.

Uexküll, Jakob von (1938) *Der unsterbliche Geist in der Natur. Gespräche*. Hamburg: Wegner.

Uexküll, Jakob von (1940) *Bedeutungslehre*. Leipzig: Verlag von J.A. Barth.

Uexküll, Jakob von (1946) *Der unsterbliche Geist in der Natur: Gespräche (4–8)*. Hamburg: Wegner.

Uexküll, Jakob von (1947) *Der unsterbliche Geist in der Natur: Gespräche (9–18)*. Hamburg: Wegner.

Uexküll, Jakob von (1950) *Das allmächtige Leben*. Hamburg: Wegner.

Uexküll, Jakob von (1982) [1940] 'Theory of meaning'. *Semiotica* 42 (1), 25–82.

Uexküll, Jakob von (2010) [1934, 1940] *A Foray into the Worlds of Animals and Humans with a Theory of Meaning*. Translated by Joseph D. O'Neil. Minneapolis/London: University of Minneapolis Press.

2 Kantian ticks, Uexküllian melodies, and the transformation of transcendental philosophy

Maurizio Esposito

1 Introduction

In one of his famous short essays, the Argentinian writer Jorge Luis Borges thought over the notion of "precursor" (Borges 1999). While considering a list of possible forerunners anticipating a great writer, Borges observed, we rarely realize that it is the author him- or herself who generates his or her ancestors: "every writer creates his own precursors. His work modifies our conception of the past, as it will modify the future" (Borges 1999, 365). In the pages of Aristotle's philosophical works, we may easily spot the revised ideas of Plato or Empedocles. There are different Platos and Aristotles in the writings of Averroes or St. Thomas. We know that Marx reframed Hegel's philosophy, and similarly, we may say that Uexküll, like Schopenhauer before him, revamped Kant (see Braga 1964; Guidetti 2013; Brentari 2015). From a Borgesian perspective, we might conclude that some of the best evidence for the success of an author is the number of his or her. reinventions, and from this perspective, Kant has been one of the most success-ful "precursors" in the history of Western thought. The advantage of adopting Borges's apparently paradoxical view is that, in inverting the historical relations of cause and effect, we might disclose unexpected hermeneutical viewpoints. Indeed, while it is certainly instructive to assess how, and to what extent, Kant changed Uexküll's mind, it might be even more revealing to understand how and why the latter enlisted the former to serve his own agenda. In other words, instead of tracing Kant's influence on Uexküll's biology, we may explore how Uexküll manufactured a new transcendental philosophy fit for his biological thought.

In order to follow Borges's suggestion seriously, we might start asking two related questions: Why did Uexküll choose Kant as one of his privileged "precursors"? and How did Uexküll transform and betray Kant's philosophy? Because Uexküll was not the only biologist of the last century who used and reframed Kant's ideas, we need to contextualize the questions within two larger questions: Why did some late 19th- and early 20th-century biologists enlist Kant among their favorite "precursors"? and What kind of Kantian contents did they usually pick up? Especially these two last questions are particularly important insofar as they help us in situating Uexküll within a heterogeneous tradition, which I have termed elsewhere "romantic biology" (Esposito 2016). Uexküll shared some fundamental principles of this tradition: among others, the rejection of mechanistic and neo-Darwinian ideas, the defense of purposiveness and wholeness in the organic world, and the idea that organisms cannot be reduced to inorganic processes. However, there is one fundamental element differentiating Uexküll from all other organicists or vitalists mainly active during the first decades of the 20th century: his distinctive devotion to Kant's *Critique of Pure Reason*. This detail is an important hermeneutical clue and should not be overlooked. After all, while Kant explicitly mentioned life sciences in the third *Critique*, the topic is largely absent in the first.[1]

Indeed, it is well known that most antimechanist thinkers appealed to Kant's *Critique of Judgment* to defuse materialist and reductionist views in biology (see Lenoir 1982). They strategically highlighted Kant's sophisticated discussion of epigenesis, the distinctiveness of *Naturzweck*, and the nature of teleological judgment in order to demonstrate the uniqueness and irreducibility of the life sciences (Esposito 2016). Although Uexküll found these discussions congenial to his overall agenda, he believed that Kant's fundamental idea, which was to transform life sciences in the 20th century, was not the supposed purposiveness of living beings but the glorious spontaneity of the transcendental subject.[2] Thus, on the one hand, biology could be related to Kant's transcendental philosophy insofar as it was a science studying the laws regulating animal subjectivity in relation to an unknowable noumenal world. On the other hand, as I argue in this chapter, organic purposiveness – what Uexküll termed conformity to a plan – could substantiate the spontaneous activity of animal subjects. Purposiveness in biology was therefore not the starting point of Uexküll's new biology, but the justification of subjectivity itself. Not surprisingly, in his seminal work, *Theoretical Biology*, he began with a transcendental subject and ended with the unavoidable presence of teleology in the living world. The idea of subjectivity and the inner teleological character of the organism constituted the two fundamental pillars sustaining Uexküll's overall anti-Darwinian program. Both subjectivity and teleology defined what biology as a discipline should be: the science of subjective entities purposively anchored to the world.

In this chapter, I first track down some of the conceptual elements that Kant and Uexküll might have shared and that make the former a natural "precursor" of the latter. In the first section, I focus my attention on the following question: Why did Uexküll's biology need Kant's first *Critique*? In the second section, I show two things: (1) the revealing and constructive distortions that Uexküll needed to make in order to squeeze Kant's philosophy into his own proposal and (2) the deep and essential relation established by Uexküll's theoretical biology between Kant's notions of subjectivity and teleology. To sum up, throughout the chapter, I argue that Uexküll strategically reshaped Kant's transcendental philosophy in order to make it congenial to his anti-Darwinian research program.

2 Why Kant?

It is widely known that behind Kant's *Critique of Pure Reason* there was a battery of fundamental epistemological issues that had tested the greatest European minds throughout the 18th century, that is, the source, origins, and limits of human knowledge. In this respect, Kant himself sets out to solve one of the most difficult problems of modern philosophy: to reconcile the rationalist with the empiricist foundation of knowledge. Gilles Deleuze famously defined Kant's solution, the first *Critique*, as an asphyxiating masterpiece hiding an astonishing architecture of new concepts (Deleuze 1978). As we will see later on, Deleuze's knowledge-ability about both Kant and Uexküll makes him an interesting interlocutor for our discussion.[3] However, what I want to highlight now is the following question: What kind of appeal did such an asphyxiating masterpiece have for a young biologist more fascinated by the physiology of the frog than by the *a priori* conditions of experience? After all, Kant's questions were not Uexküll's. Kant wanted to put scientific knowledge on a firm foundation, emphasizing the role of the subject in determining the structure of experience; Uexküll wanted to replace behaviorist, mechanistic and neo-Darwinian biology with a much more sophisticated alternative. So, what had early 20th-century biologists to do with Kant's transcendental philosophy? Perhaps we might encounter a plausible answer if we consider the intellectual background against which both Kant and Uexküll moved. In fact, despite the temporal distance, I think that an interesting point of convergence can be found in their reaction to one particular issue: the supposed passivity of the experiencing subject. We know that Kant was awoken from the dogmatic slumber of the rationalist philosophical haze of northern Prussia thanks to Hume's corrosive skepticism. Uexküll might have been similarly awoken from the dogmatic torpor of behaviorist, mechanistic, and neo-Darwinian biologies that promised much more than they could give. But, despite the different reasons for their real or alleged awakenings, we can observe that both shared the same solution: subjective spontaneity. Both Kant and Uexküll realized that the way out of empirical skepticism and behaviorism was questioning the passivity of experience. Subjects are not inert clay tablets waiting to be carved out by external forces; they are rather active and plastic entities producing a large chunk of the world, which, in the traditional picture, was supposed to spring from a series of meaningless outward stimuli.

Yet, even though they may have identified the same solution, they differed in one fundamental respect: Kant's way out of skepticism was framed uniquely for human subjects. The transcendental categories ordering intuitions, making the manifold one, were the crowning achievement of the human mind; Uexküll, instead, aimed to extend Kantian subjectivity to the entire organic world. Ticks or scallops, crabs or octopuses, moles or dogs, all had to be conceived as a spontaneous *I*, although each one perceived the world according their peculiar subjective endowments. Of course, Uexküll knew that it would have been utterly absurd to bestow all the heavy intellectual structure of the Kantian categories on the simple nervous system of a scallop. Clams, mollusks, or cats perceive the surrounding world according to the particular structure of their nervous system, not through pure concepts of understanding. Not surprisingly, Uexküll lamented that Kant had left out from his system the *a priori* elements of basic perception, considering only the relation between intuition and intellect. As he argued in one of the most idiosyncratic texts of theoretical biology written in the 20th century, before any single piece of knowledge can be received, its form must be already prepared in the mind. But these forms change in the course of experience. Kant did not concern himself with those forms of knowledge which are of such great importance biologically; he restricted himself to those which must have preceded all experience whatsoever.

(Uexküll 1926, xvi)

In other words, what about the sentient body? And what about the multiple relations between different subjects (organisms) and objects within their environments? Uexküll's biology aimed to fill the gaps Kant had left so as to extend Kant's transcendental philosophy to the entire living realm.

Yet, Uexküll was not the first in noticing those gaps. I have no evidence that Uexküll was aware of Schopenhauer's criticism of Kantian philosophy in Book One (Appendix) of his *The World as Will and Representation*. There, Schopenhauer listed a number of important mistakes that Kant had committed in his first *Critique*. And the list, in fact, includes one fundamental remark that is directly relevant to Uexküll's concerns; Kant – Schopenhauer argued – "entirely neglects the rest of knowledge of perception in which the world lies before us, and sticks solely to abstract thinking" (Schopenhauer 1966, Vol. 1, 431). What troubled Schopenhauer was the missing distinction between intuitive and perceptual knowledge versus discursive and abstract knowledge. Kant had famously claimed,

If I take all thinking (through categories) away from an empirical cognition, then no cognition of any object at all remains; for through mere intuition nothing at all is thought, and that this affection of sensibility is in me does not constitute any relation of such representation to any object at all.

(Kant 1998, B309, 349f.)

For Schopenhauer, if any kind of knowledge presupposed the use of abstract concepts, nonhuman animals could have no knowledge about things and happenings in the world. Animals would only have chaotic sensations without any possibility of organizing these sensations into structured and meaningful wholes. Of course, this was not the critique of Uexküll himself (who nevertheless asks in the conclusion of his *Theoretical Biology* why Kant did not write a Critique of Will-Power). However, as we will see, Schopenhauer's solution – that is, the distinction between intuitive and discursive forms of knowledge – was not very different from Uexküll's upgrading of Kant's transcendental philosophy.

To sum up, Uexküll claimed Kant as his principal "precursor," in spite of his different research program, because Kant was one of the first philosophers, and the most important one, to have seriously considered the subject as the center of initiatives and not as a recording black box. Yet, Kant's conceptual tools for understanding human mental activity were forged for human thinking subjects, not for any organism dealing with the environment. Uexküll had therefore to face a tremendous problem: If animals have no concepts, how do they synthesize the manifold perceptions into a unified whole? The answer to this question led to a good part, if not all, of Uexküll's ideas, including the *Umwelt*, the functional circle, and the conformity to a plan. Indeed, between the pure indeterminate spontaneous *I* and the environment, Uexküll hypothesized the existence of a whole set of processes that guaranteed the conformity between the subject and a portion of the experienced world without positing abstract concepts. For Kant, of course, the relation between the *I* and the phenomenic world was firmly established by the austere architecture of the mind, which harbored the concepts making experience meaningful. For Uexküll, in contrast, the mysterious link between subjects and worlds was established through different kinds of functional feedback interactions, which anchored the inner world of the organism to particular sections of the external world. But before expanding on the latter point, there is another fundamental element we need to elucidate, an element that clarifies the relation between Kant's transcendental philosophy and Uexküll's transcendental biology: the nature of perception and phenomena.

Here the Deleuzian interpretation of Kant's first *Critique* can be very helpful. In his first lecture given in Vincennes in 1978 on Kant's transcendental philosophy, Deleuze declared that one of the fundamental novelties of Kantian epistemology was the redefinition of the notion of "phenomenon," which, in turn, changed the meaning of the traditional dichotomy of appearance versus essence. In contrast to the traditional meaning of "phenomenon," the Kantian "phenomenon" is not synonymous with mere appearance as opposed to reality itself. It is rather what is given to the subject according to the transcendental conditions of experience. The very same appearance/essence dichotomy could therefore be reformulated as appearance and condition of appearance, the given and its transcendental conditions. In a short and revealing passage, Deleuze adds one important detail that makes his interpretation immediately relevant for our discussion:

> To make things a little more modern, I would just as well say: to the disjunctive couple appearance/essence, Kant is the first who substitutes the conjunctive couple apparition/sense, sense of the apparition, signification of the apparition. There is no longer the essence behind the appearance, there is the sense or non-sense of what appears.
>
> (Deleuze 1978)

Through a heuristic (though anachronistic) updating of Kant's language, we can transform the dichotomy appearance/reality as appearance/meaning, whereby the latter refers to the given (appearance) as semantically charged. If we have no proper access to the essence, then the philosophical task is to reveal the fundamental conditions making the frenetic phenomenic world a "meaningful" world, a world of chairs, threes, and persons against a messy bundle of colors, qualities and sounds.

Of course, Kant did not develop a semantic theory of perception, but his new notion of "phenomenon," together with the reformulated dichotomy, opened up a conceptual space where perception itself could be conceived as semiotic and not as merely causal relations between subjects and "things." In other words, far from being a simple and passive reaction to external stimuli, perception was essentially a semiotic activity, independently of the degree of subjective awareness. *Phenomena* could be redefined as meaningful stimuli for a perceiving subject, who consciously or unconsciously selects them. Uexküll profited widely from such Kantian conceptual novelty and made a fundamental contribution in "semioticizing," and then biologizing, transcendental philosophy. The way Uexküll took advantage of this "semioticization" of perception is particularly evident when we consider his attacks on reflex theory and behaviorism, from Jacque Loeb to his pupil James Watson (and, of course, the Russian physiologist Ivan Pavlov). From Uexküll's perspective, the main problem with animal tropism – physiological mechanisms and reflex conditioning – was that it overlooked the fact that receiving and processing an external stimulus was not merely a matter of causal relation between an external source and an internal physico-chemical structure. The relation between a stimulus and an organism was essentially semiotic insofar as it involved an immediate, or mediate, interpretative process, independent of the degree of complexity of the animal itself. As Uexküll stated, explicitly countering the mechanistic understanding of perception: "a stimulus has to be *noticed* [*gemerkt*] by the subject and does not appear at all in the object" (Uexküll 2010, 46). This was equivalent to arguing that subjects do not only receive physical stimuli and react accordingly; they select, assess, and interpret such stimuli before reacting, and this was precisely what a machine could not do, unless we redefine it as a hermeneutical device, which for biologists such as Uexküll would be like a Penrose polygon. In a mechanistic world, in fact, we have causes that immediately produce effects, but in a biological world, we also have, and especially, meaningful relations between subjects and things. To Uexküll, biologists may understand the animal world when they approach it in semiotic terms precisely because organisms are irreducible quasi-Kantian subjects. I emphasize the *quasi* because Uexküll's organisms lack abstract concepts that might subsume the material of experience. The simplest organism could have a meaningful experience not because it harbored, inside its nervous system, the twelve transcendental categories but because it had established a productive and constitutive net of semiotic links with its specific environments.

Now, going shortly back to the relation between subject and world, Uexküll maintained that organisms live in independent bubbles of meaningful networks. The networks are reproduced and renewed throughout the animal's life and are linked to a series of reciprocal interactions that the organism entertains with its own environment. Such a periodic, interactive, and self-contained process, which Uexküll termed the functional circle, guarantees the congruence between the inner and external world of the organism:

> Every animal is a subject, which, in virtue of the structure peculiar to it, selects stimuli from the general influence of the outer world, and to these it responds in a certain way. These responses, in their turn, consist of certain effects on the outer world, and these again influence the stimuli.
>
> (Uexküll 1926, 126)

Uexküll also broke the functional circle down into two different kinds of experience or world: the world-as-sensed (perception world, *Merkwelt*) and the world-as-action (effect world, *Wirkwelt*). The first referred to the inner perceptive world of the organism; the second denoted the set of responses and actions of the animal toward the outer world (Uexküll 1926, 127). For a simple tick, for instance, the Kantian "understanding" could not consist of subsuming a particular set of intuitions under universal concepts. Uexküll therefore had to assume, to some extent, Schopenhauer's distinction between intuitive and discursive knowledge and replace "abstract concepts" with "operation" or "activity" so that the animals' "knowledge" could be the result of receptive (passive) and operative (active) processes. The tick "understands" (or "knows" intuitively) something when the perceptive and operative organs are purposefully aligned in a functional circle within, of course, the pure forms of sensible intuitions: time and space.[4] Thus, the functional circles tied the two worlds – perception world and effect world – into a dynamic epistemic pattern that formed what Uexküll termed the *Umwelt*.[5] Uexküll understood the *Umwelt* as a material and epistemic space that constituted the beginning and the end of the organism's meaningful world. Within this context, the Uexküllian "understanding" could be reframed as an effective functional activity of the organism, which guaranteed its inclusion into its own perceived environment.

So far, we have explored one of the fundamental pillars of Uexküll's biology. We have seen how the Estonian biologist needed to transform Kant's powerful conceptual *I* into a simpler perceiving subject forming its own meaningful world. Now we need to recover the second pillar supporting the whole Uexküllian structure: the idea of the presence of teleology in the organic world. Indeed, Uexküll's subjective biology acquires its overall consistency from the fact that organisms are perfectly "fitting into" (not adapted to, in a Darwinian sense) their *Umwelten*. Yet, the concept of animal fitting into [*Einpassung*] could not be understood in terms of evolutionary adaptation but in the more traditional terms of conformity to a plan. As we will see in the next section, Kant was again strategically squeezed onto Uexküll's Procrustean bed.

3 Why purpose?

If the transcendental activity of the subject was the main pillar of the new Uexküllian biology, the question about the conditions behind this same activity needed clarification. Indeed, one could ask, if organisms interact with the external environment through a dialectics established between receptive and operative processes, how can we explain the perfect match between the two worlds of the animal and its particular environment? This was a very tricky issue for Uexküll, especially because he was determined to avoid any neo-Darwinian solution. He did not accept the idea that the correspondence between organism and environment could be explained by a gradual and progressive adaptation driven by natural selection. Not surprisingly, Uexküll believed that neo-Darwinians had reduced the dynamic and meaningful life-world to a contingent desert of physico-chemical causes and effects. Against such desert, Uexküll opposed an intricate and lux-uriant ecology of swarming subjects teleologically enclosed in their particular niches. In short, for Uexküll, the living world was pervaded by an inner purpose-fulness and meaning; it could never be reduced to a senseless mechanical universe driven by teeth and claws. It was precisely for this reason that Uexküll assumed that the very same notion of "adaptation" [*Anpassung*] needed to be supplanted by the notion of "fitting into" [*Einpassung*]. Organisms did not become adapted to their environment; they were originally and meaningfully "embedded" into it.

As I previously mentioned, many early 20th-century biologists had understood that one effective strategy to oppose neo-Darwinian and mechanistic views was to reconsider the presence and necessity of agency and teleology in the natural world. Organisms could not be equated to machines because they were highly plastic and, at the same time, autonomous entities, and for this very same reason, they could not be the mere product of a contingent process of selective forces. While a machine requires an external supervisor for working, organisms do not. Indeed, as Uexküll observed, organisms "include the activities exercised on machines by human beings. They make the machine of their own bodies them-selves, they run it themselves, and they undertake all its repairs" (Uexküll 1926, 121). In the first instance, we could recognize that the great originality of Uexküll was to associate this standard anti-Darwinian and teleological motive with the inescapable subjectivity and autonomy of the organism. But the relation between subjectivity and teleology was much more important than a strategic argument against neo-Darwinism. If we go deeper into Uexküll's philosophical proposal, we realize that the transcendental subjectivity of the organism and its essential purposiveness were two sides of the same coin. In fact, if Uexküll were to accept the neo-Darwinian hypothesis, according to which the organisms' adaptation was the outcome of natural selection, the essential subjectivity of any biological entity would have become unnecessary. Every organism would have been inexorably exposed to the same blind forces of nature that produced objects, not subjects, sharing the same material environment. In this view, organisms would have been the contingent outcome of natural laws and their supposed subjectivity could only be the latest successful adaptation to a hostile environment.

However, Uexküll understood well that without teleology it would have been very difficult to defend a transcendental biology.[6] How could a transcendental subject be the result of causal determinism? And, more generally, how could a meaningless world produce a meaningful "machine" such as an organism? The profound gap between meaningless and meaningful things could not be filled with a blind mechanism without contradicting, at the same time, the significance of a teleological entity. In fact, if the transcendental subject could be explained by natural determinist forces, it could no longer be a transcendental subject. Con-versely, if organisms were transcendental subjects, then the determinist forces could not fully explain their origins. We should bear in mind that for Uexküll the physical world (i.e., the domain of blind causes and effects or stimuli and responses) was only an abstract corollary of the much wider and more complex world of biology, an irreducible world peopled by teleological entities, meanings, and subjects. I think that it is for these reasons that Uexküll redefined life sciences in his *Theoretical Biology* as the discipline studying conformity to a plan while "the study of causality comes into the question only in so far as it contributes to that investigation" (Uexküll 1926, 130). In short, the physical world could not produce the biological world insofar the former is a subset of the latter (or its most abstract consequence).

Now, while Uexküll's notion of subjectivity was not entirely Kantian, his notion of teleology departed from Kant's original proposal too, especially as presented in his third *Critique*. Ever since Plato and Aristotle, there have been different conceptions of teleology, and these depended on the kind of phenomena or processes in need of explanation. Following André Ariew, we could roughly distinguish between cosmological and biological teleology (Arew 2002). The former refers to the explanation of the apparent order of the universe that would eventually require a final cause (the demiurge in Plato or the God in Christianity). The latter denotes the morphological and physiological order characterizing living forms. In the Aristotelian tradition, biological teleology could be further divided into "agential" teleology – which pertains to the will, desire, and activities of human or nonhuman subjects – and "organic" teleology – which includes living organisms, their development, and the relations between parts and whole. The last case directly concerns us. Indeed, both Aristotle and Kant had argued that "organic" teleology did not imply any kind of mental intentionality or awareness. Organic order was not caused by an external idea or intelligence. However, Aristotle and Kant differed on one important point: Aristotle believed that the regularity we can observe during development – and in the harmonious morphological part–whole relations – was due to an internal, immanent, goal-directed force that he termed ἐντελέχεια. Kant instead maintained that the apparent purposiveness of living things – that could in principle be related to an internal organic force informing matter – should rather be conceived as a regulative idea of reason, a heuristic assumption guiding our observations. Thus, for Kant, while we could never ascertain whether teleology is immanent in the living entities, we need the final cause as a productive presupposition if we want to have a relative, not absolute, understanding of the phenomena of life.

Uexküll's understanding of teleology oscillated between these two stances. On the one hand, he acknowledged the Kantian ontological skepticism about the actual existence and knowledgeability of goal-directed forces. On the other hand, he embraced the Aristotelian view, especially through Driesch's vitalist reinterpretation, according to which we need to posit the existence of *psychoids*, that is, function rules that determine the ordering of organic matter according to the particular species.[7] Uexküll's solution consisted in a diplomatic agreement between Aristotle (namely, Driesch's reinterpretation of Aristotle) and Kant, an agreement that, although not entirely consistent, safeguarded biology against Kant's skepticism about the possibility of having a proper science of life without giving a blank check to the Drieschian ontological vitalism.[8] Uexküll's argument runs as follows: we observe that organisms exhibit teleological properties; we may assume that these properties depend on an immanent function rule organizing the organic matter. However, we ignore the ultimate cause behind this function rule simply because we have no access to the noumenic world; we can only speculate on particular phenomenic manifestations of the function rule.

Before returning to Kant's transcendental subject, I need to add one further dimension to Uexküll's teleology: his defense of biology as an irreducible and autonomous enterprise. The issue is particularly important as it complements, in an interesting way, our understanding of Uexküll's overall project in relation to Kant's philosophy. In fact, Uexküll believed that biology was much more than a scientific discipline offering reliable knowledge about living phenomena. To him, a philosophically informed biology saved us from the abstract, inert and dead world of physics: "Biology is quite able to save the world from sinking to the low level to which blind overestimation of physics is trying to reduce it" (Uexküll 1926, 48). And again, "biology can offer the plain man an unlimited enlargement of his world, whereas the physicist would reduce him to beggary" (Uexküll 1926, 72). It is quite ironic to consider how Uexküll's downsizing of physics paralleled his overemphasis on Kant's transcendental philosophy. After all, one of the central ambitions of Kant was to protect Newtonian physics against Humean skepticism. What's more, Kant did not believe that a true science of life would really be possible. Nevertheless, Uexküll used Kant's transcendental philosophy – which was originally framed for backing the epistemic credentials of physics – in order to express the inadequacy of the physical representation of the world and, in addition, to argue in favor of biology as the science that would provide a more trustworthy experience of the world than physics. Uexküll's "betrayal" of Kant is particularly instructive here. In his view, Kant's philosophy represented the last bastion against what we would call today reductionism and crass physicalism, that is to say, the idea that biological entities are nothing but fundamental physico-chemical properties. In that sense, Uexküll felt that the Kantian transcendental subject did not necessarily reveal the precondition for having a truly scientific knowledge of the world; it rather showed the limits of a purely physical understanding of living processes. In short, the Kantian subject was a guarantee against physicists' overambitiousness to explain everything in terms of matter and movement.

Furthermore, for Uexküll, the impossibility to understand life in physico-chemical terms derived from the fact that the biological world included objects of a very peculiar kind. While physicists, chemists, and physiologists shared the same universe composed of inert bodies, biologists worked with meaningful "implements." Uexküll identified "implements" with all those entities that could not be explained by simple causality. Of course, both things and "implements" had to be understood as material entities. However, while in the former the relations between parts and wholes, as well as between structure and function, were completely symmetrical, in the latter, the same connection needed to be complemented by a plan or rule, which drove the complex morphology of the entity to a definite end. In other words, the "implements" were entities that exceeded the pure physical causality and required conformity to a plan. As Uexküll explained,

[w]hen the carpenter's axe chops up the wood into planks and pegs, and when the drill bores through the planks and the hammer drives the pegs into the

holes, there are all of them causal succession. But the structure emerging from this process, the ladder, cannot be interpreted by causality; it can be understood only from a knowledge of the designed arrangement of the rungs with relation to the main planks, and of all parts to the whole.

(Uexküll 1926, 103)

Conformity to a plan defined the basic ontology of the biologist. The "implements" could be never reduced to "causal" objects because, without teleology, we would be precluded from a real understanding of these meaningful bodies. From a broader viewpoint, Uexküll contextualized the division between a world of "things" and a world of "implements" within two larger categories: a world of "senseless" objects and a world of "meaningful" entities. The former were the objects of physics, the latter those of biology. Biological entities can be understood as meaningful things insofar as they contained, immanently, the rule of their own formation, which exhibited a teleological orientation. Such a rule, in turn, made and justified the designation of subjectivity to organic implements. In short, organic implements were subjects, and not mere objects, because they possessed the rule of their own cause in conformity to a plan. It is precisely for these reasons that the transcendental subject could not be severed from the teleological idea. In Uexküll's biology, subjectivity and teleology reinforced each other. We could even say that conformity to a plan, meaning, and subjectivity are the basic elements constituting the *Umwelt*. After all, the *Umwelt* is the creation of a subject, which perceives meaningful objects according to its morphological plan. Behind or beyond the *Umwelt*, both the world in itself and the deep causes of teleology remain forever undisclosed. As he argued in his 1920 *Theoretical Biology*, "[t]he soap-bubble of the extended constitutes for the animal the limit of what for it is finite, and therewith the limit of its world; what lies behind that is hidden in infinity" (Uexküll 1926, 42). Again the same idea in semiotic terms: "Each environment forms a self-enclosed unit, which is governed in all its parts by its meaning for the subject" (Uexküll 2010, 144).

However, these last sentences should not allow us to conclude, too easily, that Uexküll was proposing a new form of biological idealism or a scientifically clothed constructivism wrapped up by militant antirealism. If we draw this conclusion, we miss entirely the meaning and originality of his proposal. In the end, why mentioning the *Umwelt* if such a *Welt* would consist of a subjective construction? Why worry about observing organisms if everything is made up by our minds? Why present notions such as functional circle if all that we could suppose to exist is a functional noetic projection of our will? No. Uexküll was neither a traditional constructivist nor an antirealist. He maintained the Kantian distinction between phenomenon and noumenon with the proviso that the noumenon was only indirectly accessible through our semiotic interpretations. Noumena manifest themselves through signs. Our reality is the ensemble of interpreted or interpretable signs. Beyond those signs lie the unknowable or unintelligible world that we can barely conceive as the limit of our semiotic activities. From this perspective, the noumenon was not an unnecessary appendix to an overstructured philosophical system, as many post-Kantian idealists had argued against Kant himself; it was the essential and unfathomable source of those carriers of meaning constituting the environments in which we, and other organisms, live.

4 Conclusion

Throughout this chapter, we have seen why Uexküll made Kant one of his more cherished "precursors" and the amendments he needed to make in order to frame a new, partially "un-Kantian," philosophy. Indeed, Kant would have probably disapproved of a disciple who used his transcendental philosophy and teleological judgment to demonstrate the epistemic limits of Newtonian physics and, in parallel, promote the good epistemic credentials of biology. Physics was, for Kant, the model of scientific knowledge; conversely, a science of the living could never be a proper science at all. Yet, Kant's questions were not Uexküll's. The latter was not an epistemologist aiming to justify scientific knowledge against skeptics. He was a biologist trying to understand how organisms deal with the world without invoking neo-Darwinian explanations. He realized that organisms could not be conceived as black-box learning machines: they were complex subjects that produced the world in which they dwelt. And they could do that because they were "implements," that is, subjects teleologically determined. Subjectivity explained why there were no perceptions without meaning, while conformity to a plan explained why and how subjectivity existed. In coherence with the relation between subject and teleology, Uexküll also drew a fundamental line between meaningless and meaningful worlds – between the physicists and physiologists' world of atoms and machines, and the biologists' world of *harmonic* implements (signs, purposes, and subjects).

Such a sharp distinction between two worlds, one abstract and ideal and the other concrete and tangible, can remind us of the methodological distinctions between natural and human sciences. Indeed, in the spirit of Jorge Luis Borges's suggestion, we may add a further unlikely "precursor" to Uexküll's biology: Wilhelm Dilthey. After all, the Uexküllian divide between these two domains might be heuristically linked to Dilthey's famous and controversial distinction between interpretative understanding [*Verstehen*] and nomological explanation [*Erklären*], which defined the border between *Geisteswissenschaften* and *Naturwissenschaften*. For Dilthey, the difference between natural science and human science consisted, ultimately, in the irreducible presence of meaning in the human world, again a world of pure causation and laws (see Ermarth 1975). Should we be surprised that Dilthey deemed Kant one of his principal "precursors"? Like Uexküll, Dilthey considered Kant's *Critique of Pure Reason* as "the greatest philosophical work ever produced by the German mind" (Dilthey quoted in Ermarth 1975, 149). Kant was the first in questioning the idea that subjects are simple receptors of external impressions and data. For Dilthey, Kant's great achievement was the following one: in stressing the priority of mind in structuring and organizing phenomena, he had definitely challenged the copy theory of knowledge. As a consequence, the essential epistemological problem for Kant was not whether our subjective representations correspond to the external world but how the world reveals itself to us.

Thus, perhaps, Deleuze's suggestion that Kant replaced the appearance/essence dichotomy with "the conjunctive couple apparition/sense, sense of the apparition, signification of the apparition" (Deleuze 1978) pinpoints one of the most important legacies of transcendental philosophy. A legacy that could explain many aspects of neo-Kantian and post-Kantian philosophies. Kant's critical approach had indeed opened a possibility that was unavailable before him: to think the epistemic relation between subject and object in terms of meaning and not in terms of causes. From the formless and "blooming buzzing confusion" of the noumenon, the transcendental ego selects, evaluates, identifies, and represents the meaningful elements that conform to the inner structure of the "mind." As Uexküll powerfully stated,

> [m]eaning is the pole star by which biology must orient itself, not the impoverished rules of causality which can only see one step in front or behind and to which the great connections remain completely hidden.
>
> (Uexküll 2010, 160)

However, despite the fact that they were both speculating in the shadow of Kant, there are at least two revealing and fundamental differences between Dilthey and Uexküll. First, for Uexküll, the Rubicon dividing the world of sense from the world of pure causes lay not in glorious human consciousness but in the seemingly unconscious behavior of a voracious amoeba or a starving tick. In agreement with Uexküll's biophilosophy, humans could not have a monopoly on the *Geist* because even the simplest organism is a subject perceiving "meaningful" stimuli. Ants, dogs, or macaques are all hermeneutical subjects "grasping" and "understanding" their own worlds through a constant hermeneutical practice. Biology would therefore be more at home among the disciplines composing the *Geisteswissenschaften* (humanities) than those included in the *Naturwissenschaften* (sciences), insofar as Uexküll's biology put *Verstehen* before *Erklären*. Being purposely anachronistic, we could add that the Uexküllian biology would be a discipline requiring "thick descriptions" against the "thin descriptions" of the physicists insofar as the living realm was more a Geertzian ecology of "stratified hierarchy of meaningful structures" (Geertz 2017, 7) than an austere domain of stimuli and responses. In this respect, interpreting the experience of a tick would not be qualitatively different from interpreting a different human culture, although, of course, the possibility of empathic understanding would be much higher among *Homo sapiens*. In consonance with a Diltheyan humanist or Geertzian anthropologist, we might consider the Uexküllian transcendental biologist as an interpreting subject who compares his or her meaningful environment with other hermeneutical subjects. The possible intelligibility of the living world relies precisely on this simple belief: all living creatures – including humans – share a semiotic space potentially interpretable in countless different ways. The biologist's epistemic pretension to understand alternative forms of life lies in the possibility of interpreting the meaning of signs from the perspective of nonhuman organisms. In short, we may interpret Uexküll's biology as a Diltheyan *Geisteswissenschaft* extended to the entire living realm.

The second great difference between Dilthey and Uexküll is that for the former humans are essentially historical subjects, shaped by diachronic forces that constantly produce unexpected outcomes. "Kant," Dilthey observed, "was not truly critical because he was not historical" (Dilthey in Ermarth 1975, 152). Closer to Kant's perspective, Uexküll's transcendental biology was utterly ahistorical. In contrast to Dilthey and his old friend Konrad Lorenz, Uexküll maintained that Kant's transcendental philosophy could not be historicized, neither sociologically nor biologically (see Lorenz 1941). All living entities originate according to their species-specific function rule, which left no room for contingent possibilities. Furthermore, a relatively fixed plan determined the very same relation between organisms and their *Umwelten*. The original and teleological structure of the biological entity guaranteed the congruence between animal and world, in other words, the organism's "fitting into" its own niche. As Uexküll emphatically declared in the seventh letter contained in his *Biological Letters to a Lady* [*Biologische Briefe an eine Dame*]: "[T]here is no evolution, there is only origin" (Uexküll 1920, 97, my transl.). The origin was not about phylogeny but about the ontogenetic "melodies" organizing the growing body of an organism. The biological world, accordingly, was a dynamic and ahistorical process, a great symphony composed of innumerable different melodies replaying over and over again.

Thus, if you can imagine a Kantian who believes that intuitions without concepts are not always blind, a Kantian who maintains that physical sciences do not provide the best model for scientific knowledge, a Kantian who uses transcendental philosophy and teleological judgment to argue for the scientific nature of the life sciences (which eventually provide us with richer, more concrete and more appropriate access to the world), you understand how Uexküll transmuted Kant's transcendental philosophy into a new, original and powerful biophilosophical theory. In short, you can finally realize how Kant was strategically molded for being a precursor of Uexküll.

Notes

1 On Kant and biology, see Fisher (2008), Ginsborg (2007), Huneman (2008), Breitenbach (2009), and Goy and Watkins (2014).
2 Although Uexküll reinterpreted Kant's transcendental philosophy from a physiological viewpoint, as J. Müller and H. Helmholtz had similarly done before him, Uexküll went much further than presenting a coherent version of "physiological Kantianism." In fact, he not only pretended to clarify the complex relations between perception and subjectivity; he also defined the notion of life (and organism itself) in terms of subjectivity. According to Uexküll, biology had to be redefined as the science of subjectivity insofar as the living realm was essentially the realm of subjectivity (within the tradition of Kant's transcendental philosophy).
3 On Uexküll and Deleuze, see Buchanan (2008), Heredia (2011), and Brentari (2015). See also Cimatti's contribution, Chapter 10, in this volume.

4 Indeed, when the two organs are not aligned, the life of the organism is in danger. See Uexküll (1926, 127).
5 As he explained, "perception world and effect world together make a comprehensive whole, which I call the *Umwelt*" (Uexküll 1926, 127, transl. slightly modified).
6 After all, in linking purposiveness with subjectivity, he did something quite similar to Kant: carving out subjective freedom from a pure determinist universe.
7 On Uexküll and Driesch, see Driesch (1921) and Uexküll (1926). On Driesch's interpretation of Aristotle, see Driesch (1908). On Aristotle's teleology, see Bradie and Miller (1984), Cooper (1982), and Friedman (1986).
8 After all, Kant had famously stated that there could not be a Newton of a blade of grass.

References

Arew, Andre (2002) 'Platonic and Aristotelian roots of teleological arguments'. In: Andre Ariew, Robert Cummins, and Mark Perlman (eds.) *Functions: New Essays in the Philosophy of Psychology and Biology*. Oxford: Oxford University Press, 7–32.

Borges, Jorge Luis (1999) *The Total Library: Non-Fiction 1922–1985*. Translated by Ester Allen, Suzanne Jill Levine, and Eliot Weinberger. Harmondsworth: Penguin.

Bradie, Michael and Miller, Fred D., Jr. (1984) 'Teleology and necessity in Aristotle'. *History of Philosophical Quarterly* 1 (2), 133–146.

Braga, Joaquim (1964) *Críticas filosóficas: Kant e Uexküll seguidas de um discurso de justificação da crença no real*. Lisboa: Sociedade de Expansão Cultural.

Breitenbach, Angela (2009) 'Teleology in biology: A Kantian approach'. *Kant Yearbook* 1 (1), 31–56.

Brentari, Carlo (2015) *Jakob von Uexküll: The Discovery of the Umwelt between Biosemiotics and Theoretical Biology*. Dordrecht/Heidelberg/New York/London: Springer.

Buchanan, Brett (2008) *Onto-Ethologies: The Animal Environments of Uexküll, Heidegger, Merleau-Ponty, and Deleuze*. Albany, New York: SUNY Press.

Cooper, John M. (1982) 'Aristotle on natural teleology'. In: Malcolm Schofield and Martha Craven Nussbaum (eds.) *Language and Logos*. Cambridge: Cambridge University Press, 197–222.

Deleuze, Gilles (1978) *Seminar on Kant – 14/03/1978*. Translated by Melissa McMahon, www.webdeleuze.com/php/texte.php?cle=66&groupe=Kant&langue=2. (Accessed 20/08/2017).

Driesch, Hans (1908) *The Science and Philosophy of the Organism: Gifford Lectures Delivered at Aberdeen University, 1907*. London: Adam and Charles Black.

Driesch, Hans (1921) 'von Uexküll, J., Theoretische Biologie (book review)'. *Société Française de Philosophie* 26, 201–204.

Ermarth, Michael (1975) *Wilhelm Dilthey: The Critique of Historical Reason*. Chicago: Chicago University Press.

Esposito, Maurizio (2016) *Romantic Biology, 1890–1945*. London: Routledge.

Fisher, Mark (2008) *Organisms and Teleology in Kant's Natural Philosophy*. Emory University Diss.

Friedman, Robert (1986) 'Necessitarianism and teleology in Aristotle's biology'. *Biology and Philosophy* 1 (3), 355–365.

Geertz, Clifford (2017) *The Interpretation of Cultures*. New York: Basic Books.

Ginsborg, Hannah (2007) 'Kant's biological teleology and its philosophical significance'. In: Graham Bird (ed.) *A Companion to Kant*. Oxford: Oxford University Press, 455–469.

Goy, Ina and Watkins, Eric (eds.) (2014) *Kant's Theory of Biology*. Berlin/Boston: De Gruyter.

Guidetti, Luca (2013) 'Jakob von Uexküll tra Kant e Leibniz. Dalla filosofia trascendentale alla topologia del vivente'. *Rivista italiana di filosofia del linguaggio* 7 (2), 66–83.

Heredia, Juan Manuel (2011) 'Deleuze, von Uexküll y "la naturaleza como música"'. *A Parte Rei. Revista de Filosofía* 75, 1–8.

Huneman, Philippe (2008) *Métaphysique et biologie: Kant e la constitution du concept d'organisme.* Paris: Editions Kimé.

Kant, Immanuel (1998) [1781] *Critique of Pure Reason.* Translated and edited by Paul Guyer and Allen W. Wood. Cambridge: Cambridge University Press.

Lenoir, Timothy (1982) *The Strategy of Life: Teleology and Mechanics in Nineteenth Century German Biology.* Chicago: Chicago University Press.

Lorenz, Konrad (1941) 'Kants Lehre vom Apriorischen im Lichte gegenwärtiger Biologie'. *Blätter für Deutsche Philosophie* 15, 94–125.

Schopenhauer, Arthur (1966) [1818/1819] *The World as Will and Representation.* Translated by Eric F. J. Payne. New York: Dover Publications.

Uexküll, Jakob von (1920) *Biologische Briefe an eine Dame.* Berlin: Gebrüder Paetel.

Uexküll, Jakob von (1926) [1920] *Theoretical Biology.* Translated by Doris L. Mackinnon. London: K. Paul, Trench, Trubner & Co.; New York: Harcourt, Brace & Company.

Uexküll, Jakob von (2010) [1934, 1940] *A Foray into the Worlds of Animals and Humans with a Theory of Meaning.* Translated by Joseph D. O'Neil. Minneapolis/London: University of Minneapolis Press.

3 Uexküll's legacy

Biological reception and biophilosophical impact

Kristian Köchy

The following considerations are dedicated to the reception of Uexküll's work in biology and biophilosophy. In so doing, Uexküll's experimental discoveries and observations are discussed side by side with his philosophical considerations. Four closely interlinked lines of reception will be presented: (1) Uexküll's studies of invertebrates with regard to their impact on comparative physiology and the debate on the laws of organization of living beings, (2) Uexküll's behavioral studies with regard to their significance in the development of ethology and cognitive ethology, (3) Uexküll's considerations on the functional circle as a possible precursor to cybernetics, and (4) Uexküll's concept of *Umwelt* and its relation to the ecological concept of environment and to environmental ethics.

1 Studies of invertebrates, comparative physiology, and the debate on the organism

In its early years, Uexküll's research in biology is characterized above all by physiological, experimental, and anatomical studies of invertebrates. These works attest to Uexküll's experimental abilities, his innovative development of research questions and investigative methods, as well as their careful application. Uexküll was a student of Wilhelm Kühne, a famous nerve and muscle physiologist from Heidelberg (Brock 1934, 195; Mildenberger 2007, 53ff.), and he later trained in E. J. Marey's famous laboratories in Paris (G. von Uexküll 1964, 39; Mislin 1984, 46). Marey holds as a pioneer, among other things, in the automatic recording of movements in biological experiments. In Marey's laboratory, Uexküll not only perfected his experimental abilities and developed new investigative apparatuses (like the so-called *Neurokinet* – an apparatus for mechanical neuro-stimulation) but also engaged with Marey's work on chronophotography and used the latter in his own experiments (Kynast 2012). Uexküll's creative capacities were also revealed in the dummy experiments (Hassenstein 2001, 354) that made their name in the discovery of the fixed action patterns in Konrad Lorenz's and Nikolas Tinbergen's research. Furthermore, in later studies at the Hamburg Institute for Environmental Research, Uexküll developed a special aquarium observatory with moving mirrors, which notably incorporates his idea of the *Umwelt* in the designing of an observational environment.

Two aspects in Uexküll's physiologically based behavioral research largely justify its considerable impact. First, he expands his investigations to include an abundance of different species. Second, he turns away from classical physiology's investigations of the single functions of organs. He is instead interested in the interaction of elements, especially in the neuro-muscular system (G. von Uexküll 1964, 38f.; Mislin 1984, 44). Both aspects are connected by a holistic tendency. While the comparative approach aims to encompass as many areas of the animal kingdom as possible, research into interaction looks into the organic whole. These two aspects of Uexküll's work not only inspired biological research but also ignited biotheoretical and philosophical debates.

The empirical foundation of Uexküll's research is shown especially in the range of model organisms he investigated. From 1890 to 1914, he carried out research at different institutions, especially in marine research stations (Hassenstein 2001, 346; Mildenberger 2007, 55ff.). In so doing, he proves to be the advocate of a particular style of research, as Burian would put it (Burian 1993, 361). While for many researchers the pursuit of lifelong work on just one model organism is taken to be an efficient research strategy, others, among them Uexküll, pursue the opposite strategy and work on as many different species as possible. Methodological and methodical criteria, in addition to personal preferences, are always decisive factors in such an orientation. While a conservative focus on a few organisms brings the advantage of familiarity with the living beings and the investigated properties thereof, what speaks for the progressive strategy is the possibility of gaining new insights and discovering previously unknown phenomena. Above all, however, a comparative research program can only be implemented with a strategy of variation. Uexküll can thus be considered one of the founders of comparative physiology, ethology, or psychology (Brock 1934, 196). His procedure is indeed constantly dedicated to the search for general principles. Without making the claim to providing an exhaustive list here, Uexküll's research includes investigations into the squid *Eledone moschata*, the cat shark, the sea urchin, the marine medusa *Rhizostoma*, sipunculids, the brittlestar, the leech, dragonflies, the sea anemone *Anemone sulcata*, the scallop, and the crayfish. Some of these organisms play a paradigmatic role in Uexküll's research and theories and became symbolic representatives of his considerations. Uexküll's works on the sea urchin are, for instance, paradigmatic in both early investigations on the nervous and muscular system and for his claims, formulated in the *Guidelines for the Study of the Experimental Biology of Aquatic Animals* (Uexküll 1905), concerning the integration and coordination of the elements of such systems. The impact of these works in biology can be gauged by their having formed the basis and point of departure for the later formulated Uexküllian stretching rule, which also inspired later physiology (cf. Jordan 1929, 496ff.; Holst 1969, 125). The initial findings leading to the formulation of this rule refer to the spines of the sea urchin. When the shell of these animals is stimulated, the neighboring spines move in the direction of the stimulus. Here the reaction of the spines whose muscles were stretched is stronger than the reaction of the spines with contracted muscles. Uexküll found this observation on the role of the muscle tone and the relationship of stretching and flexion confirmed in other species (Uexküll 1929). Beyond their concrete physiological explanatory power, these findings also provided the occasion for reflecting on the significance of antagonisms in the coordination of organisms. The neurophysiologist and Nobel Prize winner Charles S. Sherrington had also detected similar antagonisms in the central-nervous coordination areas. With reference to these findings, Kurt Goldstein (1995, 365ff.) in his biophilosophy had declared antagonisms to be fundamental principles of life.

The sea urchin became a paradigm of biological organization also in other respects. For instance, the sea urchin's defense against its biggest enemy, the starfish, requires coordinative achievements of another kind. Special poisonous claws, the pedicellaria, are found beside the spines on the upper side of the animal. In an attack, the spines of the deterrent zone move away from the stimulus and make room for the poisonous claws. Reverse sequences of movement can be observed on the underside of the sea urchin. In flight, they must carry out coordinated movements with their long, muscular tube feet to get away from the threat. Uexküll not only observed such achievements in the intact animal in its natural environment but also developed techniques in order to record the details of the sequence. He thus investigated the movements of the legs by fixing sea urchins upside down and placing small glass beads on their legs. He examined the defense mechanisms of the pedicellaria in both intact animals as well as in isolation experiments on a shell plating with spine, pedicellaria and neuronal ganglia. Uexküll even found a metaphorical expression for the coordinative achievements of the sea urchin and for the absence of central control therein that has had a powerful influence in the reception of his work. He named the unities of reaction "reflex persons" and their organization the "reflex republic" (Uexküll 1909, 118), and he explained that

> [o]ne can therefore speak of a "reflex republic" in which, in spite of the complete autonomy of all reflex persons, a total civil peace reigns, for the tender suction feet of the sea urchin are never fallen upon by the biting, grasping pincers, which would otherwise grab any other approaching object. This civil peace is never dictated from a central location.
>
> (Uexküll 2010, 76)

Uexküll coined a phrase for these conditions: when the dog runs, the animal moves the legs, when the sea urchin runs, the legs move the animal. This phrase made waves in much research into the organizational laws of organisms (cf. Bertalanffy 1937, 53). Uexküll became a forerunner of the new biology (Portmann 1983, X). His research and theories came at a time in which philosophical reflections on biology were in great demand (Driesch 1911; Hartmann 1912; Haldane 1935, 1936) and notably promoted a discourse that postulated a special quality of organismic forms, shapes or wholes (Driesch 1908; see Ballauff 1949, 82ff.; Beckner 2006). The organism–machine comparison (Driesch 1935; see also Nicholson 2014) also belongs to this context, as does the conflict between mechanists and vitalists (Roux 1905; Driesch 1922; see also Haraway 1976). Like

John S. Haldane, Ludwig v. Bertalanffy or Julius Schaxel, Uexküll too was a leading figure for the holistic movement (see K. M. Meyer-Abich 1989) and for the organismic biology (cf. Esposito 2013, 5). As can be shown (Nicholson and Gawne 2015), this movement had an enormous influence on the development of theoretical biology and on the genesis of the philosophy of biology. Again in recent times, the significance of organisms has come to the fore, after having been threatened within the framework of evolutionary biology by its modern synthesis between the genetic level and the level of population (cf. Bateson 2005; Toepfer and Michelini 2016).

In the debate about the organizational laws and autonomy of the living, which Uexküll promoted with reflections such as those on the "reflex republic," he himself played an original role within opposing directions and schools. Furthermore, his position changed over the course of his life. Starting off as an experimental physiologist with an inclination toward proto-behaviorist considerations, he later became a theoretical biologist of the *Umwelt* who is convinced of the subjectivity of the living being and whose considerations have vitalist features. His historical position can thus also be seen in his contribution to overcoming the old conflict between mechanism and vitalism (Portmann 1983, XI). The difficulties connected to labeling Uexküll were already clear in the then contemporary biophilosophical debate. For instance, based on his studies on muscle tone, Driesch understood Uexküll as a vitalist (Driesch 1922, 205f.). Uexküll himself brushed off this appropriation by neovitalism just as much as the appropriation by holistically oriented Darwinian theories (Mildenberger 2007, 9). Nevertheless, Uexküll's reception is not free of ambiguity either. For instance, in *The Philosophy of a Biologist*, Haldane (1935, 45f.) criticizes the organismic biology in that it tends to overemphasize the autonomy of organisms and to isolate them from the environment. As the German translator Adolf Meyer – himself a significant representative of holistic biology – remarked (Meyer in Haldane 1936, 39), while this critique does apply to Driesch's vitalism, it does not apply to Uexküll's *Umwelt* doctrine. Such differentiations can also be ascertained in Uexküll's own considerations on *Planmäßigkeit* or conformity to a plan (Uexküll 1925). Accordingly, organic structures are always built up from parts that are not causally but systematically connected to each other. Uexküll emphasizes his neutrality with regard to the different interpretations of this conception, which the vitalists understood as a natural force, but the antivitalists took as an epistemic construction. Independently of such disputes, for Uexküll the heuristic fertility of the system-related view for biology is indubitable. The allegations that his presentation of conformity to a plan leads to a static understanding of organisms are also wide of the mark. Uexküll's conformity to a plan is instead dynamic and rather resembles Victor v. Weizsäcker's "theory of *Gestalten* in time" [*Zeitgestalten*] (Uexküll 1925, 7; see Magnus 2011). As Uexküll shows for the *Plasmodium vivax* (and therein refers to Fritz Schaudinn's investigations), biological systems also entail the temporal adaptation of subsequent figures (Uexküll 1922, 135ff.). The spatial sphere of the *Umwelt* thus becomes a spatiotemporal "environment-tunnel" (Uexküll 1922, 143). Uexküll also openly discusses the use of analogies or metaphors for organisms (Uexküll 1925, 8f.). He takes the analogy to melody – prominent in *Gestalt*-psychology since Christian v. Ehrenfels – to be the clearest representation of unity, although it remains purely formal. The alternative machine analogy is also taken as fruitful despite all its shortcomings. Machines are a good comparative case not only because of their material constitution but especially because of their differences to organisms. For even machines are not the mere product of mechanism. Never does the material comply with form of its own accord. A factor lying outside of mechanics must be added. The two types of system, however, are fundamentally different in their genesis – machines emerge from attaching finished parts (centripetally), but living beings always emerge from a germ (centrifugally). In both cases, material must be procured and arranged in a way that conforms to a plan. Finally, Uexküll brought together the machine and melody analogical models while explaining the organizational laws of living beings, for instance, in the example of a pipe organ (Uexküll 1930, 67ff.). His talking of "hues" [*Tönungen*], "moods" [*Stimmungen*] or "impulse melodies" [*Impulsmelodien*] belongs to this metaphorical field.

It should also be added that typically Uexküllian image language, such as for instance the idea of a "reflex republic," has reached well beyond biological reflection. Helmuth Plessner and his biophilosophy bear testimony to this (see Krüger's contribution, Chapter 5, in this volume). Plessner was also a trained biologist and had himself researched the physiology of starfish (Plessner 1913). Uexküll's name is particularly prominent in *The Levels of Organic Life and the Human* (Plessner 2019), where Plessner gives Uexküll a lecture on a large scale. Here the reflex republic stands for a basic principle of animal life; Uexküll's sea urchins notably represent one of two strands in the animal kingdom: the decentralized way of animal organization.

The reflex republic receives a similar degree of attention in Maurice Merleau-Ponty's work (2003, 167ff.). In his lectures on *Nature*, he traces transformations in our representations of nature, which he understands as ontological mutations in whose context the relationship among human being, nature, and God changes. What interests him in biology is the mutation of biological concepts. According to Merleau-Ponty, the concept of behavior, introduced by Watson, in the historical development moved out of the framework of a realistic philosophy. The schools of intentional (Edward C. Tolman) and molecular behavior (Jacob R. Kantor) show two possible directions of this movement. A third is yielded by Uexküll. The Uexküllian direction has, on the one hand, a structural reference and, on the other, it permits recourse to the psychology of mental states. This approach becomes relevant on account of its anti-Cartesian orientation. The metaphor of the reflex republic and the investigations on invertebrates thus shed significant light on machine comparisons for organisms. Uexküll shows that simple sequences of movements in jellyfish remind us of machines. The idea of a building-plan adds to this impression. According to Merleau-Ponty, however, there is no doubt that even in the case of simple life-forms, the machine analogy is inappropriate. This applies to sea urchins. They have a higher degree of complexity and integration compared to the Medusa jellyfish. But the idea of the reflex republic proves that no unity exists. The different elements of behavior are only loosely sewn together. For Merleau-Ponty as for Uexküll machines are incomplete organisms and the machinability of the living concerns its incomplete forms.

2 Animal behavior and cognition

2.1 Comparative behavioral research

These considerations show that it is especially Uexküll's anti-Cartesian orientation, which takes into account psychological aspects of behavioral expressions, what proved essential in the reception of his work. What must be taken into account here, however, is the aforementioned change in Uexküll's position from a rather behaviorist to a rather mentalist standpoint. If we follow these leads, then different strands in the reception of his work become visible. Uexküll's influence on the rise and the advancement of comparative ethology can hardly be overestimated. Even though he is often absent from established lists of well-known behavioral researchers (see, for instance, Dewsbury 1989), he can be counted as one of the founders of the discipline (see Portmann 1961, 44; Mislin 1984; Burghardt 1985, 222ff.; Wuketits 1995, 110ff.; Hassenstein 2001), who impacted its subsequent development in manifold ways (see Brock 1950, 94ff.; Portmann 1983, xvf.).

In this strand of reception too, there is a symbolic animal and a symbolic environment, notably the tick and its *Umwelt*. Uexküll's considerations on this topic, presented, for instance, in the introduction to *A Foray into the Worlds of Animals and Humans* (Uexküll 2010, 44ff.), play a central role in the physiology of the senses and in behavioral research (see Leyhausen 1973b, 292ff.). Uexküll's relevance here is also proved by the influential monograph *Animal Behavior* by Robert A. Hinde. Hinde (1966, 43) mentions Uexküll in connection with explanations of effective stimulus. Uexküll's tick example proves the high stimulus selectivity of animals that, from the many stimuli of their surroundings, seek out only those that are relevant, "like the gourmand [who singles out] the raisins from the cake." The organism's agency, based on the high degree of selectivity in the sensory organs, corroborates Uexküll's findings and this strand of his work opened up the discussion of the subjective capacities of animals. At the same time, Uexküll provides the occasion for Hinde to reflect on the role of human observers and their ability to comprehend animal capacities (Hinde 1966, 64). One last reference proving Uexküll's key role for behavioral research is to be found in Konrad Lorenz and the Lorenz school (Hinde 1966, 88). Hinde refers to the connection between stimulus selectivity and fixed action patterns that play a central role in comparative behavioral research (see Hassenstein 2001, 353).

With this reference, we arrive at a personal relationship in this reception, that between Uexküll and Lorenz (see Brentari 2015, 217ff.). The ambivalences of Uexküll's reception are also spelled out in this long and intense relationship. It is clear that Uexküll influenced the development of behavioral research in two ways. On the positive side, he is held to be the model of excellent experimental research; on the negative, he provided an occasion for warning against far-fetched vitalist speculation and an anti-Darwinian regression. In the early phase of his work, which finds its expression in the *Companion as Factors in the Bird's Environment* (1935), Lorenz is especially interested in the modalities with which animals perceive objects of their *Umwelt* and in the differences between the cognitive capacities of human beings and animals. There are points of overlap with Uexküll in both areas right up to their philosophical framework, that in both instances is provided by Kant's philosophy. Because of these overlaps, Lorenz uses and modifies Uexküll's theoretical conceptions, for instance the assumption of a projection of the sensation of perceived stimuli onto outward 'things' in the external space [*Hinausverlegung*] or the functional circle (see the following discussion). However, significant differences quickly surfaced. These may be said to be stirred up directly by the desire to use the *Umwelt* doctrine as the basis for a natural-scientific biology. Uexküll's notion that all animal species have their own respective *Umwelt* that differs fundamentally from the *Umwelten* of others tends toward a monadology (Lassen 1939). However, considering Leibniz's idea of windowless monads leads to the question of intersubjectivity in the sense of scientific objectivity (a central critical point raised by Hans Blumenberg, see Borck's contribution, Chapter 11, in this volume).

In "The Innate Forms of Possible Experiences" (Lorenz 1943, 353), Lorenz emphasizes that the aim of his comparative behavioral research is based on the acceptance of an intersubjective reality. This latter seems to be impossible in Uexküll's *Umwelt* doctrine. In this respect, the image of the *Umwelt* as a bubble can be somewhat misleading (Warkentin 2009, 24). This simmering conflict breaks to the surface at the latest in a talk given by Lorenz at a conference on Uexküll (Brentari 2015, 221). Lorenz here describes Uexküll as an enemy of science. But in this stark settling of accounts too, old moments of recognition acknowledge Uexküll as an accurate physiologist and a brilliant scientist. The underlying discrepancy here points us toward the different theoretical presuppositions of the two researchers. According to Paul Leyhausen, a student of Lorenz, Uexküll's doctrine of *Umwelt* is a phenomenology, whereas Lorenz's behavioral research is a genetic causal analysis (Leyhausen 1973a, 95). Accordingly, Uexküll's concept of "functional circle" is phenomenological, whereas Lorenz's concept of "instinct movements" is inductively acquired. A further discrepancy lies in their respective relation to the theory of evolution. While Uexküll's stance has anti-Darwinian features and remains indebted to the embryologist Karl Ernst von Baer (see Mildenberger 2007, 19ff.), Lorenz wants a biological behavioral doctrine based on the evolutionary theory. Finally, what is to be taken into account is Lorenz's demarcation from vitalist approaches, which, however, had inspired his own research. In the "Russian manuscript" of *The Natural Science of the Human Species* (Lorenz 1996), Lorenz acknowledges the vitalists' admiration and awe of nature. He, like Tinbergen, holds precisely this lay admiration of nature as the beginning of behavioral research. However, he criticizes the corollary of renouncing further scientific investigation because of this admiration. This despondent resignation is not for Lorenz, who seeks a holistically directed method. But in the cold light of day, one cannot accuse Uexküll of such a resignation; on the contrary, he puts the challenges of the living to experiment time and again. Uexküll even does justice to Lorenz's insight that as a researcher the vitalist must be a mechanist (Lorenz 1996, 195f.). Finally, Uexküll's assumption of harmonizing teleological natural forces has also been the subject of harsh critique from several corners. On this central critical point, however, Lorenz recognizes the basic antiperfectionist nature of Uexküll's theory and the prominence, there within of the idea according to which all kinds of living beings are in each case completely adapted to their *Umwelten*. As Brentari summarizes it,

> [i]n other words, once the vitalist faith in the total teleology of nature is over come, the Uexküllian idea of an adaptive insertion [*Einpassung*] between organism and environment therefore stays valid.
>
> (Brentari 2015, 223)

To this very idea, in fact, regardless of its epistemological assumptions, recent niche-construction approaches have given new impetus (West-Eberhard 2003; Laland 2014). In this respect, Uexküll's notion that animals are not passively formed by external forces but on their part actively form their *Umwelt* has proved absolutely topical.

2.2 Cognitive ethology

Let us recall Uexküll's change from an experimental physiologist to a theoretical environmental biologist and the associated change in his understanding of organisms (see Brock 1934, 198). This switch between two different fundamental perspectives explains a further strand in the reception of Uexküll's thought that leads to cognitive ethology. Two divergent programmatic papers on terminology, written thirty-six years apart from one another, illustrate the matter quite clearly. These papers are the "Proposals toward an Objective Nomenclature in the Physiology of the Nervous System," written together with Th. Beer und A. Bethe (Beer, Bethe, and Uexküll 1899) and its later complementary outline, the "Proposals toward a Subject-Related Nomenclature in Biology" (1935) (Uexküll 1980). The reference in the two titles changes from physiology to biology and brings a change in direction to the fore. As Uexküll in *A Foray* clearly explains,

> [f]or the physiologist, every living thing is an object that is located in his human world. He investigates the organs of living things and the way they work together just as a technician would examine an unfamiliar machine. The biologist, on the other hand, takes into account that each and every living thing is a subject that lives in its own world, of which it is the center. It cannot, therefore, be compared to a machine, only to the machine operator who guides the machine.
>
> (Uexküll 2010, 45)

This Copernican turn can be vividly outlined based on a close reading of the two programmatic articles (for context, see Mildenberger 2007, 59ff.). The authors begin the earlier essay on objective nomenclature by pointing out that everyone always only knows sensation from their own experience. By analogy we arrive at the assumption of psychic qualities in other humans or higher animals as well. According to this fundamentally proto-behaviorist position, the question of inter-subjectivity is solved through inference by analogy in ordinary life, but in science this solution leads to serious problems. The stamp of the subjective is the cause of the misunderstanding commonly known as anthropomorphism. The authors advocate introducing an objective nomenclature as to avoid this. This is of differing importance for different disciplines. Psychology, which is only preoccupied with the subjective, is not affected. Human sensory physiology, however, must adapt its terminology, according to whether it is speaking of objective stimulus, of physiological processes, or of sensations. Finally, comparative physiology especially must consider whether it is dealing with objective stimulus or physiological events. What is problematic here is that "stimulus" has a double meaning in ordinary language. For example, "light" can mean an objective quantity (the wavelength of light) or a subjective quality (color sensation). On this basis, the authors develop technical concepts that are today obsolete. However, the behaviorist aim behind them is not. The determination of biologically relevant events in terms of physically measurable results is constitutive of experimental neurobiology and behavioral research up to the cognitive turn. Subjective concepts, introspection, or inferences by analogy are bracketed there, and said bracketing occurs in reference to the programmatic writing of the three authors (see Kandel 1976, 3ff.). Inherited from the "three man manifesto" are likewise the names, proposed by Bethe and still customary today in physiology, for sense organs ("reception-organs" or "receptors"), sense-nerves ("receptor nerves" or "afferent nerves"), motor nerves ("effector nerves" or "efferent nerves") and motor organs ("effectors").

The "Proposals toward a Subject-Related Nomenclature" (Uexküll 1980) begins instead with the example of a musical apparatus (here, the barrel organ) and the question concerning its appraisal by a deaf person. This person either ascribes a building-plan to it but supposes a false meaning, or denies the building-plan because he does not recognize any meaning in the instrument. This is comparable to human access to the vital expressions of animals. In retrospect, the article by Beer, Bethe, and Uexküll – Uexküll explains – can be understood as an attempt to avoid anthropomorphic interpretations without denying the existence of conformity to a plan. The behaviorist conclusion of dispensing with the question of the meaning of animal activity and concentrating on the exploration of building-plans is appealing. However, the ensuing result, namely, declaring everything that cannot be proved to exist as a material process in the organism to be nonexistent, is way less so. Furthermore, the experimentally developed psychology of empathy in the animal soul – Uexküll continues – still bears heavy anthropomorphic marks inasmuch as it understands the animal soul as a modified or minimal human soul. Uexküll's solution is based on certainties of the life-world. It does not, however, concern isolated inner sensations but overall contexts of activity. When observing human beings utilizing objects, the meaning of the observed activity should not be placed in the acting subject but rather in the object on which the activity is performed. A chair becomes an object for sitting. The significance of the object for the acting subject in whose environment it emerges as the bearer of significance [*Bedeutungsträger*] is inferred from the activity. Uexküll assumes something analogous to the human interpretation of animal behavior. There is an active forming of the bearer of significance by the animal subject here too – a subject- and species-related structuring of the milieu, which thus becomes the *Umwelt*. The bearers of significance are to be accounted for with appropriate expressions in the relationships to the animal subject; concepts from which the kinds and manners of subject-relatedness unambiguously emerge. To this extent, the new terminological approach stands for the role of the subject in biology (Uexküll 1931a).

In the face of the above outlined development, Dzendolet's conspicuously misleading interpretation of the Beer–Bethe–Uexküll paper can be easily explained. Dzendolet (1967, 256) characterizes this early approach, which, for the authors, was a contribution to comparative *physiology* as having an important position in the development of animal or comparative *psychology* (see Mildenberger 2006). Dzendolet's perspective only suffices up to the year 1967, the year in which his contribution was made. Accordingly, it only has the innovation to Watson's behaviorism in view (see also Mildenberger 2007, 85). However, recent developments in neuroscience and behavioral research have led to the repealing of previous restrictions and now the investigation of internal representations or mental events is regarded as the main concern of cognitive neuroscience (Kandel, Schwartz, and Jessell 2013, 371). This cognitive turn in ethology opens up a wide field of behavioral elements beyond behaviorist restrictions (Menzel and Fischer 2011). It is obvious that Uexküll's thematic expansion to the subjective is carried out here, albeit with different methical means. As seldom as there are references to him (nevertheless, see Saidel 2002, 55; Timberlake 2002, 105), so seldom too is there a sense for the methical, methodological, or terminological problems discussed by Uexküll. In this respect, the reference made to Uexküll by Dorothy L. Cheney and Robert M. Seyfarth (1990, 8f.) in their book, *How Monkeys See the World*, provide good evidence for the presence of Uexküll in this debate too. The authors, pointing to the literary proximity of the title of their book to Uexküll's *A Foray*, reflect on how both approaches are connected by the emphasis on the unique perspective that forms the subjective world of a species. More closely considered, there are further family resemblances. Cheney and Seyfarth reference the turning away from uncritical mentalism with Lorenz and Tinbergen and the focus on experimental research. Despite the methods shared with behaviorism, they stress that they are rather agnostic with regard to the mind and its causal effects. Like the later Uexküll, then, they employ a behaviorist method combined with a mentalist terminology.

3 Functional circle and cybernetics

According to the "Proposal toward an Objective Nomenclature," comparative physiology encompasses the physiological process from the entrance of the stimulus to the completion of the reaction. This sequence corresponds to the typical schema of the course of a reflex arc. It was frequently represented in a linear fashion in Uexküll's illustrations and allocated to physiology (Uexküll 2010, 46). Within the framework of the *Umwelt* doctrine, the linear schema then develops into a circle that, as "functional circle," has had an enormous impact.

According to the *Theoretical Biology* (1926, 70), the difference to the unified view of reality in physics consists of biology's maintaining that there are as many worlds as there are subjects. When the observers find themselves facing an animal whose world they want to investigate, then they must distinguish between their own fixing of the "indications" (perception marks; *Merkmale*) of this world and the "mark-signs" (perception signs; *Merkzeichen*) of the animal subject (Uexküll 1926, 78). Under the *ignorabimus* that the sphere of animal sensation remains forever closed to the biological method, Uexküll's concern is the identification, through experiment and observation, of those perception marks to which the animal reacts. The starting point for the concept of a functional circle (Uexküll 1926, 126ff.) is the insight that every animal, due to its peculiar design, only selects a few stimuli from the effects of the external world, which it responds to in a specific way. In the context of the functional circle, the stimuli form the indications that the animal can rely on to have control of its movements ("as the signs at sea enable the sailor to steer his ship"; Uexküll 1926, 126). The sum of the indications is called the "world-as-sensed" (perception world; *Merkwelt*). The animal has effects on the external world, thus yielding the "world of actions" (effect world; *Wirkwelt*). Perception world and effect world form a coherent whole, the *Umwelt*. As the second edition of *Theoretical Biology* explains,

> [t]he properties of the object that act on the receptors are the perception-mark carrier [*Merkmalträger*] for the subject. Under their influence, it sets its effectors into action which in turn imprint their effect marks [*Wirkmale*] on the properties of objects and so they become the subject's effect-mark carriers [*Wirkmalträger*].
>
> (Uexküll 1973, 158, my transl.)

Perception-mark carrier and effect-mark carrier are held together by the counter-structure [*Gegengefüge*] of the object. As with the two parts of a pair of pliers for noticing and effecting, the animal subject comprises every object (Uexküll 2010, 46ff.). There is thus a meaningful coherence between animal and *Umwelt* (Uexküll 1931a, 389; see also Hassenstein 2001, 349ff.).

As Robert McClintock's inspiring study (1966) shows, twenty years before cybernetics could do the same, Uexküll's functional circle connected the feedback principle to the question concerning purposive behavior (see also Hassenstein 2001, 354f.; Lagerspetz 2001; Mildenberger 2007, 219ff.). Arguably, the same principles underpinning Uexküll's *Umwelt* doctrine lie at the heart of Nobert Wiener's mechanical cybernetics. There is not only a conceptual family resemblance, as Uexküll even applies schematic signs for the representation of the feedback mechanism and diagrams that are similar to those later devised in cybernetics (McClintock 1966, 252). However, some ambivalent relations can be pointed out between cybernetics and Uexküll's approach. On the one hand, astonishing similarities are opened up in the supposed organizational principles of living beings and their relation to the *Umwelt*. On the other hand, the disjunctive frameworks of mechanism and vitalism have here, once more, an impact. McClintock's impression is that one can make these differences fruitful, as to make up for the possible deficits of cybernetics, which consist in its being only directed toward quantifiable magnitudes and functional aspects. Questions concerning the genesis of systems described cybernetically are excluded. Uexküll's approach, however, includes a theory of development and can comprise phenomena such as self-repair, self-reproduction, and self-improvement. Also, it is not limited to quantitative relations, but, due to the purposiveness of organisms and the information of their vital plans, it is directed toward qualitative elements of significance and content. Uexküll's approach thus allows the distinction between quantitative information and qualitative informativeness. For McClintock, there is no question:

> in the light of Uexküll's rule of function, information theory should have only one duty, to enter into the design of cybernetic systems; the function circle, replete with its teleology, is the proper way to describe the functioning of brains – be they neuronic or electronic.
>
> (McClintock 1966, 253)

If this assessment is correct, then the biophilosophical significance of Uexküll's reflection comes out reinforced by what is emphasized, for instance, in Hans Jonas's criticism of cybernetics (Jonas 2001, 108ff.). Jonas devotes philosophical attention to cybernetics because of its antidualistic prospects. At the same time, it challenges the previous human self-understanding. Jonas focuses then on the programmatic contributions of Rosenblueth, Wiener, and Bigelow. These had dichotomously classified behavior in a logical διαίρεσις and distinguished between active and passive forms. In so doing, these determinations remain, as McClintock's takes them, an external relation in space and time that can be studied by behavioristic methods. However, by comparing organisms with three typical artifacts (a roulette, a clock, and a homing missile), Jonas shows that the understanding of target-oriented behavior presupposed here is inadequate and unable to grasp the conditions of internal purposiveness. The intrinsic purposefulness of living beings presupposes a subject of action who is interested in achieving the goal or purpose. For Jonas and, before him, for Uexküll, feedback in the receptor-effector system only leads to a purposeful action if it is *more* than a feedback mechanism. This is the case – so also Plessner and Merleau-Ponty – when the two elements (receptor and effector) are not directly coupled, but an interest or concern occurs between them as a third link. The duality between sentience and movement in the cybernetic model therefore is replaced by a triad (perception, motility, and emotion) in Jonas's model. As a result, Jonas lays emphasis on the subject status of animals, as Uexküll did before him, and thus exposes the inability of the cybernetic model to account for the qualitative dimension of interest.

4 The conception of *Umwelt*, ecological models, and environmental ethics

As previously made clear, from the outset a distinctive feature of Uexküll's program was its tendency toward holism. Notably, his comparative approach in physiology and ethology entails an expansion to the whole of genera and species; while his studies of the coordination of movements bring into focus the interaction of parts in the organism. Similarly, the key concept of conformity to a plan does not stand, as in classical anatomy, for a typical fundamental pattern in the morphological composition of a group of organisms, but rather for the relation between the structures and functions of living beings and their respective surroundings. These relationships to the wholes, which are expressed in the functional circles, become prominent in the elaboration of the doctrine of *Umwelt* (Uexküll 1931b). Here, not only the boundaries of the organisms are opened up to the *Umwelt* in the relational system of noticing and effecting, but also the inner world of sensation is understood as projected or transferred outwards. The interaction between parts and wholes in organisms, understood by Kant as inner purposiveness, is extended by Uexküll to the purposive order between the organism and its *Umwelt*. Uexküll's holism is therefore not limited to organismic units but also refers to organism–environment–units. A clear distinction should then be introduced between mere surroundings and the *Umwelt*, particularly when including the human observer (Uexküll 1909, 249, 1930, 129ff.). Needless to say, Uexküll's understanding of the *Umwelt* has attracted considerable attention (Tønnessen, Magnus, and Brentari 2016), especially in the philosophical discussions about the ecological crisis (Langthaler 1992, 35ff.). Uexküll's doctrine of *Umwelt* is even taken as an essential contribution to the elucidation of the fundamental natural-philosophical problems of ecology and is considered to be the basis for recent developments in ecology (K. M. Meyer-Abich 1989, 327; see also Teherani-Krönner 1996).

However, things are not always that easy (see already Haldane 1936, 32ff.). One might take into account, for instance, the manifold definitions and theories of "ecology" and *Umwelt* (Toepfer 2011a, 2011b). A great classic in ecology, such as August Friedrich Thienemann's book, *Leben und Umwelt* (1956), although it starts off with a reference to the *Umwelt* doctrine (Thienemann 1956, 8) and closes with a citation from Uexküll (Thienemann 1956, 131), also underlines the different usages of the theory of *Umwelt*. While Uexküll understands the *Umwelt* in a narrow sense, Thienemann understands by it the complex of relationships a living unity has to its surroundings. It is therefore helpful to recall Canguilhem's (2008) reconstruction of the conception of "milieu" with reference to the history of ideas. He notably concludes that Uexküll's concept expresses a paradigmatic shift. Organisms are no longer interpreted as externally determined mechanisms. For the first time, it is acknowledged that not everything can be imposed on the organism because its existence *as* an organism consists precisely in relating itself to things according to its own orientations. The organism becomes the center of all references to the milieu (see Ostachuck, Chapter 9, in this volume).

Already this philosophical and scientific-historical characterization shows that the concept of *Umwelt* has other connotations than that of ecology (see Weber

1939). Contributions to the journal *Studium Generale* have made this clear in the 1950s. The well-known German ecologist, Karl Friedrichs (1950), even attempted to reconcile the possible discrepancies between the ecological concept of *Umwelt* and the Uexküllian concept, the latter standing for the subjective taking on relations in the sense of a world of one's own [*Eigenwelt*]. To this end, he advocates a stratified concept in which Uexküll's psychological *Umwelt* constitutes one level. Friedrichs also stresses that reality itself does not know of any opposition between organism and *Umwelt*. An external point of view would then be vital to ecology, inasmuch as a biological observer objectifies both living being and environment. In the same volume, Uexküll's student, Friedrich Brock (1950), takes an alternative stance on this question. He advocates, with Uexküll, a "biological research of the world of one's own" [*biologische Eigenweltforschung*] as an element of a doctrine of life that is one of understanding rather than one of explanation. Here the methodological access to the living and its relatedness to the *Umwelt* takes a different turn. In addition to the causal order of the natural sciences, a hermeneutic system of interpretation is also legitimized. Further specificities to this approach concern the meaning of experiential phenomena, where the relationship between observer and the organism–*Umwelt* relation is structured anew. In general terms, as also Thure von Uexküll would have it (1983, xxx), attention given to the standpoint of the observer in the relational system of research is defined as characteristic of the doctrine of *Umwelt*. As an observant, I experience the animal subject and its surroundings in the manner of an outside that can be detached from me. As the experiencer, however, I am at the same time a subject myself with my outside as counterpart. Like all other subjects, it is impossible for me as an observer to be directly experientially involved in the dynamic of alien subjects. The access remains secondarily analogous – in the sense of a scientific interpretation of the building-plan.

It is therefore clear that Uexküll brings into play another understanding of ecological contexts, one in which material relationships are dissolved into epistemic or symbolic ones (Tønnessen 2009). The Uexküllian program also requires a new stance of the observer toward nature, a stance that can be characterized as participation and that has bioethical implications (Altner 1991, 140ff.). One can find here a family resemblance to Maturana's ontology of observation (Maturana 1998, 26, 156f., 169f.) in the assumed commonality between observer and observed. Finally, with respect to the relationship of the organism to the *Umwelt* understood in Uexküllian terms, the aforementioned conception of "niche construction" (Laland *et al.* 2014) has been able to grasp the core of the advocated change. As a result, the organism becomes the determining center of all relatedness to the *Umwelt*. The main lesson derived from this alternative approach amounts to always keeping in view the other living beings who are beside us, which is more than just an epistemological task. As Brett Buchanan (2008, 187) sums up, "[Uexküll's] stroll through the environments of animals asked us to step out of ourselves [...] so as to view our surroundings from perspectives other than our own." It is also clear that Uexküll's considerations take on new significance in the current debates about the agency of animals (Warkentin 2009). From the wider perspective of environmental ethics, Uexküll would ultimately prompt thinking about how we have to behave toward other living beings in the superordinate ecological community (Gens 2013; Beever 2016).

References

Altner, Günter (1991) *Naturvergessenheit. Grundlagen einer umfassenden Bioethik*. Darmstadt: Wissenschaftliche Buchgesellschaft.

Ballauff, Theodor (1949) *Das Problem des Lebendigen*. Bonn: Humboldt Verlag.

Bateson, Patrick (2005) 'The return of the whole organism'. *Journal Biosciences* 30, 31–39.

Beckner, Morton O. (2006) 'Organismic biology'. In: Donald M. Borchert (ed.) *Encyclopedia of Philosophy*. Vol. 7. Detroit/New York: Macmillan, 36–39.

Beer, Theodor, Bethe, Albrecht, and Uexküll, Jakob von (1899) 'Vorschläge zu einer objektivierenden Nomenklatur in der Physiologie des Nervensystems'. *Zoologischer Anzeiger* 22, 275–280 (Engl. in Dzendolet 1967).

Beever, Jonathan (2016) 'The mountain and the wolf: Aldo Leopold's Uexküllian influence'. *Resilience, A Journal of the Environmental Humanities* 4 (1), 85–109.

Bertalanffy, Ludwig (1937) *Das Gefüge des Lebens*. Leipzig/Berlin: Teubner.

Brentari, Carlo (2015) *Jakob von Uexküll: The Discovery of the Umwelt between Biosemiotics and Theoretical Biology*. Dordrecht/Heidelberg/New York/London: Springer.

Brock, Friedrich (1934) 'Jakob Johann Baron von Uexküll'. *Sudhoffs Archiv* 27 (3–4), 193–203.

Brock, Friedrich (1950) 'Biologische Eigenweltforschung'. *Studium generale* 3 (2–3), 88–101.

Buchanan, Brett (2008) *Onto-Ethologies: The Animal Environments of Uexküll, Heidegger, Merleau-Ponty, and Deleuze*. Albany, NY: SUNY Press.

Burghardt, Gordon M. (ed.) (1985) *The Foundations of Comparative Ethology*. New York: Van Nostrand Reinhold.

Burian, Richard M. (1993) 'How the choice of experimental organism matters'. *Journal of the History of Biology* 26 (2), 351–367.

Canguilhem, Georges (2008) [1952] 'The living and its milieu'. In: Georges Canguilhem (ed.) *Knowledge of Life*. Translated by Stefanos Geroulanos and Daniela Ginsburg. New York: Fordham University Press, 98–120.

Cheney, Dorothy L. and Seyfarth, Robert M. (1990) *How Monkeys See the World*. Chicago/London: University of Chicago Press.

Dewsbury, Donald A. (ed.) (1989) *Studying Animal Behavior: Autobiographies of the Founders*. Chicago/London: University of Chicago Press.

Driesch, Hans (1908) *The Science and Philosophy of the Organism*. 2 Vols. London: A. and C. Black.

Driesch, Hans (1911) *Die Biologie als selbständige Grundwissenschaft und das System der Biologie*. Leipzig: W. Engelmann.

Driesch, Hans (1922) *Geschichte des Vitalismus*. Leipzig: J. A. Barth.

Driesch, Hans (1935) *Die Maschine und der Organismus*. Leipzig: J. A. Barth.

Dzendolet, Ernst (1967) 'Behaviorism and sensation in the paper by Beer, Bethe, and von Uexküll (1899)'. *Journal of the History of the Behavioral Sciences* 3, 256–262.

Esposito, Maurizio (2013) *Romantic Biology 1890–1945*. London: Pickering & Chatto.

Friedrichs, Karl (1950) 'Umwelt als Stufenbegriff und als Wirklichkeit'. *Studium generale* 3 (2–3), 70–75.

Gens, Jean-Claude (2013) 'Uexküll's "Kompositionslehre" and Leopold's "land ethic" in dialogue: On the concept of meaning'. *Sign Systems Studies* 41 (1), 69–81.

Goldstein, Kurt (1995) [1934] *The Organism: A Holistic Approach to Biology Derived from Pathological Data in Man*. New York: Zone Books.

Haldane, John B. S. (1935) *The Philosophy of a Biologist*. Oxford: Oxford and the Clarendon Press.

Haldane, John B. S. (1936) *Die Philosophie eines Biologen*. Translated by Adolf Meyer. Jena: Fischer.

Haraway, Donna (1976) *Crystals, Fabrics, and Fields: Metaphors of Organicism in Twentieth-Century Development Biology*. New Haven/London: Yale University Press.

Hartmann, Nicolai (1912) *Philosophische Grundfragen der Biologie*. Göttingen: Vandenhoeck & Ruprecht.

Hassenstein, Bernhard (2001) 'Jakob von Uexküll (1864–1944)'. In: Ilse Jahn and Michael Schmitt (eds.) *Darwin & Co. Eine Geschichte der Biologie in Portraits*. München: Beck, 344–364.

Hinde, Robert A. (1966) *Animal Behaviour: A Synthesis of Ethology and Comparative Psychology*. London/New York: McGraw-Hill.

Holst, Erich von (1969) [1939] 'Die relative Koordination als Phänomen und als Methode zentralnervöser Funktionsanalyse'. In: Erich von Holst. *Zur Verhaltensphysiologie bei Tieren und Menschen*. Vol. 1. München: Piper, 33–132.

Jonas, Hans (2001) [1966] 'Cybernetics and purpose: A critique'. In: Hans Jonas. *The Phenomenon of Life*. Evanston, IL: Northwestern University Press, 108–127.

Jordan, Hermann (1929) *Allgemeine Vergleichende Physiologie der Tiere*. Berlin/Leipzig: De Gruyter.

Kandel, Eric R. (1976) *Cellular Basis of Behavior: An Introduction to Behavioral Neurobiology*. San Francisco: W. H. Freeman.

Kandel, Eric R., Schwartz, James H., and Jessell, Thomas M. (eds.) (2013) *Principles of Neural Science*. New York: McGraw-Hill.

Kynast, Katja (2012) 'Jakob von Uexküll's Umweltlehre between cinematography, perception and philosophy'. *Philosophy of Photography* 3 (2), 272–284.

Lagerspetz, Kari J. H. (2001) 'Jakob von Uexküll and the origins of cybernetics'. *Semiotica* 134, 643–651.

Laland, Kevin, *et al.* (2014) 'Does evolutionary theory need a rethink? Yes, urgently'. *Nature News* 514, 161–164.

Langthaler, Rudolf (1992) *Organismus und Umwelt*. Hildesheim/Zürich/New York: Olms.

Lassen, Harald (1939) 'Leibniz'sche Gedanken in der Uexküll'schen Umweltlehre'. *Acta Biotheoretica* 5 (1), 41–50.

Leyhausen, Paul (1973a) [1954] 'The discovery of relative coordination'. In: Konrad Lorenz and Paul Leyhausen. *Motivation of Human and Animal Behavior*. New York: Van Nostrand Reinhold, 70–97.

Leyhausen, Paul (1973b) [1967] 'The biology of expression and impression'. In: Konrad Lorenz and Paul Leyhausen. *Motivation of Human and Animal Behavior*. New York: Van Nostrand Reinhold, 272–380.

Lorenz, Konrad (1943) 'Die angeborenen Formen möglicher Erfahrung'. *Zeitschrift für Tierpsychologie* 5 (2), 235–409.

Lorenz, Konrad (1996) [1944–48] *The Natural Science of the Human Species: An Introduction to Comparative Behavioral Research*. Cambridge MA: MIT Press.

Magnus, Rijn (2011) 'Time-plans of the organism'. *Sign Systems Studies* 39 (2/4), 37–56.

Maturana, Humberto R. (1998) *Biologie der Realität*. Frankfurt a. M.: Suhrkamp.

McClintock, Robert (1966) 'Machines and vitalists: Reflections on the ideology of cybernetics'. *The American Scholar* 35 (2), 249–257.

Menzel, Randolf and Fischer, Julia (eds.) (2011) *Animal Thinking: Contemporary Issues in Comparative Cognition*. Cambridge MA: MIT Press.

Merleau-Ponty, Maurice (2003) [1956–1960] *Nature: Course Notes from the College de France*. Translated by Robert Vallier. Evanston, IL: Northwestern University Press.

Meyer-Abich, Klaus Michael (1989) 'Der Holismus im 20. Jahrhundert'. In: Gernot Böhme (ed.) *Klassiker der Naturphilosophie*. München: Beck, 313–329.

Mildenberger, Florian (2006) 'The Beer/Bethe/Uexküll Paper (1899) and misinterpretations surrounding "Vitalistic Behaviorism"'. *History and Philosophy of the Life Sciences* 28 (2), 175–189.

Mildenberger, Florian (2007) *Umwelt als Vision. Leben und Werk Jakob von Uexkülls (1864–1944)*. Stuttgart: Franz Steiner.

Mislin, Hans (1984) 'Jakob von Uexküll'. In: Roger A. Stamm (ed.) *Tierpsychologie*. Weinheim/Basel: Beltz, 44–52.

Nicholson, Daniel (2014) 'The machine conception of the organism in development and evolution'. *Studies in History and Philosophy of Biological and Biomedical Sciences* 48, 162–174.

Nicholson, Daniel and Gawne, Richard (2015) 'Neither logical empiricism nor vitalism, but organicism: What the philosophy of biology was'. *History and Philosophy of the Life Science* 37 (4), 345–381.

Plessner, Helmuth (1913) 'Untersuchungen über die Physiologie der Seesterne'. *Zoologische Jahrbücher* 33, 361–386.

Plessner, Helmuth (2019) [1928] *The Levels of Organic Life and the Human: An Introduction to Philosophical Anthropology*. Translated by Millay Hyatt. New York: Fordham University Press.

Portmann, Adolf (1961) [1953] *Animals as Social Beings*. Translated by Oliver Coburn. London: Hutchinson.

Portmann, Adolf (1983) [1956] 'Ein Wegbereiter der neuen Biologie'. In: Jakob von Uexküll and Georg Kriszat. *Streifzüge/Bedeutungslehre*. Frankfurt a. M.: Fischer, IX–XXI.

Roux, Wilhelm (1905) *Die Entwicklungsmechanik ein neuer Zweig der biologischen Wissenschaft*. Leipzig: Engelmann.

Saidel, Eric (2002) 'Animal minds, human minds'. In: Marc Bekoff, Colin Allen, and Gordon M. Burghardt (eds.) *The Cognitive Animal*. Cambridge, MA: MIT Press, 53–58.

Teherani-Krönner, Parto (1996) 'Die Uexküllsche Umweltlehre als Ausgangspunkt für die Human- und Kulturökologie'. *Zeitschrift für Semiotik* 18, 41–53.

Thienemann, August F. (1956) *Leben und Umwelt*. Hamburg: Rowohlt.

Timberlake, William (2002) 'Constructing animal cognition'. In: Marc Bekoff, Colin Allen, and Gordon M. Burghardt (eds.) *The Cognitive Animal*. Cambridge, MA: MIT Press, 105–113.

Toepfer, Georg (2011a) 'Ökologie'. In: Georg Toepfer. *Historisches Wörterbuch der Biologie*. Vol. 2. Darmstadt: Wissenschaftliche Buchgesellschaft, 681–714.

Toepfer, Georg (2011b) 'Umwelt'. In: Georg Toepfer. *Historisches Wörterbuch der Biologie*. Vol. 3. Darmstadt: Wissenschaftliche Buchgesellschaft, 566–607.

Toepfer, Georg and Michelini, Francesca (eds.) (2016) *Organismus. Die Erklärung der Lebendigkeit*. Freiburg/München: Verlag Karl Alber.

Tønnessen, Morten (2009) 'Umwelt transitions: Uexküll and environmental change'. *Biosemiotics* 2 (1), 47–64.

Tønnessen, Morten, Magnus, Rijn, and Brentari, Carlo (2016) 'The Biosemiotic Glossary Project: Umwelt'. *Biosemiotics* 9 (1), 129–149.

Uexküll, Gudrun von (1964) *Jakob von Uexküll. Seine Welt und seine Umwelt*. Hamburg: Wegner.

Uexküll, Jakob von (1905) *Leitfaden in das Studium der experimentellen Biologie der Wassertiere*. Wiesbaden: Bergmann.

Uexküll, Jakob von (1909) *Umwelt und Innenwelt der Tiere* (1st ed.). Berlin: Springer.

Uexküll, Jakob von (1922) 'Technische und mechanische Biologie'. *Ergebnisse der Physiologie* 20, 129–161.

Uexküll, Jakob von (1925) 'Die Bedeutung der Planmäßigkeit für die Fragestellung in der Biologie'. *Wilhelm Roux' Archiv für Entwicklungsmechanik der Organismen* 106, 6–10.

Uexküll, Jakob von (1926) [1920] *Theoretical Biology*. Translated by Doris L. Mackinnon. London: K. Paul, Trench, Trubner & Co.; New York: Harcourt, Brace & Company.

Uexküll, Jakob von (1929) 'Das Gesetz der gedehnten Muskeln'. In: Albrecht Bethe (ed.) *Handbuch der normalen und pathologischen Physiologie*. Vol. 9. Berlin: Springer, 741–754.

Uexküll, Jakob von (1930) *Die Lebenslehre*. Potsdam: Müller & Kiepenheuer.

Uexküll, Jakob von (1931a) 'Die Rolle des Subjekts in der Biologie'. *Die Naturwissenschaften* 19, 385–391.

Uexküll, Jakob von (1931b) 'Der Organismus und die Umwelt'. In: Hans Driesch and Hans Woltereck (eds.) *Das Lebensproblem im Lichte der modernen Forschung*. Leipzig: Quelle & Meyer, 189–227.

Uexküll, Jakob von (1973) [1928] *Theoretische Biologie*. Frankfurt a. M.: Suhrkamp.

Uexküll, Jakob von (1980) [1935] 'Vorschläge zu einer subjektbezogenen Nomenklatur der Biologie'. In: Jakob von Uexküll. *Kompositionslehre der Natur*. Frankfurt a. M./Berlin/Wien: Propyläen, 129–142.

Uexküll, Thure von (1983) 'Die Umweltforschung als subjekt- und objektumgreifende Naturforschung'. In: Jakob von Uexküll and Georg Kriszat. *Streifzüge/Bedeutungslehre*. Frankfurt a. M.: Fischer, XXIII–XLVIII.

Uexküll, Jakob von (2010) [1934, 1940] *A Foray into the Worlds of Animals and Humans with a Theory of Meaning*. Translated by Joseph D. O'Neil. Minneapolis/London: University of Minneapolis Press.

Warkentin, Traci (2009) 'Whale agency: Affordance and acts of resistance in captive environments'. In: Sarah E. McFarland and Ryan Hediger (eds.) *Animals and Agency*. Leiden/Boston: Brill Academic Press, 23–42.

Weber, Hermann (1939) 'Der Umweltbegriff der Biologie und seine Anwendung'. *Der Biologe* 8, 245–261.

West-Eberhard, Mary Jane (2003) *Developmental Plasticity and Evolution*. Oxford: Oxford University Press.

Wuketits, Franz M. (1995) *Die Entdeckung des Verhaltens*. Darmstadt: Wissenschaftliche Buchgesellschaft.

Part II

Jakob von Uexküll's relevance for philosophy

4 Creative life and the ressentiment of *Homo faber*

How Max Scheler integrates Uexküll's theory of environment

Ralf Becker

Introduction

The far-reaching philosophical implications of Uexküll's ideas in biology were first instrumentally grasped by Max Scheler. He was indeed the first to interpret and implement the biologist's environment theory philosophically. Given the nature of Scheler's theoretical agenda, the core assumptions of Uexküll's *Umwelt* were investigated mainly on an ethical and political level and tied up with selected notions stemming from Scheler's readings of Nietzsche and Bergson. In his paper "Ressentiment," released in 1912 in the first edition of the *Zeitschrift für Psychopathologie*, Scheler refers to Uexküll's early work, Umwelt *and Inner-world of Animals* [*Umwelt und Innenwelt der Tiere*, Uexküll 1921], while in his later writings the same text is quoted as "Innenwelt und Umwelt der Tiere," thus allocating the first position to the living being and its related instincts and needs. Uexküll puts forward, according to Scheler, "more appropriate ideas" than the "mechanical theory of life" endorsed by Charles Darwin and Herbert Spencer (Scheler 1912, 364, my transl.). It can be argued, all in all, that Scheler sees in the biologist, just as in Henri Bergson, an ally against the selectionism of Darwin's descent theory. Alongside Bergson, whose work, *Creative Evolution*, was later awarded the Nobel Prize for Literature and translated into German in the year of publication of Scheler's work on *Ressentiment*, Scheler conceives evolution as an active, creative process of increase of life. From Uexküll, he gains key additional insight that there is not *one* single environment, to whose changes all living creatures have to adapt, but that the creatures' abilities of perception and behavior are what initially constitute the structure of their milieu, consequently influencing the organism. Scheler could then argue that there are as many environments as existing animal species and that before the animal environment can ever exert selective pressure, a virtual selection of this environment is made by the organism itself. As a result, Scheler entrusts biology with the task of establishing that one "must always begin with the *basic relation of an organism to its environment*." This leads to claim that neither is any form of life the result of environmental factors nor is the environment a construct of the organism, and consequently that an animal and its correlative environment form a "unit of life" (Scheler 1973, 154f.).

Further contact with Uexküll's ideas comes from Scheler's interest in the *Ideas for a Biological Conception of the World* [*Bausteine zu einer biologischen*

Weltanschauung, Uexküll 1913]. Already in 1914, Scheler wrote a broad audience review of this collection of essays for the monthly expressionist journal, *Die weißen Blätter*. As is well known, Uexküll's *Ideas* start with a bang: "We are on the eve of a scientific bankruptcy, whose consequences are still not assessable. *Darwinism* is to be swept away from the set of scientific theories" (Uexküll 1913, 17, my transl.). Clearly, Uexküll is once more grist for Scheler's anti-Darwinian mill. He is quick to remark that the "modern theory of inheritance founded on Mendel's laws" acts similarly to "Darwinian genealogical tables as does chemistry to alchemy" (Scheler 1993, 394, my transl.) but, more importantly, that what rediscovered Mendel is for genetics, Uexküll is for morphology. Alchemy is here the unflattering correlate of anthropomorphism in "Darwin's and Spencer's opinion of nature." This latter, Scheler tells us, takes "our special human environment as the base" of its studies of adaptation and not the species-specific "world of perception of the animal, with which the animal carves out its environment within the abundance of the universe." Darwin and Spencer "hypostatized" then the human environment and "falsely" turned it into "the world per se" (Scheler 1993, 395, my transl.). In this regard, Uexküll's biological input helps Scheler construct a criticism of this kind of model and claim that "the mechanistic theory and Darwinism are images of life" based on an "artificial dominion of the dead world" via "labor and fabrication," whereas in "Uexküll, according to disposition, character and origin belong fully to the world, where everything develops and grows" (Scheler 1993, 396, my transl.).

The reader should be aware that after 1915, exception made for one cursory reference in two small footnotes of *The Forms of Knowledge and the Society* [*Die Wissensformen und die Gesellschaft*] (1926), Uexküll stops being mentioned in Scheler's works, at least those published during his lifetime (Scheler 1960, 259, 341). Astonishingly, there is no reference to the biologist in *The Human Place in the Cosmos* (1927/28). Nevertheless, one can easily assume that Uexküll is Scheler's clandestine source for his introductory diagnosis: "Darwin's solution of the origin of the human being" is "severely shaken today" (Scheler 2009, 5). What's more, behind the decisive introduction of the notion of world-openness (Scheler 2009, 27ff.), it is very tempting to assume some close debate with Uexküll producing the need to somehow overcome the limits of his "environment." It is true, however, that nothing *in the texts* explicitly indicates that Scheler perceived it this way himself. Wherever he makes explicit reference to Uexküll, his remarks are especially praising of his being close to his own anthropology. The human being is there introduced as "*that X who can comport himself, in unlimited degrees*, as 'world-open'"[1] (Scheler 2009, 28) – with special emphasis on *can*! Scheler's solution consists in claiming that, as "living beings," humans know an instinct-bound environment with respect to the structure of the milieu, and only as "persons" they are intellectual and world-open beings. In my opinion, the fact that Uexküll did not acknowledge this layer, which animals lack, is not of great importance. Arguably, Scheler was especially engaged with the biologist's writings only in the years between 1912 and 1915, such an episodic reception perfectly corresponding to his temper as an author and reader. Based solely on the available texts in his *Collected Works*, one could say that Scheler was only familiar with Uexküll's early writings Umwelt *and Inner-World of Animals* and *Ideas for a Biological Conception of the World*. Finally, it should not be forgotten that the description of "the drama of animal behavior" (Scheler 2009, 28), as a full account of the structure of instincts in the inner world, is not a pure recital of Uexküll's theory. Many sources actually influenced Scheler's philosophical biology, not least Uexküll. What is certain, moreover, is that Scheler's anthropology never had Uexküll as its polemical target.

What follows aims to chronologically reconstruct Scheler's interest in Uexküll based, philologically, on his writings. The reader will learn how Uexküll suitably served Scheler's cultural-philosophical, ethical, and political interests: (1) first of all, Uexküll confirms Scheler's inquiries on *ressentiment* as distinctive feature of the spirit of modern culture; (2) consequently, Uexküll's ideas, while contributing to Scheler's critique of Spencer, provide solid ground also to his materialist ethics of values; (3) finally, Scheler does not consider himself too good for jumping on the bandwagon of anti-English rhetoric and declares war against English biology, under the banner "*Los von England.*" Regardless of whether cultural-philosophical, ethical, or political – all these strands in Scheler's intellectual personality rely on one decisive philosophical distinction concerning life and one decisive idea: the difference between development and preservation of life and the idea that preservation is in the service of development. It is Scheler's opinion that only lack of development leads to that "struggle for life or existence," on which Darwin and Spencer place special emphasis. For Scheler, life is much more solidarity and affection than competition and egoism; it is active increase and not just passive adaptation, not only self-preservation but also creative evolution.

For modern readers, Scheler's rejection of Darwinism may well seem obsolete, even revisionist. That is why some clarification seems to be necessary. First of all, it should be clear that doubts about Darwin's "theory" do not necessarily imply doubts about "evolution." For Scheler, beyond a shadow of a doubt species living today have formed over the course of a million-year-long process of development and shift, survival and extinction. His own theory diverges from the Darwinian – decades before the synthesis of "neo-Darwinism" – concerning species-forming mechanisms. Unlike Darwin, or more emphatically than Darwin, Scheler hinges this mechanism on life itself: organisms are neither the product of a creator nor the product of coincidence (as both are external factors) but the product of an (internal) abundance; life is creative by itself. Scheler's criticism attacks the whole "model" of life as struggle and competition over a shortage of resources, based on which the individuals that are best adapted to their environment emerge as winners. Behind such a model Scheler perceives the transfer of Manchester Capitalism to natural circumstances. In this regard, his criticism of Darwinian "selectionism" and "adaptivism" mirrors a criticism of a civilization that sets its life values on (economical-functional) utility.

1 Cultural criticism (1912/15)

Uexküll is quoted by Scheler for the first time in the essay "Über Ressentiment und moralisches Werturteil. Ein Beitrag zur Pathopsychologie der Kultur," republished under the new title *Das Ressentiment im Aufbau der Moralen* in 1915 (English translation: *Ressentiment*). Since Uexküll is not mentioned until the last part of the essay, the context has to be defined more closely. The essay represents Scheler's answer to Nietzsche's *On the Genealogy of Morality* and is his attempt to defend Christian ethics against the accusation of being a "slave morality" born out of "ressentiment." Devaluation of Christian values, especially of altruism, is, according to Scheler, in itself a manifestation of a profound, modern "ressentiment" of the weak against the strong, which leads to a revaluation of the merely pleasant and useful opposed to vital, spiritual, and holy values. Nietzsche and his psychology of *ressentiment*, as Scheler sees it, end up throwing out, so to speak, the baby with the bathwater. For it is not Christianity, according to Scheler, the symbol of the "slave revolt in morality" but, rather, the "mechanistic world view" of industrial modernity (Scheler 1994, 124). The "spirit of modern civilization" constitutes

> [a] *decline* in the evolution of mankind. It represents the rule of the weak over the strong, of the intelligent over the noble, the rule of mere quantity over quality. It is a phenomenon decadence, as is provided by the fact that everywhere it implies a *weakening of man's central, guiding forces* as against the anarchy of his automatic impulses. The mere means are developed and the goals are forgotten. And that precisely is decadence!
>
> (Scheler 1994, 125)

Scheler's polemical target, then, is not the religion of incarnate God but, rather, the ideology of the apotheosis of humans as *Homo faber*. As he understands it, the triumphal march of instrumental reason subordinates everything under the diktat of utility, its ethics being utilitarianism.

A few more words should be added here on Scheler's ideas on technology. According to him, it is not creative genius what creates technology, but rather the hindered living creature:

> The near-sighted man will praise his eyeglasses, the lame man his stick, the bad mountain climber will extol the rope and climbing irons which the better one holds for him with his arms.
>
> (Scheler 1994, 123)

In a remarkable anticipation of Gehlen, Scheler emphasizes organic deficiencies, which humanity has to relieve itself from, as the trigger of technical civilization. Tools are only poor "compensations" of organic features, not their "extension." Furthermore, the value attribution within the relationship of organs and tools is identified by Scheler as the result of "ressentiment" as implied by utilitarian "slave morality": "The positive valuation of tools is not due to *ressentiment* – only the assumption that tools are as valuable as organs!" (Scheler 1994, 123).

Scheler's criticism of technology is not a critique of technology as a "matter" but a critique of the "value" we attach to it:

Man, as the biologically *most stable* species, must create civilizations, and he should do so – provided that the subordinate forces, and the forces of dead nature, are employed in order to relieve nobler forces. But he should remain within these *limits*, i.e., the tool should *serve* life and its expansion.

(Scheler 1994, 123)

However, it is exactly this master–slave ratio what industrial modernity reverses:

With the development of modern civilization, *nature* (which man had tried to reduce to a mechanism for the purpose of ruling it) and *objects* have become *man's lord and master*, and *the machine* has come to dominate *life*. The 'objects' have progressively grown in vigor and intelligence, in size and beauty – while man, who created them, has more and more become a cog in his own machine.

(Scheler 1994, 123)

According to Scheler, this is mainly

[d]ue to a fundamental *subversion of values. Its source is ressentiment*, the victory of the value judgements of those who are vitally inferior, of the lowest, the pariahs of the human race!

(Scheler 1994, 124)

Finally, human civilization is a *"surrogate* for the formation of organs." (Scheler 1994, 122). Scheler sees "the formation of organs" in the "formative activity of an agent" (Scheler 1994, 121) and not in the adaptation via tools. Looking back to Nietzsche, he retrieves a notion of life as chiefly "activity." Modern biology pushes forward "the theory of 'adaptation'" "into the van of the argument, exploited; adaptation – that means to say, a second-class activity, a mere capacity for 'reacting' [...]. This definition, however, fails to realize the real essence of life, its will to power" (Nietzsche 1913, 92; cf. Heinen 2012). Although Scheler does not endorse Nietzsche's notion of "will to power," which he considers a faux pas, he takes on board the ranking of actual activity (self-activity) and mere reactivity (adaptation). In this respect the lower-ranking paradigm of adaptation is directly linked to anthropomorphism, and detected already in the sphere of language. Philology tells us, for instance, that *organon* is the ancient Greek word for *tool*. And since the value of tools lies in their utility, when we conceive organs as tools, produced, that is, analogously to functional human artifacts, despite their not being crafted, we actually tend to pit their value against how successfully they can help the living being in dealing with the demands of its environment. That is how, in Scheler's analysis, the success of adaptation becomes the alleged agent of the formation of organs.

This helps us understand in what respects Nietzsche's and Scheler's notions of life radically differ from Darwin's and Spencer's understanding of the same issues. To an idea of life understood mainly as preservation, the German philosophers oppose an inquiry on life in terms of increase of being; whereas Darwin and Spencer explain the diversity of species as the result of lack of resources, Nietzsche and Scheler refer to an abundance (of vital energy), and where the ones focus on the organism adapting to its niche, the others bring attention to how the organism expands its milieu. According to the latter perspective, only when "the vital activity *stagnates* and can *no longer extend* the milieu by the formation of new organs" (Scheler 1994, 122) does adaptation take place. In this respect, adaptation is a reaction to lacking activity. And this is exactly, according to Scheler, what happens with humans. Since we do not extend our milieu through the development of new organs, we shift to the invention of tools. The stagnation of our physical evolution forces us to do so. From this viewpoint, theorists of adaptation mistakenly apply "our" strategy of coping with life to the evolution of "all" species:

> Guided by *ressentiment*, the modern world view reverses the true state of affairs. [...] It interprets life as such as an accident in a universal mechanical process, the living organism as a fortuitous adaptation to a fixed dead milieu. The eye is explained by analogy with spectacles, the hand by analogy with the spade, the organ by analogy with the tool.
>
> (Scheler 1994, 122f.)

This amounts to declaring prostheses the measure of nature. According to Scheler, adjustment mechanisms are only effective when a creature's vigor is not sufficient to shape its environment. In any other scenario, adaptation is unnecessary. Besides Nietzsche, Scheler's biophilosophy finds support also in Bergson and in Uexküll (in 1915 Driesch joins this party). According to Bergson, the *élan vital* cannot be comprehended and explained

> [b]y applying concepts and forms of perception that are proper to an 'intellect' which has itself originated as *an instrument* of the specifically *human* vital activity and is completely dependent on its tendencies.
>
> (Scheler 1994, 140)

Similarly, Uexküll teaches us that the natural *Umwelt*

> [i]s not a mere 'datum' to which [man's] vital activity must passively adapt itself. It has been *selected* from an abundance of phenomena by the course and direction of this vital activity.
>
> (Scheler 1994, 122)

By integrating Bergson's terminology and Uexküll's emphasis on activity, Scheler is then able to outline a bi-directional understanding of evolution: "The same process which forms the organ also determines the character and structure of the 'milieu' or 'nature' to which a species tries to adapt by means of tools" (Scheler 1994, 122). Hence, organisms are not merely passive objects but active subjects of evolution. Our "vital organism," Scheler claims, selects "that corner [...] in the

universe" to which we humans adapt our tools (Scheler 1994, 122). On the opposite theoretical corner, Darwin and Spencer are said to expand our nook to "one"
environment for "all" living beings and then consequently subjugate them under
the compulsion to adapt themselves.

Within the framework of Scheler's criticism of Herbert Spencer's (theoretical)
biology and sociology, Uexküll provides him with a powerful argument. In a few
words, one can say that Spencer mistakes "an *image* [...] for the *thing itself*"
(Scheler 1994, 121). As the reference to *Ideas* – included in the second edition –
proves, Uexküll allows Scheler to argue that Spencer ignores the species-specific
dependency of environments and succumbs to anthropomorphism, as he understands the different milieus of the several organisms as different branches of
human nature, which is instead specifically dependent on human perception and
conceptualization. Clearly, the core of Scheler's criticism does not target empirical biology but, rather, an underlying philosophical stance. In this regard, Spencer
finds himself on the firing line much more often than Darwin, as the relative fre-
quency of their respective names in Scheler's texts shows.

As previously anticipated, the criticism of this scientific model mirrors the
criticism of a cultural self-conception as *Homo faber*. This is the powerful image
that is being confused with the thing itself. Technology, Scheler would argue,
does not derive from nature but from a culture of coping with life. Such a strategy has then led to the utilitarian ideology, which declares utility its determining
value: to be good is to be useful. As a result, Scheler can argue, modern culture
mistakes the bottommost for the topmost. The modern attribution of values – the
appreciation of the *bonum utile* at the expense of values of life and intellectual and
cultural values – is derived, Scheler concludes, from the *ressentiment* felt by the
organically disadvantaged human creature against the organically creative life:
since humans have to get by via technology in those cases where life can usually
find a way for itself, mankind devalues organic vital processes and revalues its
own technical achievements – up to the point that the vivid side of mankind is
subordinated to the technical mastery of nature in our industrial modern world.

The next section looks closely at Scheler's theory of milieu and his value ethics. At this point, it has to be noted that at the core of Scheler's reception of
Uexküll lies the idea that animal environments are not subsets of the human environment because the "essence and structure" of any environment are dependent
on the respective "vital organization." If instrumental behavior is part of the "vital
organization" of humans, that does not imply that every animal milieu is coined
by the essential structure of utility.

2 Ethics of values (1913/16)

In *Formalism in Ethics and Non-Formal Ethics of Values*, Scheler grasps the
opportunity to expand his theory of milieu (for a concise account, see Gurwitsch
1977, 82, 95) to the realm of ethics and key elements of his arguments are indeed
clearly reminiscent of Uexküll's lesson. Scheler distinguishes four types of values: (1) values of the pleasant and unpleasant (including the useful), (2) values

of life and vitality (the noble and common), (3) spiritual values (aesthetic values of the beautiful and the ugly, values of the fair and wrong as well as values of the pure cognition of truth), and (4) values of the holy and unholy. All these share a common reference, inasmuch as something is, for instance, pleasant, unpleasant or useful, only "for" a living being. The value of life itself is not a physical value but an intellectual one and, as such, is rooted in infinite spirit [*Geist*]. While the holy and unholy values are the highest, the ones of the pleasant and unpleasant score the lowest in an argumentatively determined hierarchy of values. It is clear, then, that Scheler aims to demonstrate that "it is impossible to reduce vital values to the *useful*" (Scheler 1973, 277) and thus dismisses utilitarian ethics altogether. I will come back to Scheler's criticism of utilitarian and deontological ethics' common premise later on.

Scheler understands milieu as "the value-world as effectively experienced in practice" (Scheler 1973, 142). His idea is then that, depending on our attitudes toward the values we experience, our environment is shaped by certain qualities of values. Practical experience, in this respect, precedes conscious perception and attention-guiding interest alike. "[W]e also frequently experience the *effectiveness* of *something* that we do *not* perceive" (Scheler 1973, 140), Scheler argues. "The milieu is *not* the sum of all that we sensibly perceive; rather, we can only sensibly perceive *what belongs to the 'milieu'*" (Scheler 1973, 148):

> To a forester, a hunter, and someone talking a walk, the same forest represents different 'milieux.' In principle, the forest provides a milieu for a deer which differs from that for a human and, again, for a lizard living in it.
>
> (Scheler 1973, 143)

This is, clearly, where Uexküll comes into play: just like a "dragonfly's world" knows nothing but "dragonfly things" (Uexküll 1921, 45, my transl.), so does a hunter's world know nothing but hunter things and so on. Scheler then understands the difference between social human milieus in the light of the specific difference – effectively elucidated by Uexküll – between various animal environments, as we can see in the following remark:

> A Philistine remains a Philistine; a Bohemian remains a Bohemian. Only that which carries with it the value-complexes of their attitudes becomes part of their 'milieu.' Human beings belonging to a specific social rank, race, ethnic group, or occupation, and even individuals, carry with them the structure of their own milieu.
>
> (Scheler 1973, 143)

As to nuance the species-specific dimension of milieus, a "variation" in social regard is not to be excluded. This can come about in the form of a shift of what is actually experienced – for example, a ranger can go through a forest not as ranger, but as stroller on a Sunday stroll – thus implying a shift in our attitude toward values but not a shift of the whole perspective (cf. Scheler 1973, 140). Environment is still a milieu dependent on evaluating attitudes.

Aron Gurwitsch (1977, 95) has rightfully objected that Scheler fails to clarify what allows him to speak of the "same" forest, that ranger and hunter, roebuck and lizard experience differently. Only his anthropology actually provides an answer to the problem of perspectivism via the differentiation of environment-dependence and world-openness. His early remarks also point in that direction, as Scheler distinguishes between two kinds of transcendence: "the primary tendency of life," "in the *transcending* of a given milieu, in extending it and conquering something new in it" (Scheler 1973, 283) and the other tendency of human beings to "*transcend [their] own life and all life*" (Scheler 1973, 289). While the first kind of transcendence does not repeal the dependence of the milieu on our perception, the second kind allows grasping the idea of an object, that is, independent of attitudes, something in itself. This transcendence does not turn humans into a different kind of beings but makes them "intellectual" creatures, persons. This particular primary tendency of life to expand and conquer milieus spawns new species of living creatures, not the enhanced adaptation to a milieu. This latter rejected claim is "not only an error of observation but also one of philosophical relevance" (Scheler 1973, 155, fn. 46), as it relies, according to Scheler, on the wrong "idea of environment."

The environment of a living being is the "totality, or the *uniform whole*, of the world [...] which is effectively experienced" (Scheler 1973, 154). What is not being experienced as effectively does not belong to the environment of a living being:

> One who fails to see this point will be led into a false *anthropomorphism* by considering man's environment *basic* to all other organizations and by examining their adaptations to *man's* environment, which is not *theirs*. A worm's or a fish's environment is not at all 'contained' in the human environment. The environment of different animals is always fixed by special procedures. [...] It is only between the environment of an organization and its *members* that there are different adaptations. Spencer's primary mistake in biology and the theory of knowledge is his idea that the world of organizations is related to the environment of *man* and that changes in higher organizations are to be reduced to a mere adaptation of organisms to this 'environment.' The *activities* of life (and its *directions* and their changes) that alone *determine* environment fall by the wayside.
>
> (Scheler 1973, 155f., fn. 46; cf. 282, 291)

The theory of adaptation, it is argued, is unable to account for the conquest of new habitats. It interprets the relationship between organism and environment as unilateral, instead of conceiving "*each* as a dependent variable of the *processes of life* in its uniform occurrence" (Scheler 1973, 155). Scheler does not deny that living beings adapt to environmental circumstances, but he distinguishes "traits of adaptation" (e.g., different types of leaves of aquatic and desert plants) from "traits of organization" that cannot be linked to features of adaptations. Thus, one cannot explain the difference between botanical and animal organization or between invertebrates and vertebrates via processes of adaptation.

Nor does the principle of preservation suit the explanation of the diversity in the forms of organization of the living. For Scheler, the "reproductive drive in all living things is stronger and more original than the drive of self-preservation" (Scheler 1973, 158, ft. 50):

> The drive of reproduction and propagation *precedes* the drive of preservation; and only to the degree that the drive of propagation meets certain obstructions is an increased drive of preservation formed in individuals.
>
> (Scheler 1973, 280)

What applies to adaptation holds true also for preservation. Preservation becomes the predominant force only when the vital force stagnates. The tendency of spatial extension corresponds to the reproductive drive in temporal regard. Alongside the references to the couplets milieu/adaptation and reproduction/preservation, Scheler adds one more hierarchy-instituting conceptual pair, that of sympathy/egoism: "love, sympathy, devotion, and striving for sacrifice essentially belong to life as much, as does the tendency to growth, development, and power." Egoism is not an "original vital tendency"; rather, it is based on "a *loss*, on a *removal* of the *feelings of sympathy* that belong *originally* and naturally to all life" (Scheler 1973, 278). Egoism, the instinct of self-preservation, and adaptation dominate only provided that living beings lack the vigor of appeal, reproduction, and transcendence (of the milieu). Only then the so-called struggle for life – that Darwin and Spencer claim is the main motor of selection – takes place.

Struggle is not representative, according to Scheler, of the norm in the realm of the living:

> If we begin with the most marked differences among the living beings that we know, i.e., between plants and animals, the struggle is in this case totally *sub*ordinated to the principle of solidarity. The different forms of nutrition for each realm preclude struggle, or at least make it quite secondary in importance.
>
> (Scheler 1973, 282)

Plants and animals do not compete with each other, because they live in "different milieus":

> [A] basic condition for the possibility of a *competitive struggle* is that this struggle take[s] place in a given milieu-structure which is *common* to the units of life in this struggle, or, in other words, that this milieu-structure present[s] *common elements*. When this is not the case, there is of course no common ground for the struggle.
>
> (Scheler 1973, 283)

Whether or not the notion of solidarity as an antonym for struggle was a wise choice is debatable. It could indeed be argued that the idea of a competition about scarce resources is not less anthropomorphic than the one of solidarity between forms of life. Rather one-sidedly, Scheler is only keen on remarking that Darwin's and Spencer's concept of a "struggle for life" is the result of

> [a] transfer of concepts derived from human civilization to the extra-human world. [...] Haunted on the one hand by a vision of poverty of the masses and the accompanying struggle of the workers for higher wages and more food in an industrialized England, and prepared on the other hand by his Calvinistic dogmatism to regard nature as 'poor' and 'needy,' as too small for the powers of vital drives, an orthodox preacher came up with the idea of man's necessarily competitive struggle for food and other well-known theories connected with this, which are expressed in Malthus' laws of population. Darwin extended this idea to all organic life.
>
> (Scheler 1973, 281)

The science of the struggle for life seems to reflect a bit too closely the special circumstances of the industrial modern age, back then particularly prominent in Great Britain. There, the tendency to adapt to a stagnating milieu via hard labor would prevail. Fierce competition would then ensue as necessary effect, given the competing of many individuals for the limited resources in a "single milieu" – provided no change takes place within the working-class environment. The expansion of the milieu would make, for instance, competition unnecessary. The underlying claim, here, being that only the ones who are not able to change their living conditions adapt to these.

These remarks are to be found in the second part of *Formalism*, published in 1916, and here Uexküll is never mentioned. His milieu-pluralistic ecology is recognizably overshadowed by the critique of the primacy of preservation and the principle of competition, which Scheler combines with his critique of modern ethics: rationalists and empiricists, according to Scheler, base their different ethical concepts on the same psychology, which applies the "basic concepts and principles of mechanics, especially the *principles of preservation*, to the phenomena of life" (Scheler 1973, 277). According to this genealogy, the concept of preservation – whether as physical being (Spencer) or rational being (Kant) – traces back to the classical mechanics of the preservation of impulses and is transferred into biology and psychology. The physicalistic view on life, according to this viewpoint, plays a central role even in Kant's deontological ethics: perfect duties (e.g., prohibition of suicide and forbiddance of lying) are duties of preservation. A full discussion of the eligibility of this critique would clearly exceed the scope of this chapter. It is nevertheless worth emphasizing to what extent Scheler was able to trace ethics back to an anthropological self-conception. The manner in which we think of ourselves as corporal beings or, more precisely, as living entities made of individual being and environment, is mirrored by our "valuation of values." If we assume that we have to adapt to a given milieu, egoism, preservation, and utility rank first. On the other hand, if we understand our milieu as something dependent on individuals, social groups, and species, then appreciation, growth, and prosperity receive a higher status in the hierarchy of values. One has to distinguish the afore-mentioned objective order of values (according to circumstances of foundation) from the subjective (sociocultural) "appreciation" of values. Where higher values are esteemed as something low, there morality is supported by "ressentiment."

3 Politics (1914/15)

The third writing that documents Scheler's engagement with Uexküll utilizes his theory of environments for political purposes. Scheler claims that war is more than a mere "struggle for existence" and the continuation of self-preservation by different means. Self-preservation in itself is in service of the creative devel-opment of life. Especially the fusion with Bergson's *élan vital* reveals where Uexküll inspired Scheler's thinking: the "diversity" of the animate world is not due to the adaptation to an interspecies "common" milieu in the struggle for the "same" scarce resources in the "same" territory. In fact, the extension of different lifeforms has perfectibility as "super-mechanical cause." The pure competitive struggle is therefore the exception because it presupposes that the competitors' environments meet the "same" requirements. But if the milieu is dependent on criteria of relevance that are "different" from species to species that would mean that competition would only happen among beings with a similar vital organiza-tion and accordingly very similar criteria of relevance. What is new about the fol-lowing context under reconstruction is that Scheler, with his ally Uexküll, carries the "battle for ideas" to the battlefields of the Great War.

War unleashes rhetoric. In October 1914, Scheler released his pamphlet "The Genius of War." This war is for Scheler "first and foremost a *German-English* war" (Scheler 1914, 1348, my transl.). It is not a war of economic and political interests but a war of philosophies: metaphysics versus positivism, ethics of con-viction versus utilitarianism, communitarianism versus contractualism, creative mind versus rational intellect, vitalism versus mechanism, culture versus mere civilization. It is a war against capitalism and the "motherland" of the "positivist doctrine of interests" (Scheler 1914, 1348, 1333, my transl.). War takes place in human life whenever life increases its power through the development of organs and the expansion of milieus – it has a "*vital* root" (Scheler 1914, 1342, my transl.), but it is spiritually reshaped in the human sphere. "Real culture is greater than power; the genius is of a higher echelon than the hero!" (Scheler 1914, 1342, my transl.) This German–English war, according to Scheler, is not just a matter of two nations fighting each other but, rather, of two "cultures" of valuation of values opposing one another: the English preference of utilitarian values versus the German preference of "higher," spiritual, (and holy) values. The pamphlet is a continuation of the critique of Darwin's and Spencer's biology on the battle-ground of politics and declares war as the rebuttal of competitive struggle by dif-ferent means. Scheler identifies "two roots of all human 'struggles'":

> For every economic competition between individuals and nations, the root is the same as the one that governs the animal struggle for nourishment and prey; this root is the principle of the increasing 'adaptation' to a given, steady

environment for the development of technology and ecological forms of organization. [...] However, the root of *war* is [...] the deeper principle – that is more substantial to life – of original increase of power by extension and shaping of the environment of the *nobler* and more mannered human groups.
(Scheler 1914, 1336f., my transl.)

The former root has the self-interest of the individual as a precondition, while the latter has the common interests of the collective as precondition. Economical competition and war correspond in the human realm to similar life tendencies such as self-preservation and the "extension and shaping of the 'environment'" (Scheler 1914, 1335, my transl.). Once more Scheler refers here to Uexküll, who "very instructively" studies milieus with a species-specific approach (Scheler 1914, 1336, my transl.). The critique of Spencer's anthropomorphic theory of evolution, which equates the human environment to the general environment of all living beings and the struggle for existence to the vital factor in the development of new forms of life, is turned into a political affair by Scheler, thereby making an inglorious contribution to the warmongering against England.

Darwin and Spencer represent the "English biology" that "is the mere projection and universalization of the *liberal and utilitarian principles of the English philosophy of businessmen on the whole realm of organic life*" (Scheler 1914, 1337f., my transl.). For the "English thinking," every fight is a competition; for Scheler, every competition is a sign of decadence. This goes along with "a lack in development of power of life" and the stagnation of organic formation as well as the formation of environment. War, on the other hand, is a "*milieu-extending force*," an expression of the original tendency of life toward "increase, growth and development of its form and functions" (Scheler 1914, 1334, my transl.).

In conclusion, Scheler's apologetics of war combines the theory of milieu and the theory of values with a Nietzsche-inspired life philosophy: The "Darwinian and Spencerian conception of life [...] strips [...] life of its nature: 'activity.'" The different forms of life are not due to the successful but ultimately random adaptation to variable environmental conditions. All processes of adaptation and mere self-preservation are preceded by "the tendency of extension and the active shaping of the environment – Nietzsche called it unilaterally and inappropriately the 'will to power'" (Scheler 1914, 1335, my transl.). On this basis, Scheler legitimizes war as the human variant of the extension of the milieu and the active shaping of the environment. Without war, "the mighty and noble, located within the minority," would succumb to the ordinary mediocrity, "the owners of the virtues and vices of adaptation such as cunning, slinkiness, industriousness, but also mendacity, subservience, egoistic righteousness" would survive and outlive the "owners of the contrary 'heroic' properties" (Scheler 1914, 1338f., my transl.).

Nevertheless, weapons do not have the final say. The "last objective *telos*," which serves war, is the

reign of the mind on earth and above all: the development and extension of any of the many forms of entities of life, that formed as people, nations and so on, represent the opposite to the mere factual and legally formed communities of interests.

(Scheler 1914, 1328, my transl.)

Thus, war is as much a milieu-extending force as the "*strongest* force of human unification" (Scheler 1914, 1349); the "genius of war" is the development of unity. First, the entity of the different war parties for themselves: war binds a nation together in an "action group." But history does not stand still at this point. Scheler is definitely aware of the idea of perpetual peace, which is only possible when egoism and mere "solidarity of interests" are replaced by the "attitude and solidarity of will." The ideal condition is the unity of a "*comprising community of love* of all rational beings, which is precisely the *opposite* of the positivist ideal of humankind merely unified by solidarity of interests and contracts." The role model for such a broad community of love of all humans is the "Christian idea of the Kingdom of God." The historical mission of war is to "make war dispensable!" (Scheler 1914, 1351f., my transl.) In order to do so, it is at first necessary to defeat the utilitarian valuation and the interest-driven competition – wherein Scheler sees the historical mission of Germany and Austria.

However, Scheler's final word is not one of vitalistic bellicism. He dissociates quite promptly from his enthusiasm for war. That is why the reconstruction of Scheler's engagement with Uexküll should conclude with a remark that guides our view away from death and back to life. Once more, Scheler quotes Uexküll in order to disagree with Darwin's and Spencer's adaptativism. Two "basic kinds" of values of life are to be distinguished: the "pure" values of life as *values of development* as well as the "technical values of life or *values of preservation*. The latter serve the former" (Scheler 1957, 311, my transl.). Mere preservation and adaptation are, according to Scheler, no explanation for why evolution spawned ever so complex types or organizations. The development from "jellyfish to human" cannot "be traced back to the mere value of benefit of always heaping adaptations to one milieu common to all living beings." Here, Scheler recommends Uexküll again as further reading and continues: the extension of space has "an over-mechanic cause (urge of perfection, 'élan vital', entelechy of real species and so on)" (Scheler 1957, 312, my transl.). If life would be pure preservation – it would have stopped at the stage of protists. The diversity and increasing complexity of forms of life must have another cause: Scheler sees this cause in evolution itself, that is not only the effect (of preservation and adaptation) but also the driving force. Life is a waste of itself. For Scheler there is no other way to explain the exuberant diversity of living beings. Uexküll promises Scheler the beginning of a new biology, which fits this diversity in a better way than Darwin's descent theory.

The history of biology passed over Uexküll's and Scheler's cancelation of Darwinism from the set of scientific theories. At the beginning of the 20th century, naturally, one could not anticipate how modern evolutionary biology would differentiate theoretically. If, from a distance of meanwhile over hundred years, one reconstructs the discourse in which Scheler participated, it becomes evident that not the validity claim [*Geltungsanspruch*] of empirical knowledge is debatable but the validity claim of a normative assertion. We are dealing with a "practical" discourse, which is concerned with the question of what "value" we assign to life as such and especially to pre-personal life. Scheler, with his analysis of the "ressentiment," wants to draw attention to the fact that life is more than mere coping with it and that the instrumental-functional utility must be subordinated to the actual values of life. In his *Formalism*, he emphasizes the principle of solidarity and the role of emotions of sympathy even in prehuman nature. And free of all apologetics of war, one can subscribe to the assumption that self-preservation is important but that self-development is of higher value.

As was shown in this chapter, Scheler adopts Uexküll especially in the systematic and temporal context of his ethics of values and of his order of values. His first anthropology is the "ethical personalism," whose foundation should be built by his *Formalism*. According to the terminology of *The Human Place in the Cosmos*, utilitarianism, which is the philosophy behind the "Darwinian-Spencerian notion of life," detaches the level of practical intelligence from the level of the vital force as well as from the sphere of the mind (and thereby from personality). *Homo faber* (as well as his youngest offspring, the *Homo deus*) is a torso – without legs and without a head. Uexküll's theory of environment provides Scheler with an antidote to anthropomorphisms and to the "biologist's fallacy": the biologist must not project his own environment into animal environments. He has to gain knowledge of the species-specific environment from the animal behavior. Animals "shape" their environments; they are not just passively exposed to its changes. Selective pressure of adaptation could turn out to be a genuine human experience that does not explain the abundance of life forms necessarily. That is what Scheler has learned from Uexküll.

Note

1 The use of italics in Scheler's quotations, if not otherwise indicated, always follows Scheler's original text.

References

Gurwitsch, Aron (1977) *Die mitmenschlichen Begegnungen in der Milieuwelt*. Berlin/New York: Walter de Gruyter.

Heinen, René (2012) 'Darwinismus als Mythos und Ideologie. Nietzsches Kritik an Darwin und ihre Fortsetzung bei Scheler und Adorno'. *Nietzscheforschung. Jahrbuch der Nietzsche-Gesellschaft* 19 (1), 353–373.

Nietzsche, Friedrich (1913) [1887] *The Genealogy of Morals*. Translated by Horace B. Samuel. In: *The Complete Works of Friedrich Nietzsche*. Vol. 13. Edinburgh/London: T. N. Foulis.

Scheler, Max (1912) 'Über Ressentiment und moralisches Werturteil. Ein Beitrag zur Pathopsychologie der Kultur'. *Zeitschrift für Psychopathologie* 1, 268–368.

Scheler, Max (1914) 'Der Genius des Krieges'. *Die neue Rundschau* 25 (10), 1327–1352.

Scheler, Max (1957) [1911–1921] 'Vorbilder und Führer'. In: Max Scheler. *Gesammelte Werke*. Vol. 10: *Schriften aus dem Nachlass*. Band I. *Zur Ethik und Erkenntnislehre*. Bern: Francke Verlag, 255–318.

Scheler, Max (1960) *Gesammelte Werke*. Vol. 8. *Die Wissensformen und die Gesellschaft*. Bern: Francke Verlag.

Scheler, Max (1973) [1913–1916] *Formalism in Ethics and Non-Formal Ethics of Values: A New Attempt toward the Foundation of an Ethical Personalism*. Translated by Manfred S. Frings and Roger L. Funk. Evanston, IL: Northwestern University Press.

Scheler, Max (1993) [1914] 'Jakob Baron von Uexküll: Bausteine zu einer biologischen Weltanschauung' (book's review). In: Max Scheler. *Gesammelte Werke*. Vol. 14. Bern/München: Francke Verlag, 394–397.

Scheler, Max (1994) [1915] *Ressentiment*. Translated by Lewis B. Coser and William W. Holzheim. Milwaukee, WI: Marquette University Press.

Scheler, Max (2009) [1928] *The Human Place in the Cosmos*. Translated by Manfred S. Frings. Evanston, IL: Northwestern University Press.

Uexküll, Jakob von (1913) *Bausteine zu einer biologischen Weltanschauung. Gesammelte Aufsätze*. München: Bruckmann.

Uexküll, Jakob von (1921) *Umwelt und Innenwelt der Tiere* (2nd ed.). Berlin: Springer.

5   Closed environment and open world

On the significance of Uexküll's biology for Helmuth Plessner's natural philosophy

Hans-Peter Krüger

1 Three systematic focal points of Plessner's engagement with Uexküll

As early as the preface to the first edition of *The Levels of Organic Life and the Human. Introduction to Philosophical Anthropology* (1928), Helmuth Plessner refers to the "new biology" and names (alongside Hans Driesch) Jakob Johann von Uexküll as its key proponent (Plessner 2019, 15). As can be seen throughout the whole book, Plessner understands the "new" biology to be an experimental, and hence modern, empirical science, which, however, unlike physics and chemistry, seeks to do justice to the specific nature of the living. Plessner recognizes the essential validity of the theory of evolution stemming from Darwin but criticizes the Darwinian one-sidedness of the theory. This one-sidedness, Plessner argues, consists in conceptualizing the relation between the organism and its environment primarily as a struggle for existence, in which the particular organism plays a primarily passive role toward the environment, since it must adapt to its environment and gets selected by it. A less one-sided approach to the theory of evolution, in general, would instead grant equal rights of consideration, alongside the struggle for existence, to the harmony, the concordance, or the emergent attunement between organism and environment. Moreover, both adaptation and selection can arguably be conceived not just as requested by the environment but also as actively proceeding from the organism as well. Under these respects, Plessner finds precisely in Uexküll's new biology a powerful counterweight to the narrow constraints of Darwinism. Much to Plessner's appreciation, Uexküll would, namely, uncover the "blueprint" [building-plan; *Bauplan*] of the individual organism and thereby the specific environment in which it can become active. However, one should also add that, on closer inspection, Uexküll's counterposition starts appearing to Plessner just as one-sided as the Darwinian focus, only in the opposite direction. When it comes to Darwinism and evolution, then, the complex relation of contact and distance between the two authors here under investigation finds its first exemplification. Equally close to Plessner's chords, inasmuch as it stems from an idea of biology as reviewable experimental science (see Plessner 1982, 80f.), Uexküll's fundamental criticism of the anthropomorphic projection of human self-understanding onto animals is invariably deemed as valid. Rejecting "anthropocentrism" in general, Uexküll's criticism is deployed positively on behalf of a "mechanical biology" focusing on the structures and functional capacities of organisms (Uexküll 2014, 42, 62f., my transl.). His positions result from the preliminary expansion of the stimulus–response explanatory pattern beyond behaviorism, as to include behavioral functions in the organisms' specific "blueprints" [building-plans; *Baupläne*], and thereby place emphasis on the unity of the organism with its environment. At the same time, Uexküll also shows full awareness of the *supra*mechanical character of living unities, even though he would still relegate protoplasm to the domain of "technical biology" (Uexküll 2014, 63, my transl.).

In this respect, what Plessner criticizes – despite all the points in common in their respective understanding of the problem – is that Uexküll misses entirely the intermediate layer of the lived body [*Leib*] between the organism and its environment. Not only a viable topic but also a necessary one, this intermediate layer of lived bodily experience is introduced by Plessner above all in order to differentiate the behavioral activities of lower and higher, more developed animals, in particular the intelligence of the anthropoids. As a result, by placing emphasis on the *Bauplan* element, Plessner shows an inclination to read Uexküll's research program merely as a minimal program and not as a maximal program. Unlike Uexküll, concerning the more developed animals, Plessner would take up elements of animal psychology from Wolfgang Köhler (1887–1967) and Frederik Jacobus Johannes Buytendijk (1887–1974). Differently, Uexküll would consider their approach to animal psychology as pertaining to the realm of psychology but not at all to that of biology (Uexküll 1973, 215).

Finally, Plessner reproaches to Uexküll an insufficient distinction between the notion of world and the notion of environment, which leads to the parallel positing of different environments. In reaction to Uexküll's leveling of all environments, Plessner supports a clear philosophical-anthropological distinction between the intelligent way of living seen in the great apes, whose intelligence was demonstrated without a doubt by Wolfgang Köhler, and the co-wordly life of the mind seen in the personal sphere of human life (Köhler 1925). Furthermore, he also criticizes the converse mistake, consisting in the mere reversal of Uexküll's biological pluralization of various environments all equal in nature through rigid categorical distinction. The sharp separation of "environment" and "world" would lead indeed to an equally erroneous contrast, assuming that only animals are bound to a certain environment while human life plays out in the openness of a world, as, for instance, it is the case according to Martin Heidegger's theory of human existence in *Being and Time* (1927). Plessner insists that the personal life of humans requires a sociocultural shaping of the biotic environment. From a systematic point of view, the personal sphere of life is made possible by an opening of the world, which can be created, organized, and changed only historically. As a consequence, the manner in which animals are bound to their environments [*Umwelten*] cannot be equated to our relation to sociocultural environments [*Mitwelten*]. After an intense discussion of foundational principles in the 1920s, the issue of false equivalences emerged quite dramatically in the racist ideology of National Socialism (Plessner 2001a) and then once more throughout the entire postwar era in which this categorical error persisted in shaping ideologies (Plessner 1983a, 2001b). The interweaving of environment and world in personal life, as opposed to the dependence of animals on their environment, was at the center of the interdisciplinary philosophical congress organized by Plessner in 1950, seeing the participation of biologists and physicians (Plessner 1983b) and leading to the great interdisciplinary work, *Propyläen Weltgeschichte* (1961), edited by Golo Mann and Alfred Heuß, who asked Plessner to write the philosophical-anthropological introduction (Plessner 1983c).

In what follows, the above outlined systematic focal points are discussed in relation to the most important passages in Plessner's writings on Uexküll. The thoroughly systematic nature of these focal points results from Plessner's continuously philosophical concerns. As an assistant to Buytendijk, Plessner also worked as an empirical and theoretical biologist; he was therefore able to complement theoretical discussions with direct observations. However, it is not in this field that he engages with Uexküll. Plessner responds to Uexküll's biology not as a biologist but as a philosopher of living nature (Plessner 2019, 21, 22, 71). He thus aims to establish a "philosophical biology" (Plessner 2019, xv, 61, 71), en route to which, in order to avoid the traps of anthropological circular biases, he also puts together a "philosophical anthropology" (Plessner 2019, xvi, 26–32). Plessner's main aim is to categorically reconstruct the qualities in the personal conduct of life that laypeople and experts can share within an enlightened "common sense" (Plessner 2019, xxxi, xxxiv). In this respect, he believes that the biologically necessary reduction of these qualities to the experimentally observable and calculable is a real gain that can be productively put to service, but it cannot replace the experience of these qualities in the conduct of life by free persons. The philosophy of living existence is thereby supposed to bring to light those "presuppositions" and "preconditions" that make biology and its application to the life-world possible, just as, conversely, biology is supposed to free these phenomena from their life-world contexts via their methodical and theoretical reduction for the sake of experimental and calculable reproducibility (Plessner 2019, 67, 84, 106, 279). Plessner thus establishes that both sides, biology and philosophy, require a collaboration (Plessner 2019, 60f., 109), in which no side can replace the other.

2 Plessner's critical engagement with Uexküll's biology in *The Levels of Organic Life and the Human* (1928)

Plessner's primary natural-philosophical work hosts the most comprehensive account on Uexküll's biology. Plessner once more presents Uexküll's critique of "anthropomorphisms," in general, as something enabling modern biology to operate as a strictly "experimental investigation." But under the *dualistic* presupposition that other bodies are only accessible to us from outside through sensory perception alone, while the direct access to the psychic requires an "introspective self-observation," this represents a very limited solution:

> Uexküll in particular argued for replacing animal psychology with biology, a science that is able to determine the objectively controllable correlations

between stimulus and reaction in the makeup of a particular animal. The scientific program of the 'animal psychologist,' according to these scholars, should concern not the eternally hidden world of animals with its feelings and perceptions, inaccessible to us, but rather their environment – that is, the various unified forms of the elements that affect them and upon which they can have an effect. Instead of crypto-psychology, what was called for was a phenology of living behavior: the explanation of animal behavior visible to us in factors perceivable by the senses.

(Plessner 2019, 58)

Plessner himself aimed at a "revision of the Cartesian dichotomy in the interest of a science of life" (Plessner 2019, 58). Hence, he asks whether the Uexküllian program represents a maximal or a minimal program:

If it were a maximum program, all the questions raised by so-called animal psychology would be resolved in terms of the physiology of irritability or of movement. In fact, however, Uexküll's biology (life plan research) does not pursue its program in this sense. It is instead restricted to the study of the stimuli and reactions characteristic of the organization type of the animal in question – not because it rejects the aim of a comparative psychology (as would dogmatic mechanism), but because it considers them to be unattainable due to a lack of means. It in no way denies that the notion of a 'life plan' might contain another side; it only contests the possibility of conducting *empirical* research on it, without, however, negating from the outset the possibility of a *non*empirical approach to the matter.

(Plessner 2019, 59, emphasis in the text)

This results from the fact that the "idea of the plan" includes a *whole* representing "more than the sum of the factors realizing it," although this plan does not necessarily have "the character of a goal or end":

A life plan as a unity of stimuli to which the organism recognizably responds *and* these responses can thus not be identical to the sum of these perceivable processes. This unity of the sphere that is the given framework for stimuli *and* reactions and that is itself invisible but becomes visible *in* the processes belongs neither to the body of the organism nor to the world surrounding it alone. If such 'plans' exist, sense organs and organs of locomotion cannot 'precede' the world of things *for* which they are there – and vice versa.

(Plessner 2019, 59, emphasis in the text)

This not sensibly perceptible wholeness in the individual life plan clears the way for Plessner to those categories of a "philosophical" biology that explicate the implicit presuppositions of "empirical" biology. Where Uexküll's program comes to an end because of its dualistic presuppositions, Plessner's philosophy of living nature has only just started. Philosophical "categories" of life pave the way to empirical "concepts" in which the presupposed wholeness of the living can be analyzed without breaking it up into the results of the analysis (Plessner 2019, 108f.):

> Any empirically ascertainable conformity of the organism to its environment, any adaptedness of the environment to the organism points to overarching laws equally governing living subject and world. That fact that neither of the two members of this reciprocal relationship has priority over the other cannot, however, be grasped empirically.
>
> (Plessner 2019, 60)

"Laws" is not meant here in the sense of a causal explanation but, rather, as the laws of understanding of the perceptible facts of the matter as a whole: "A category, then, is a form that belongs neither to the subject, nor to the object alone but allows these to come together by virtue of its neutrality" (Plessner 2019, 60). The intelligence of other animals, as well as the pre-reflective consciousness of humans, shows us that consciousness does not need to be self-consciousness. "Actually, things are exactly the opposite" of the reification of consciousness in a chamber that is supposed to reside in the head: "consciousness is not in us, but we are rather 'in' consciousness – that is, we relate to our surroundings as motile, lived bodies" (Plessner 2019, 62). By positioning pre-reflective consciousness in the conduct of the living body, according to Plessner, we are liberating animal psychology from its erroneous orientation of having to get to the animal psyche through introspection:

> Accordingly, the young science of animal psychology is not concerned with lived experiences, but is rather interested in developing a theory of animal behavior, of its forms and factors. Its narrow path runs between the Scylla of anthropomorphic descriptions of the intelligence, loyalty, and love in an animal's soul and the Charybdis of Uexküll's program and its attempt to discredit all research on consciousness. In this way it avoids both the 'behaviorism' cultivated by some Americans, whose overanxious compulsion to objectivity leads to the structure of a physiological stimulus-reaction schema [...], and the laypersons' uncritically romantic panpsychism and anthropomorphism.
>
> (Plessner 2019, 63)

In short, Plessner's philosophical transformation of Uexküll's research program is mostly about extending in the direction outlined in this passage:

> Uexküll was the first to declare the relationship between the organism and its environment to be the domain of an animal psychology (biology, life plan research) brought to reason. This young science has gone beyond him, however, in the sense that (unlike the 'Kantian' Uexküll) it strives to understand this relationship in its vitality and intelligibility, no longer identifying the physiological conditions of its realization with the overall habitus of animal

conduct. Wolfgang Köhler, thanks to whom we have excellent studies of anthropoid intelligence, David Katz and F. J. J. Buytendijk seem to be the most decisive in their approach to this kind of animal psychology: they are fully conscious of the purely image-based nature of this science, of the purely phenomenal character of 'behavior'.

(Plessner 2019, 63f.)

Since living behavior has at least *Gestalt*-character and is "only given in the habitus picture", it is not necessarily directly understandable, above all in the case of "lower" animals. The experimental method includes "physiological observation controls" but directed at the "unified *Gestalt* of living behavior" (Plessner 2019, 64). In the wake of this assessment of Uexküll's contribution to the state of research in biology, Plessner develops his own natural philosophy. Since in carrying this out he refers back to Uexküll here and there, we must first clarify the natural-philosophical context of these passages in order to be able to understand his remarks on Uexküll.

Plessner's categorical distinction between living bodies and physical bodies is independent of Uexküll. It starts with an intuition about behavior. According to this distinction, physical bodies live "insofar" as they "execute" their own borders in their behavior (Plessner 2019, 97f.). In behaving, they go out of themselves into a "medium" or environment and from there they return into themselves. The performance of this boundary-crossing can be understood in intuition through qualitative dimensions of sense that are "between" space and time in the physical sense and the spacelessness and timelessness of purely mental contents. The directional sense of a living movement unfolds in its own "spatiality"; it fulfills its sense in its own "temporality" (e.g., catching prey, fleeing, etc.). Since, however, this intuition could be mere appearance; it is not just phenomenologically investigated but also hermeneutically understood in the ontology of the living being. Plessner, therefore, wonders: under what ontological and ontic conditions can this intuition represent the true self-positioning of the body (Plessner 2019, 113, 121)?

To answer this question, Plessner works on a deduction of philosophical categories, which are able to both explain the qualitative mode of the intuitions of living organisms and simultaneously support the empirical concepts of biology. This includes, first of all, the essential structural and functional conditions of the dynamic realization of the organism's own boundary, that is, the processual character and typed nature (Plessner 2019, 123) of living existence, the developmental character of the living processes (Plessner 2019, 129). This is then followed by the conditions of the "static realization" of the organism's own boundary in the systematic character of the living individual thing (Plessner 2019, 144f.), its "self-regulation" (Plessner 2019, 149), and its "organized" nature (Plessner 2019, 154).

Plessner argues that we must distinguish, among the static conditions of realization, between two ways in which the whole of a living individual could be represented in the parts. As to the first, "[t]he whole of the living body is immediately potentially present in its parts." This form of representation is called the "harmonious equipotential system" (Plessner 2019, 157). Such a system is described by Plessner with reference to the phenomena of self-regulation studied by Hans Driesch. Here, however, he still holds Driesch's assumption that the entelechy is a natural factor to be completely superfluous (as Uexküll also does):

Entelechy as natural factor can be replaced by entelechy as a mode of being in corresponding to the boundary condition, which can be understood despite the fact that it cannot be characterized physically ('explained').

(Plessner 2019, 137)

As to the second mode of representation, the whole is not just "immediately" present as "potency" (capacity) in the parts but also in a mediated form as "actuality": "This form of representation exists in the harmonious divergence of specialized organs" (Plessner 2019, 157). Plessner discusses this form with reference to Uexküll, who understands "organization" as "the combination of varied elements according to a unitary plan for common effect" (Plessner 2019, 158; cf. Uexküll 2014, 184–187). As it actually entails the realization of the whole in the organs, namely, through the mediation of the "blueprint" of behavioral functions, Plessner finds Uexküll's systematic conceptualization of the problem to be suitable for explaining the static conditions of realization of the organism's own boundary:

The whole of the organism is not only logically but also ontologically able to set itself apart from the body that it is as a physical body; indeed, it even constitutes itself in and with this setting apart that is captured by the words 'according to' [*nach*] and 'for' [*zu*]. Only as a unity of means and end is the living body whole or an autonomous system.

(Plessner 2019, 159)

In this respect, it is all the more important to ask how the organization "according" to a unified plan and "for" a common effect can be understood within the dimensions of spatiality and temporality. Unlike the reversibility of directions in physical space and physical time, irreversibility characterizes the orientations of the lived body in its spatiality and temporality. For a particular living body, its front and back, its left and right, its below and above, its stages of aging as a whole are not reversible but rather irreversible. When it comes to the dimension of its temporality, irreversibility goes from the future into the present and past, hence the inversion of the causal sequence from the past by way of the present into the future (see Plessner 2019, 164). In other words, in its potencies the being of the living body is ahead of itself, such that it can be actually realized in specific situations, in the fulfillment of its relation to the mode of the present being conditioned by the fulfillment of its relation to the mode of the future (Plessner 2019, 165):

Connected back from the future the living body is ahead of itself – that is, is its own end; it stands *over against* its constant passing over from the not-yet into the no-longer, or persists. The abstract now between future and past is no longer suitable as the schema of the living body's existence, but only concrete presence, whose differential is the instant, or *blink of an eye*, the unity of future and past. This is why the living body, which, connected back, is accorded to or after itself [*das ihm selber Nachseiende*] still *has* a past.

(Plessner 2019, 167f., emphasis in the text)

While engaging with the theory of natural evolution, Plessner is concerned to uncover the categorical assumptions underpinning the biological concepts of "adaptation" and "selection." He thereby claims that in order to speak about a real adaptation in a particular situation here and now, we have to presuppose a structural and functional "adaptedness" in the relation between the organism and its environment (Plessner 2019, 186f.). And the actual selection here and now does not occur *ex nihilo* but, rather, only under the premise of a selectivity in structural and functional regards (Plessner 2019, 196f.). Moreover, adaptation/adaptedness and selection/selectivity, Plessner argues, can be understood from both sides, that is to say from the standpoint of the organism or of its environment. By placing emphasis on the categorical network and the two-way directions of fulfillment, Plessner intends to make the limitations in Darwin's model visible, inasmuch as Darwin clearly implies that the organism must adapt itself to its environment (and not vice versa) and that the organism is selected by the environment (and not vice versa; Plessner 2019, 191). Differently, Plessner conceives the framework of his philosophical categorical network as a spectrum of structural and functional potentials for play, that is, for an attuning and emerging attunement in both directions.

In the context of this categorical network, Plessner grasps the opportunity to make reference to Uexküll's biology as well. In plain Uexküllian terms, he speaks about the "conformity of the organism to its environment" and the "adaptedness of the environment to the organism" (Plessner 2019, 60; cf. Uexküll 1973, 319–321, 2014, 169). Uexküll notably teaches how to see the "primary harmonies" of adaptedness to what he presents as forms of "life plan" but "not [...] without tending toward the other extreme of espousing the absolute adaptedness of life systems, giving way, as it were [...] to a biological monadology" (Plessner 2019, 192). Although inspired by Uexküll's approach, Plessner is careful to remind the reader that the structural and functional adaptedness in the form of behavior does not supplant the actual adaptation in behavior here and now in view of a specific content. The organism remains "endangered, regardless of how secure it is" (Plessner 2019, 192):

The life of the organism plays out in the relation to its surrounding field in a way that is anticipatory in its form, that seeks contact to the medium in the concrete living act, that is adapted and adapting. The form, of course, cannot be separated from the concrete individual content.

(Plessner 2019, 192)

Moreover, it is not possible to develop "the fullness of the construction plans [...] from their underlying laws," rather this abundance presupposes a "playful capriciousness" in life: "The living physical thing carries in itself the order-creating conditions that come into play, themselves hampering and furthering, once the game has begun" (Plessner 2019, 159). At variance with Plessner's ideas, then, Uexküll's theoretical biology ended in a "overwhelmingly large accordance with a plan" that is thought to include "all plans" (Uexküll 1973, 342, 2014, 237).

Similar assimilation and distancing take place around the topic of the animal's closed form of organization. As is well known, according to Plessner, the closed form of organization of the animal, as opposed to the open form of organization of plants, is characterized by the nervous centers mediating the contrast between sensory and motor organs into a unity. Uexküll explains how the real existence of this closed form is possible in terms of a "functional circle," both morphologically, hence ontogenetically, and physiologically (Plessner 2019, 212). In this respect, Plessner readily adopts Uexküll's schema, not without some modifications though (Plessner 2019, 230). Uexküll's distinction between perceiving and effecting, Plessner specifies, categorically presupposes a lived body than can notice and effectuate something (Plessner 2019, 213). And since he holds the description of the "psychic" inner world of animals empirically impossible, Uexküll arguably stops at the "physiological" doubling of the body, which emerges from the representation of the organs in the central organ. However, according to Plessner, it has to be acknowledged that the fact that the "lived body is the positional equivalent of the physical separation into a bodily zone that contains the center and the bodily zone bound by the center" (Plessner 2019, 220) must be found in the doubled aspect of animal behavior from outside (not in the psyche believed to be inside). Not as a total body (including the central organ) but, rather, as a lived body (*Leib* – as the bodily zone dependent on the central organ) can the animal become the "subject of having" or a "self," that is, act (Plessner 2019, 220f.), and thus perceive and effectuate. As a result, in his modification of Uexküll's schema, Plessner systematically avoids the term "inner world," since he is not concerned with the study of an "inner" psyche. A clearly Cartesian assumption leads to the inner localization of the psyche (Köchy 2015, 43f.). However, for Plessner, the psyche is not situated inside the organism but, rather, in the expressive dimension of behavior toward the environment. In addition, he replaces Uexküll's terms "world of observation" and "world of effect" with "sphere of observation" and "sphere of effect." He thus intends to preserve Uexküll's notion of the unity of subject and object but without Uexküll's conflation of "world" and "environment" (Plessner 2019, 230).

Further problematization of Uexküll's assumptions leads Plessner to deal with the topic of what Uexküll calls the "counter-structure" in the "object" of the environment (Uexküll 2014, 45), based on which the vehicle of effects and the vehicle of properties can be said to belong to one and the same object. In simple terms, in order to say with Uexküll that the worm only encounters worm things in its environment, one needs, according to Plessner, a biological observer who can ascertain the unity of the object as the bearer of various properties and effects. Uexküll could not derive the unified nature of this object from the "counter-world" of the animal in which, as he himself admits, one can detect the mirroring not of the environment but rather of the counter-world itself. what is at stake is the mere "projection" of the animal "counter-world" on the "structure" of objects in the environment (Uexküll 2014, 169). Differently, Plessner argues for a relation of "coexistence" of organism and environment: "In this relation no member outweighs the other" (Plessner 2019, 240).

Inasmuch as he adheres to the "ignorance" in neurophysiology of his time, according to Plessner, Uexküll underestimates the psychic performances of the consciousness of primates (see Uexküll 2014, 308). To be fair, in his *Theoretical Biology*, Uexküll does indeed account for the type of "monitored action" that is key to learning processes: "Here the necessity arises of monitoring the activity of one's own effectors through one's own receptors. The monitored action is an action of experience that becomes a reflex" (Uexküll 1973, 307, my transl.). But while Uexküll explains this type of action in reference to humans only, for Plessner, it has long since been demonstrated for the great apes as well. According to him, the central nervous connection of an animal's actions to its own sensorium, which allows the monitoring of its actions, is the "closed loop of the sensory-motor functional play," which in behavior "corresponds to the appearance of things in the field of noticing" (Plessner 2019, 237). This closing of the circle in the central organ, that is to say, the brain, represents the physiological condition for an animal's ability to perceive "its own movements in its environment," hence to take note of itself as a lived body in the zone it occupies, whereby the environment separates off from the lived body with its own boundary and receives a "structure," namely, the structure of things bearing properties and effects. The environment of the animal hence becomes "a signal field and an action field in one" (Plessner 2019, 234). In this respect, it could be easily argued that Uexküll's schema of the functional circle actually presupposes the same scenario, without, however, being able to apply it to the higher animals. The very hypothesis that "the circle of sensory-motor functions [...] is closed again in the central organ" (Plessner 2019, 234) is assessed also by Uexküll under the assumption that it would link sensory and motor functions better as they would be more malleably connected during learning processes. For Plessner, however, even just in neurophysiological terms, the brain is not just a better possibility for connecting sensory and motor functions here and now, as he believes to be already the case for lower decentralized animals. For Plessner, the brain is, above all, the potential to interrupt this connection as well, as one can see in the more-developed mammals. If we consistently think through the "centralization" (instead of decentralization of the nervous system) in the closing of this form of organization toward the environment, then the brain (neocortex) takes on an autonomy relative to the sensory and motor functions: from a neurophysical point of view the brain is the correlate of precisely that "*interruption, inhibition, interval* (between stimuli and response) *which in positional terms is* the being of a self in the central position – that is its being 'against something in the surrounding field' or its intuition of something" (Plessner 2019, 242, emphasis in the text). Precisely this interruption or pause in the actual performance of conduct enables the animal to filter its current impressions, to imagine something that is not yet there (Plessner 2019, 249), and to form a memory from the fragments of its past that are relevant to it (Plessner 2019, 264). In this respect, Köchy has convincingly reconstructed Plessner's demonstration of higher animals' consciousness of things, of their graspability, constancy, and solidity, as stemming from a contrastive confrontation with Uexküll (Köchy 2015, 54–56).

On the whole, Plessner is convinced that a brain physiology of the pauses in the connection of sensory and motor functions will contribute to an "objective disciplining of interpretations" in animal psychology:

> The more differentiated the receptors and the brain, the more varied the evoked excitations and the more manifold the intervals and thus the structure of the positional field. Nervous excitations of the sensory (and motor) apparatus merely afford the living being the opportunities to occupy the central position as which and in which its conscious life plays out.
>
> (Plessner 2019, 242)

For us observers, this conscious life is the mediation of the relation between organism and environment. But for the animal, it is experienced in an unmediated manner. Plessner calls such a relation a "mediated immediacy" in life (Plessner 2019, 241).

In closing forms, Plessner develops his distinction between the environment of animals, which he calls "central positionality," and the world of people or other personal creatures, which he describes categorically as the "excentric positionality," without particular reference to Uexküll. Biology takes place in an excentric positionality, which is the presupposition that biologists must make without being able to integrate it as biologists. Uexküll shows awareness of this problem in the closing chapter of his book, Umwelt *and the Inner-World of Animals* [*Umwelt und Innenwelt der Tiere*], titled "The Observer" – referring to the biologist positioned outside the animals' environment (Uexküll 2014, 233). On the one hand, the biologist's understanding of the world seems to be a precondition for recognizing the environment of animals as a selection of said world: "Every environment of an animal forms a delimited part, both spatially and temporally as well as in content, of the phenomenal world of the observer" (Uexküll 2014, 236, my transl.). On the other hand, the world of humans, including biologists, is also alleged to be just the environment of an organism: "The phenomenal world of every human is also comparable to a solid casing," just like the environment of an animal is an "impermeable casing" for it, "which encloses it continuously from its birth to its death" (Uexküll 2014, 237, my transl.). "The same holds for the perceived world of the observer, this also wholly closes him off from the universe, since it represents his environment" (Uexküll 2014, 237, my transl.).

Uexküll leaves it open how this comparability of environment and world is to be understood: whether it is the philosophical confession of an analogy between humans and other animals that are all equal before nature, but which can clearly only be drawn by humans and not other animals, or is the biologist really supposed to develop a standard for this comparison of environment and world that can be empirically and arithmetically realized, even if Uexküll is not able to accomplish this himself? For Plessner, the idea of this type of biological standard of comparison is a clear categorical "cardinal error," by measure of which the human could at best appear as a "healthy" or a "sick" animal, as a "predator" or a "domestic animal," without being able to comprehend the specifics of our personal sphere of life (Plessner 2019, 293, 295; see also Plessner 1983b).

According to Plessner, a sphere of life is an excentric positionality precisely when an ex-centering of the concentrics of the lived body becomes truly possible within it (Plessner 2019, 285). The concentrics of the lived body are directed at the "convergence" with something in its environment (Plessner 2019, 270f.). The center of centric positionality is precisely wherever and whenever the conduct of the lived body fulfills itself in its encounter with something in the environment, for example, when the dog catches its ball. Ex-centering this center requires attaining some distance from it, a distance sufficient to recognize this center as an object. Plessner calls this standpoint, located "outside" the unity of the lived body in between the organism and its environment, that of the "person" (Plessner 2019, 272). The person can use his or her organism as a physical body, that is to say, instrumentalize it and use it as a medium in order to bring about the unity of his or her lived body with something in the environment. For this purpose, however, the person has to be able to stand in a co-world with other persons that share his or her mind. The mind represents the "nothing" of determinations in physical space-time and the "nowhere" and "no-when" of determinations in the dimensions of the lived body. This allows for a contrast between that which is perceptually given and what it means conceptually, a contrast that gets interpreted symbolically in cultural connections. From the standpoint of the mentally shared co-world, persons can interpret how they should, can, and must conduct themselves in the world.

Since persons as members of a mentally shared co-world do not cease to be organisms, they have to arrange the world they have opened mentally in an artificially and historically mediated fashion. For the personal sphere of life, the character of a historical task is essential, which Plessner elucidates in three fundamental anthropological laws: first, the necessary habits of second nature need to be artificially produced through culture and technology. Plessner calls this the path of "natural artificiality" (Plessner 2019, 287). At the same time, this artificiality must be capable of being lived naturally in the sense of the concentrics of the lived body, which leads to a continual tension. That which intentionally satisfies such living creatures here and now as a matter, of course, has a long and circuitous historical road behind and ahead of itself. Plessner calls this the path of "mediated immediacy" or "indirect directness" (Plessner 2019, 298). After all, despite all artificial and historical successes, living creatures in excentric positionality are never spared the adversities of life, from illness to death, even if they can be better treated and postponed. What the affirmation of an adverse and mortal life ultimately allows is a "utopian standpoint" (Plessner 2019, 316; for a thorough interpretation of Plessner's *Levels*, see Krüger 2017).

3 The historical interweaving of sociocultural environment and world in personal life and the opening of biotically closed environments (1938–1961)

Plessner often repeats his fundamental claim in the most various discussions from 1938 to 1961: "Every biological interpretation of an environment must ultimately rest on an extra-biological concept of the world" (Plessner 1983a, 59, my transl.). He thus intends to claim that the biological knowledge of animal behavior in certain environments does not occur within this animal behavior but, rather, presupposes a distance from said behavior in order to treat it as an object. In this respect, since as a biologist Uexküll describes how he sees the world through the eyes of a fly, or of a spider, how he sees a dog with fly-interests, spider-interests, and dog-interests, he must make more than one "concession," Plessner argues, to his notion of human "environment" in the direction of Plessner's understanding of the human "world." He famously claims then that "[o]nly because it is a world and not just an environment does the human world provide the constant background" (Plessner 1983a, 59, my transl.). Reaching here a clear friction with Uexküllian biology, Plessner goes further in criticizing the implications of Uexküll's positions on the matter:

> With this notion of a pluralism of environments, which introduces a pluralism of biological standards of value, which renders impossible the idea of progress, of increasing adaptation to a single environment, and which relativizes intelligence, of course also vitiates humanity's special position within nature, if the world in which humans live is nothing more than the environment of the human blueprint.
>
> (Plessner 1983a, 59, my transl.)

For Plessner, humanity's special position in nature rests on the fact that the human species takes part in *logos*, that is, in a shared mental world that allows the "reasonable" formation of personal modes of behavior (Plessner 1983a, 55) within relations of love that are ends in themselves with no ulterior purposes (Plessner 2001a, 167). We should not conflate this mental self-understanding of humans in a world with the "instinct-bound intelligence" of animals, that is, with a behavior that "is open to correction by experience" but only extends "to a specific environment," "granting insight into the constellations" of the environment specific to that animal (Plessner 1983a, 56, my transl.). In contrast to the instinct-bound intelligence for biotically specific environments, the mental and general distance from bodies and lived bodily experience opens a world of potentials representable linguistically and symbolically, within which and against the background of which it is necessary "to create an artificial environment" (Plessner 1983a, 63, my transl.). This historical task presupposes that "humans are unstuck," that we are no longer pre-adapted to any specific biotic environment, and requires "the ability to shut off and to interrupt the biological connections" (Plessner 1983a,

64, my transl.). This special position of humanity in nature is also suggested by biological insights into humanity. Along these lines, Louis Bolk emphasizes how humans maintain similarities to the fetal stages of other anthropoids and have a long stage of childhood and youth. Paul Alsberg explores the supplementation and replacement of natural organs by artificial organs, particularly speech, by humans (Plessner 1983a, 60, 62f.). "Seen fundamentally," Plessner claims, "humans are emigrants of nature, who have no home by nature but rather only insofar as we conquer it and hold on to it with all of the mental powers of thought and heart" (Plessner 1983a, 64, my transl.).

Plessner also tries to lead us out of the false opposition between closed environment and open world: "The usual correlation nowadays of animals with the boundedness to a closed environment and of humans to the openness of a world makes the matter too simple" (Plessner 1983c, 182, my transl.). The opening of the world cannot be stabilized without the establishment of an artificial environment, which remains dependent on earthly conditions of life. Personal beings remain dependent on their physical bodies and their lived bodily experience, by representing in these something and someone, hence mental content – in contrast to angels, who might be able to exist in pure world-openness (Plessner 1983c, 187).

> [The compulsion in personal life] [t]o face the open reality and master its unpredictability produces everywhere an artificial narrowing of the horizon, which encloses the whole of human life like an environment, but does not close it off. This artificial narrowing of the horizon is rather simply the mode of mediated immediacy, which characterizes all human behavior.
>
> (Plessner 1983c, 189, my transl.)

Rather than simply speaking of world-openness, Plessner refers to, as he already did in his *Levels* (Plessner 2019, 272), the "fragmentary character of human world-openness" (Plessner 1983c, 188, my transl.). But it is not just the category of world that gets mediated in itself through the category of the sociocultural environment – the environment in the biological sense does not always have to be assessed as closed. The environment is only closed insofar as the physiological blueprint corresponds to Uexküll's closed functional circle and determines the behavior of the animal accordingly. Only then it is true for the animal in question that its environment is not transposable; its blueprint requires this environment for its selection, and the relativity of the animal's actions isolates it against everything irrelevant to its environment (Plessner 1983c, 182f.). However, in the case of mammals and in particular of anthropoids, the biotic environment displays numerous instances of opening connected to intelligence, tool production, and social co-relations, all the more prominent given the assumed evolutionary history of humans.

As far as the biological possibility of attaining some kind of "independence" from the environment it is concerned (Plessner 1983c, 166), Plessner emphasizes the following: among mammals, the "line of increasing cerebralization" and intelligent processing of experience also include the growth of a "youth phase" as a "time of learning and play." Buytendijk shows, according to Plessner, how in this early ontogenetic stage an excess of drives is formed that presses for release but coincides with an unready motoric ability and sexuality: "This results in a play, a behavior free of purpose, between bond and release, which gives rise to a self-fulfilled relation unburdened by fear and avarice" (Plessner 1983c, 167, my transl.). For humans, this gain in freedom relative to the biotic environment increases with the creation of a sociocultural environment as the horizon of the world. The end of embryonic development takes place outside the uterus in the sociocultural environment (Plessner refers in this regard to Adolf Portmann; see Tolone 2015), in which, in the last third of the first year of life, acquired motoric activity (as opposed to inherited motor activity) and the acquisition of language and transposable intentionality of action begin to mature at the same time (Plessner 1983c, 166, 182). The embedding of speech into the upright gait and hand–eye coordination integrates a "virtual organ" (Plessner 1983c, 177) into the composition of human behavior, the mental performance of which is no longer subject to any biological function (Plessner 1983c, 181):

> Imitation and reification, which the acquisition and use of language live from, have the same human root: the sense for reciprocity and perspectives in the relation of my own lived bodily existence to that of the other.
>
> (Plessner 1983c, 179, my transl.)

At the same time, this opening of the world through symbolic forms requires a stable reproducibility of behaviors in the artificial forming of physical bodies [*Körper*] and lived bodies [*Leiber*] – which become media in which people represent factual matters and interpersonal relations. Plessner calls these media of representation and performance "roles" of persons, in which and with which one plays community and society (Plessner 1983c, 193, 205, 209). I have thoroughly described his historical theory of performance of personal roles elsewhere, in particular the doubling of the person into private and public (Krüger 1999). Concerning the relation of the sociocultural environment and world, it is worth drawing here attention to Plessner's criticism of Erich Rothacker's cultural anthropology, in which the sociocultural environment is again drawn closer to a biotic "real" environment:

> Just as a mental framework consisting of language, values, goods and customs in all its closure and intransferability remains at the same time open outwardly – between languages there is the possibility of translation – and forms bridges to other mental frameworks of history and the contemporary world and allows for a view into foreign mental life, it sets itself clearly apart from the environmental bonds of a purely vital and emotional character in which we humans with our deep person live preconsciously, affectively, instinctively.
>
> (Plessner 1983c, 187, my transl.)

The "whole formation of the environment for humans" has "an acquired and conserved essence" that "is not simply given with the nature of the lived body, but rather – because this latter leaves it open – is made and only grows naturally in a metaphorical sense" (Plessner 1983c, 184, my transl.).

At the end of his attempt to find his way out of the false dichotomy of closed environments of animals and the world-openness of humans by drawing new distinctions, Plessner clarifies his own standpoint and, in contrast to other philosophical anthropologists such as Max Scheler and Arnold Gehlen, he claims that:

> Neither the classical nor the pragmatist anthropology reckon with the possibility that our boundness to the environment and the openness of the world might collide and only obtain in a reciprocal interweaving that cannot be brought to any kind of balance.
>
> (Plessner 1983c, 182, my transl.)

The surprises of history consist of such collisions in which the balance no longer works, even if one believes oneself to have seen the necessity of reciprocally interweaving environment and world (on Plessner's philosophy of history, see Krüger 2013): the ecological problem is an example of this.

In 1965, in the appendix to the second edition of the *Levels*, Plessner noted that Uexküll's "dual animosity toward animal psychology, which in his time operated with anthropomorphic analogies, and American behaviorism, which was guided by chain reflex models, had lost all currency in modern ethology" (Plessner 2019, 329). His biology had too little differentiating potential for the phenomena "between" animal psychology and chain reflexes. In contrast, Plessner's differentiations among intelligence, consciousness, and mind in various spheres of life and among cooperation, imitation, and emulation of interconnections of things and persons in the reciprocity of roles have taken on new relevance for the comparative study of brain and behavior (Krüger 2010, 2014).

In conclusion, I agree with Köchy that Plessner has broadened and developed Uexküll's program (Köchy 2015, 31), but I do not contend, for all the points of divergence stated earlier, that Uexküll's doctrine has provided the foundation to Plessner's thinking nor that Uexküll's ideas have served as decisive model for some fundamental positions of Philosophical Anthropology (Köchy 2015, 25–27; regarding the many sources of Plessner's Philosophical Anthropology, see Krüger 2017).

References

Köchy, Kristian (2015) 'Helmuth Plessners Biophilosophie als Erweiterung des Uexküll-Programms'. In: Kristian Köchy and Francesca Michelini (eds.) *Zwischen den Kulturen. Plessners* Stufen des Organischen *im zeithistorischen Kontext*. Freiburg/München: Verlag Karl Alber, 25–64.

Köhler, Wolfgang (1925) [1921] *The Mentality of Apes*. Translated from the second German edition by Ella Winter. New York: Harcourt, Brace and World.

Krüger, Hans-Peter (1999) *Zwischen Lachen und Weinen.* Vol. 1. *Das Spektrum menschlicher Phänomene.* Berlin: Akademie Verlag.

Krüger, Hans-Peter (2010) *Gehirn, Verhalten und Zeit. Philosophische Anthropologie als Forschungsrahmen.* Berlin: Akademie Verlag.

Krüger, Hans-Peter (2013) 'Die säkulare Fraglichkeit des Menschen im globalen Hochkapitalismus – Zur Philosophie der Geschichte in der Philosophischen Anthropologie Helmuth Plessners'. In: Christian Schmidt (ed.) *Können wir der Geschichte entkommen? Geschichtsphilosophie am Beginn des 21. Jahrhunderts.* Frankfurt/New York: Campus Verlag, 150–180.

Krüger, Hans-Peter (2014) 'Mitmachen, Nachmachen und Nachahmen. Philosophische Anthropologie als Rahmen für die heutige Hirn- und Verhaltensforschung'. In: Olivia Mitscherlich-Schönherr and Matthias Schloßberger (eds.) *Das Glück des Glücks* (Internationales Jahrbuch für Philosophische Anthropologie, Vol. 4). Berlin: De Gruyter Verlag, 225–243.

Krüger, Hans-Peter (ed.) (2017) *Helmuth Plessner: Die Stufen des Organischen und der Mensch. Einleitung in die philosophische Anthropologie* (Klassiker Auslegen, Band 65). Berlin/New York: De Gruyter Verlag.

Plessner, Helmuth (together with F. J. J. Buytendijk) (1982) [1925] 'Die Deutung des mimischen Ausdrucks. Ein Beitrag zur Lehre vom Bewusstsein des anderen Ichs'. In: Helmuth Plessner. *Gesammelte Schriften.* Vol. 7. *Ausdruck und menschliche Natur.* Frankfurt a. M.: Suhrkamp, 67–129.

Plessner, Helmuth (1983a) [1946] 'Mensch und Tier'. In: Helmuth Plessner *Gesammelte Schriften.* Vol. 8. *Conditio humana.* Frankfurt a. M.: Suhrkamp, 52–65.

Plessner, Helmuth (1983b) [1950] 'Über das Welt-Umweltverhältnis des Menschen'. In: Helmuth Plessner. *Gesammelte Schriften.* Vol. 8. *Conditio humana.* Frankfurt a. M.: Suhrkamp, 77–87.

Plessner, Helmuth (1983c) [1961] 'Die Frage nach der Conditio humana'. In: Helmuth Plessner. *Gesammelte Schriften.* Vol. 8. *Conditio humana.* Frankfurt a. M.: Suhrkamp, 136–217.

Plessner, Helmuth (together with F. J. J. Buytendijk) (2001a) [1938] 'Tier und Mensch'. In: Helmuth Plessner. *Politik – Anthropologie – Philosophie. Aufsätze und Vorträge.* München: Wilhelm Fink Verlag, 144–167.

Plessner, Helmuth (2001b) [1951] 'Das Problem der menschlichen Umwelt. Vortrag im Sender RIAS'. In: Helmuth Plessner. *Politik – Anthropologie – Philosophie. Aufsätze und Vorträge.* München: Wilhelm Fink Verlag, 168–175.

Plessner, Helmuth (2019) [1928] *The Levels of Organic Life and the Human: An Introduction to Philosophical Anthropology.* Translated by Millay Hyatt. New York: Fordham University Press.

Tolone, Oreste (2015) 'Helmuth Plessner und Adolf Portmann. Zur philosophischen Bestimmung des Menschen durch Exzentrizität und Frühgeburt'. In: Kristian Köchy and Francesca Michelini (eds.) *Zwischen den Kulturen. Plessners* Stufen des Organischen *im zeithistorischen Kontext.* Freiburg/München: Verlag Karl Alber, 141–159.

Uexküll, Jakob von (1973) [1928] *Theoretische Biologie* (2nd ed.). Frankfurt a. M.: Suhrkamp.

Uexküll, Jakob von (2014) [1928] *Umwelt und Innenwelt der Tiere* (2nd ed.). Berlin/ Heidelberg: Springer Verlag.

6 Ernst Cassirer's reading of Jakob von Uexküll

Between natural teleology and anthropology

Carlo Brentari

1 Introduction

Two main philosophical issues led Ernst Cassirer to plunge into Jakob von Uexküll's work. The first one is the problem of teleology in living beings, which Cassirer discusses in *The Problem of Knowledge* inside a critical survey of the epistemological debate about biology in the 19th and early 20th centuries (Cassirer 1969). In this context, Cassirer shows appreciation for Uexküll's idea of natural teleology understood as the correspondence of the organism to an inner plan that determines it both anatomically and functionally. According to Cassirer, as we shall see in Section 2 of this chapter, Uexküll's position makes a good case for the disciplinary autonomy of biology without falling into the opposite poles of the metaphysical vitalism à *la* Driesch or of the mechanistic reductionism. This issue is still of the utmost actuality for the discussion on the topic of teleology developed by a contemporary philosophy of biology. Whereas the most severe risks of vitalism – notably the undesirable consequences ensuing from postulating the direct action of spaceless and timeless entelechies, or vital forces, on the organism, which means, ultimately, to accept that essential factors of biological processes cannot be the object of empirical investigation – had been identified already in Cassirer's time, the subsequent debate exposed the disadvantages also of the opposite pole (for a review, see Perlman 2004). For the sake of eliminating every trace of metaphysics, in fact, reductionist positions tend to understand the living being as nothing more than a set of physical-chemical processes; pushed to the extreme, this stance prevents us from making any reference to a whole range or "family" of concepts, such as the "purpose" of a behavioral sequence or the "function" of an organ (see Mayr 1992; Cummins 2002; Perlman 2010; Cummins and Roth 2010; Godfrey-Smith 2014, 59–65).

As to the second point of Cassirer's confrontation with Uexküll, the focus shifts from the general theory of life to that particular living being that is man. Cassirer does not abandon the biological-theoretical perspective, but he feels the need to develop specific theoretical tools in order to identify, describe, and, if possible, explain the peculiarities of human beings inside the field of living beings, and particularly through the comparison with nonhuman animals. This leads to the definition of the best-known notions of Cassirer's thought, namely, the concept of symbol, the idea of man as *animal symbolicum*, and the remarks on culture and its forms. This stage of his philosophical production is mirrored by two texts: "The Problem of the Symbol as the Fundamental Problem of Philosophical Anthropology," written in 1928 but published posthumously in *The Philosophy of Symbolic Forms* (Cassirer 1996), and the second chapter ("A Clue to the Nature of Man: The Symbol") of *An Essay on Man* (Cassirer 1944). As we shall see in Section 3 of this contribution, within this second context Cassirer seeks the handhold of Uexküll's theoretical biology with a different aim: the definition of the specificity of man in relation to animals. In the final part of both Sections 2 and 3, as well as in the concluding remarks, I then offer an overall evaluation of Cassirer's reading of Uexküll's theoretical biology.

2 Uexküll and a key problem in philosophy of biology: how to think about purposiveness without postulating the direct action of final causes

2.1 The place of Uexküll in Cassirer's history of theoretical biology

In Cassirer's *The Problem of Knowledge*, Uexküll makes his first appearance in the section devoted to vitalism. Together with Ludwig von Bertalanffy (1901–1972), Emil Ungerer (1888–1976), and John Haldane (1892–1964), Uexküll is positively seen as a biologist engaged in the task of preserving biology from the many forms of one-sidedness that threaten it, especially, after the arrival on the scene of Darwin's theory, the latter's metaphysical extension with Ernst Haeckel (1834–1919), and the *Entwicklungsmechanik* of Wilhelm Roux (1850–1924; see Köchy's contribution, Chapter 3, in this volume).

From a widely philosophical point of view, Cassirer is ready to acknowledge the valuable contribution of Darwinism to the elimination of the metaphysics of final causes, replaced instead by "a strictly unitary causal explanation, with no assumption of any special type of causality equal or superior to the psychochemical" (Cassirer 1969, 166). However, the elimination of metaphysics and final causes, maintains Cassirer, should not be seen as the permanent removal of teleology from the philosophy of biology. The actual problem is not the elimination of "purposes," but "whether the 'category' of purposiveness could assert its role as a peculiar 'principle of order' in scientific description and presentation or whether it had become superfluous" (Cassirer 1969, 166).

Cassirer's ultimate goal is to restore the Kantian distinction between constitutive and regulative uses of the idea of purposiveness. This distinction has the advantage of providing orientation to the search for mechanical and antecedent causes, on the basis of considerations that concern not only a wider sphere than the single process but also a qualitatively different one. The (dangerously "romantic") holistic assessment of the organism, for example, might suggest the function of an organ or process, thus directing the more sober biochemical or genetic analysis. Moreover, Cassirer emphasizes that Darwinism itself cannot help but make use of concepts such as fitness, selection, struggle for existence, or survival of the fittest, which "have all plainly a purposive character," and this fact "would alone suffice to show that in opposing a definite form of metaphysical teleology it in no way renounces 'critical teleology'" (Cassirer 1969, 166; see also Krois 2004, 284–290).

Although a full account of Cassirer's insightful analysis of the disputes in theoretical biology at the turn of the 19th and 20th centuries certainly exceeds the scope of this contribution, emphasis should be placed on the two opposing fronts Cassirer identifies. Against the supporters of the elimination of the purposiveness in all its forms stand those who propose various combinations of experimental research on antecedent causes, on the one side, and teleonomic and/or holistic heuristic patterns, on the other. As aptly put by Ernest Nagel,

> he [Cassirer] rejects all forms of substantival vitalism, such as that of Hans Driesch (1867–1941) [...]; but he finds merit in holism because it confirms in a significant manner Kant's conception of biological form as a heuristic rule.
>
> (Nagel 1951, 149f.)

In Cassirer's search for a biological knowledge that does not reject holism and purposefulness but, at the same time, does not make a constitutive use of final causes, vital forces, souls, or entelechies, Uexküll's theoretical biology appears to tick most boxes. Less positive is, conversely, Cassirer's evaluation of Driesch's solution for the problem of the purposiveness of organic processes. Although Driesch and Uexküll share the opposition to Darwinism and Roux's developmental mechanics, how their goals are achieved is very different. Based on Cassirer's reading, Driesch tries to support vitalist principles by postulating the direct action of a spaceless and timeless force, the entelechy, on the material substrate of the organism and, in particular, on the embryo. Uexküll, on the other hand, would avoid the slippery ground of embryology and ontogenesis altogether, as it is too exposed to the influence of evolutionary mechanics – the reader might recall here the extraordinary impact of Haeckel's hypothesis that ontogeny recapitulates phylogeny. In Cassirer's view, Uexküll opts for a synchronic approach to the organism, based on the notions of form and structure: "Driesch started from *physiology*" – writes Cassirer – "but Uexküll was above all an *anatomist*. [...] The concept of 'anatomical type' [...] permeated and dominated all his thinking" (Cassirer 1969, 199). As we shall see in what follows, however, Uexküll does not neglect the realm of physiology; many of his best-known books include, indeed, a wealth of examples and case studies of precise behaviors based on physiological processes.

If this is true, Cassirer's reading is justified insofar as what determines and organizes the physiology, perception, and behavior of each animal species is the "plan of construction" (building-plan; *Bauplan*), to use one of Uexküll's keywords. In the notion of *Bauplan* (freely translated in the English edition of Cassirer's work as 'structure' or 'structural plan' [Cassirer 1969, 129, 199]), Cassirer sees the continuation of a tradition of biological thinking based on the careful observation of the overall shape, of the holistic structure, or of the unity of an organism as *Gestalt*. According to him, in modern times, this tradition finds its paramount representatives in the morphology of Georges Cuvier (1769–1832) and Karl Ernst von Baer (1792–1876) and in the "idealistic morphology" developed by Goethe (Cassirer 1969, 137–150). Within this approach, the immediate grasp of the *Bauplan* would allow for a regulative and not constitutive use of teleological notions: if not their antecedent causes, it allows to understand the "sense" of an organ, as well as the "function" of a tissue, or the "purpose" of a behavioral sequence (on the nature and function of the *Bauplan*, see also Kull 2004, 105–107).

2.2 Some critical remarks on Cassirer's view of Uexküll's vitalism

Cassirer's choice of presenting Uexküll primarily as an anatomist leads him to start from one of Uexküll's less quoted texts, *Theory of Life* [*Die Lebenslehre*, Uexküll 1930], in which the basic coordinates to the exposition are provided by the concepts of *Bauplan* and "universal conformity to a plan [*universelle Planmäßigkeit*]" (Uexküll 1930, 156, my transl.). The other texts mentioned by Cassirer (Uexküll 1909, 1921, 1928) are less easy to relate to Cassirer's reading of Uexküll; the same is true, as we shall see, also of some key passages in *Theory of Life* that are omitted by Cassirer's account. In general terms, some objections might be raised concerning what Cassirer presents as a main distinctive feature in Uexküll's theoretical biology, that is to say, the rejection of any direct or immediate intervention of teleological principles (e.g., Driesch's entelechy) in organic processes. It could hardly be maintained, in fact, that Uexküll rules out such ultimate principles. According to Cassirer, he just chooses to limit himself to postulating the existence of a natural factor [*Naturfaktor*] with harmonizing effects (it is "a nonmaterial ordering, a rule of the living process" [Cassirer 1969, 202]). Yet, even this confinement of the teleological principle to the role of a distant harmonizing factor is not always respected by Uexküll himself. Quite surprisingly, evidence of this can be found precisely in the very texts that Cassirer takes into account in *The Problem of Knowledge*. Here we find, on the one hand, the general definition of teleological principles as distant and unknowable ("rule and plan are only the form in which we know the effects of that natural factor [*Naturfaktor*]. In itself, it is completely unknown to us" [Uexküll 1909, 13, my transl.]); on the other hand, however, Uexküll expressly likens the natural factor to similar notions playing a leading role in other forms of vitalism ("Driesch, referring to Aristotle, calls it 'entelechy', Karl Ernst von Baer calls it 'goal-directedness' [*Zielstrebigkeit*]" [Uexküll 1909, 13, my transl.]).

More importantly, we find ourselves faced with at least one organic phenomenon in which the *Naturfaktor* directly enters into play, notably the physiology of the cell protoplasm, whose peculiarities push Uexküll to the bold statement that its study offers us the possibility to "make a closer experience of this enigmatic natural factor" (Uexküll 1909, 13, my transl.; see also Cheung 2004). Uexküll is certainly not the only biologist of his time who entrusted to the protoplasm a key role in the understanding of the living. After an initial phase in which it focused on the cellular components of the membrane and the nucleus, from the 1860s onward cytology developed a strong interest in the protoplasm. What's more, the question of the nature and functions of the protoplasm was nested from the outset in the dispute opposing mechanists and vitalists. The nodal point of the dispute is the question of the structure of the protoplasm. In this regard, the mechanists (e.g., Julian S. Huxley, 1887–1975) believed that the composition of the protoplasm followed the ordinary laws of chemistry and physics and, hence, could be further studied in terms of deeper molecular structures (as it happened later). Differently, the vitalists postulated the presence of a vital force in the protoplasm that is not reducible to physicochemical factors. Close to Uexküll, Claude Bernard (1813–1878) stated, for instance, that the protoplasm was "life in the naked state" (Welch and Clegg 2010, 1281; see also Geison 1969).

Uexküll shows a clear adherence to the vitalist field. He defines protoplasm as "the original designer [*Urgestalter*] of life," whose "work is the fullness of animal and plant forms through the ages of the Earth" (Uexküll 1930, 16). Moreover, the teleological action of the protoplasm is understood by Uexküll in terms of the construction of structures [*Strukturbildung*]. In most cases, this action only occurs in the embryonic stage, as the adult organism, states Uexküll, is ruled by the principle that "the structure inhibits the construction of structures" (Uexküll 1909, 13). However, in some cases, such as that of the *Amoeba terricola* and other unicellular organisms, the action of the protoplasm lasts for the entire life of the animal, leading to the formation of temporary organs, as and when necessary. In these cases, but also in cases of the regeneration of organs in certain amphibians, Uexküll claims that the original plasticity of the cellular protoplasm is not channeled into a definitive *Bauplan* but persists in the adult stage.

Understandably, Cassirer devotes to the problem of protoplasm only a brief quote (Cassirer 1969, 201). In truth, the emphasis given by Uexkull to protoplasm is simply not compatible with Cassirer's idea that he was not interested in the level of the ultimate organizing factors or of the "special purposive forces" (Cassirer 1969, 202). In Cassirer's defense, one should say that in the four texts by Uexküll mentioned by Cassirer, if taken in their entirety, the topic of the protoplasm is not always given the same importance. Whereas in the first edition of Umwelt *and Inner-World of Animals* [*Umwelt und Innenwelt der Tiere*, see Uexküll 1921] and in *Theory of Life* it has an opening position and is given a foundational role, in both the first and the second edition of *Theoretical Biology* the issue is confined to an internal section of the chapter "Object and Living Organism" [*Gegenstand und Lebewesen*] (Uexküll 1920, 93f., 1928, 97f.). It cannot be argued, however, that the problem of protoplasm as well as, in general, the ontogenetic perspective ever disappear from Uexküll's works. *Theoretical Biology* goes as far as including some proposals on how to interpret, from a vitalist perspective, the new discoveries concerning the composition of the cell nucleus (such as the identification of genes by Wilhelm L. Johannsen [1857–1927]; see Uexküll 1928, 160, 178) and the chromosomes by Thomas H. Morgan (1866–1945; see Uexküll 1928, 167). In the same book, the reader even finds a cautious acceptance of the variability of the species, although traced back to teleological factors and not to Darwinian selection (Uexküll 1928, 183).

Despite their rather weak scientific value, Uexküll's claims on protoplasm bear testimony to his lasting effort to understand how form arises. Hence, more than one doubt can be cast on Cassirer's vision of Uexküll as a biologist who is only interested in a synchronic study of living forms. Behind his evaluation, one could easily see Cassirer's personal agenda playing a role, notably: his search for elements of continuity stemming from Goethe's idealistic morphology, the correlated criticism of the exaggerated interest of the evolutionists in a diachronic analysis of the living being, the limitation of the methodological hegemony of mechanistic schemes, and the criticism of naive forms of vitalism based on the direct intervention of final causes.

In conclusion, one could certainly see Cassirer's reading of Uexküll's theoretical biology as partial, insofar as it ignores the biologist's desire for scientific completeness and his awareness of the necessity to confront his theories with the recent findings from experimental biology, which ultimately forces him to move onto the ground of ultimate causal factors. Nevertheless, concerning Cassirer's account, what we must now emphasize is that he aptly grasps how thoroughly Uexküll's approach to biology is permeated by the Kantian idea that "[mechanistic] causality is not the only rule at our disposal for construing the world" (Uexküll 1928, 81; quoted in Cassirer 1969, 203). In so doing, Cassirer gives the right prominence to the leading notion of an overall conformity to a plan [*Planmäßigkeit*] characterizing the living being, for which "we may use indifferently the terms 'structural character [*Planmäßigkeit*]', 'functional character' [*Funktionsmäßigkeit*], 'harmony' or 'wisdom'" (Uexküll 1928, 144; quoted in Cassirer 1969, 202f.).

Moreover, as Cassirer rightly observes, Uexküll is very attentive not to turn this "inner purposiveness" into some form of "outer purposiveness" (Cassirer 1969, 203). This point can be understood in at least two different ways. As to the first, as Cassirer explicitly points out:

> the living world is not so ordered that the various kinds of animals [...] relate to each other in such a way that the whole series finds its end in mankind – "end" in the two senses of a terminus and a purpose. [...] Every organism has its center of gravity in itself.
>
> (Cassirer 1969, 202f.)

The second way to understand Uexküll's rejection of outer purposiveness is to remember that, in all his major works, Uexküll emphasizes the fact that the plans of nature are blind. The wisdom of nature is not based on knowledge, *Weisheit* is not *Wissen*; when faced, in particular, with complex animal behavior, human observers are "all too inclined to search in a knowledge of the subject the source of the coincidence of an action with the advantage for the agent" (Uexküll 1928, 143, my transl.).

Beside the already-detected emphasis on morphology, the consideration of each organism as a center of gravity in its own right, the idea of the blindness of the structure, and the recurrent conceptual distinction between *Planmäßigkeit* (conformity to an inner plan) and *Zweckmäßigkeit* (correspondence to an external goal or order of goals), all bear witness to Uexküll's critical approach to the problem of teleology; this is why the neo-Kantian Cassirer, while investigating in *The Problem of Knowledge* the methodological assumptions of biology, can find in Uexküll's vision of organic life multiple elements of great relevance and usefulness.

3 Symbolic *Welt* versus Uexküll's *Umwelt*: Cassirer's view of the qualitative novelty of man

As anticipated in the introduction, Uexküll also helps Cassirer focus on the emergence of human beings, culture, and symbolic thought. In this context, Cassirer relies on what is probably the most influential concept of Uexküll's thought: the concept of *Umwelt*. *Umwelt* is, for Uexküll, the species-specific subjective environment, or "world," developed by all living beings that are endowed with a central nervous system connecting sense perception and operative organs. In the central nervous system, Uexküll maintains, there are quanta of excitation, stimulated by contacts with the external reality, that the organism immediately transforms in environmental "marks." Such marks are spontaneously outward-transposed [*hinausverlegt*] and experienced by the animal as objective reality. The resulting *Umwelt*, which according to the species manifests different degrees of complexity, is divided into perception world and effect world [*Merkwelt* and *Wirkwelt*] and – from the behavioral perspective – articulated in functional circles [*Funktionskreise*], such as the prey circle or the reproductive circle.

As long as Cassirer's focus is the problem of teleology and its legitimate usage in biology, the concept of *Umwelt* is considered exclusively as one of the many areas in which the goal-directedness of the organism unfolds itself: the *Bauplan* "is the inner structure that creates the environment by its own activity" (Cassirer 1969, 201). Cassirer does not misrepresent the notion of *Umwelt*; nevertheless, only brief summaries are given of it, in order to quickly return to the problem of the plan-based teleology of the living beings (see Cassirer 1969, 202). Moreover, in *The Problem of Knowledge*, Cassirer never implies that the notion of *Umwelt* can be contrasted to that of human life context. This last point should be preliminary emphasized in order to assess the radical change in Cassirer's perspective in the other two texts here under consideration (Cassirer 1944, 1996). In Uexküll's works, neither in the overall discussion of purposiveness in organic beings nor in the exposition of the unfolding of the species-specific *Umwelten* can any hint be found of an eventual qualitative novelty of man. The only reference concerning the relations between man and other animal species stresses, rather, the risks of anthropocentrism undermining the Haeckelian vision of evolution as culminating in man.

This point is important because, in *The Philosophy of Symbolic Forms*, Cassirer relies on Uexküll precisely to establish the qualitative novelty of man in relation to animals, in other words, to outline "the basic problem of 'philosophical anthropology'" (Cassirer 1996, 43). The novelty of man consists in the symbolic elaboration of the world: a form of consciousness that is free from the obligation to respond to stimuli with overt behavior, with "instinctive compulsion" (Cassirer 1996, 41). The issue of symbolic forms – language, art, myth, religion, philosophy, science, and even technique – is solved by Cassirer by making reference to the basic mediatedness of the relationship between man and the natural world. With tools, for instance, "the immediate grasping of an object is replaced by a mediated relation. [...] Technology sees them through a kind of refracting medium" (Cassirer 1996, 41).[1]

It is important, at this point, to underline that, according to Cassirer, the insertion of a whole range of mediation processes between man and world does not jeopardize the basic naturalness of man:

> a world shaped by myth could not be grasped if we conceive it exclusively as a form of thought or if we take it purely as a form of life; only the interconnection of both these determinations can provide its true constituent principle.
> (Cassirer 1996, 41)

Cassirer shares with the German philosophical anthropology the quest for a dynamic balance between *Geist* – human superior intellectual faculties – and *Leben* – the substratum of powerful, but incomplete drives and behavior. The German philosophical anthropology does not reiterate the scholastic explanatory model according to which man would have a core of animality to which higher faculties are added. As is particularly clear in Gehlen, the animality in man is instead deficient and fragmented and could never direct the action of humans without the institution of a new relationship with the outer reality, based on symbolic mediation and culture.[2] In this regard, however, Uexküll's theory is not of much use, even though – as we are now going to see – this is precisely what Cassirer is reluctant to admit in the two texts here under investigation.

3.1 Man as the only truly semiotic being

According to Cassirer, the first dividing line between man and the animals places on one side the immediacy of animals' reactions to the *Umwelt* and, on the other side, the mediated nature of human responses to the outer reality. To illustrate this opposition, Cassirer relies on semiotic vocabulary. In *The Philosophy of Symbolic Forms*, Cassirer writes that man "has directly to do no longer with 'things', but with 'signs' that he has created" (Cassirer 1996, 59f.). Although, as we shall see in the next section, the "things" to which animals relate are, rather, bundles of stimuli, the meaning of Cassirer's statement is clear: animals are immediately part of their life sphere and seem to react to the stimuli from the environment rather than to respond to them with flexibility.

In *An Essay on Man*, a more sophisticated semiotic toolkit is put into play. An important distinction between signs and signals, on the one hand, and symbols, on the other, is notably introduced. Although signs and signals are said to belong also to animals, they do not break, Cassirer argues, the basic immediateness of animal reactions: they are "part of the physical world of being" and have "a sort of physical or substantial being; symbols [instead] have only a functional value" (Cassirer 1944, 51). As Cassirer's pupil, Susanne K. Langer (1895–1985), puts it, signs are only a substitute, a "proxy" of the material thing (Langer 1948, 48f.), and they limit themselves to evoke a reaction that would be adequate to the material thing itself. According to Cassirer's preferred terminology, signals are "operators," while symbols are "designators" (Cassirer 1944, 51); animals have heavily limited semiotics skills.

For the purposes of this inquiry, what should be stressed is that the alterity of man as symbolic being pivots on the modification of Uexküll's concept of *Funktionskreis*:

> The functional circle of man is not only quantitatively enlarged; it has also undergone a qualitative change. [...] Between the receptor system and the effector system, which are to be found in all animal species, we find in man a third link which we may describe as the *symbolic system*.
>
> (Cassirer 1944, 43)

Language, art, myths, religions, and even science are rooted in this break with animal immediateness: whereas, in animals, "a direct [...] answer is given to an outward stimulus," in man, "the answer is delayed, [...] interrupted and retarded by a slow and complicated process of thought" (Cassirer 1944, 43). As a consequence of the "interposition of this artificial medium," man does not live "in a world of hard facts, or according to his immediate needs or desires. He lives rather in the midst of imaginary emotions, of hopes and fears, in illusions and disillusions, in his fantasies and dreams" (Cassirer 1944, 43).[3]

One may wonder whether Uexküll's thought is actually compatible with this approach to philosophical anthropology. As we shall see in the concluding remarks, no intention to separate human and nonhuman organisms and *Umwelten* can be legitimately ascribed to Uexküll. Consequently, any attempt to use his ideas to this aim entails some kind of forcing action. An example of this is how Cassirer uses the Uexküllian concept of counter-world [*Gegenwelt*]. In Uexküll's early texts, the counter-world is presented as the dynamic reproduction of the outer reality through an organism's complex set of inner physiological structures and processes. With this notion, Uexküll tries to explain how external reality can be "mirrored" on a physiological level (through the differential activation of the nerve fibers, the threshold value of excitation, etc.), in a way that ensures a close and functional interaction between organism and environment (see Uexküll 1909, 196). Coming back to Cassirer, he clearly borrows Uexküll's concept of counter-world – a physiological notion regarding all organisms endowed with a nervous system – to explain a distinctively human phenomenon, that is to say, the interposition of symbolic forms between perception and action: "*man* has built an 'opposing-world [*Gegenwelt*]' which now is added to the surrounding world of immediate existence" (Cassirer 1996, 61; emphasis mine).

Another example of Cassirer's interpretative forcing is the omission of the semiotic aspects of Uexküll's theory of *Umwelt*. As we have seen, the immediacy of the relationship between animal and physical world is the logical presupposition of Cassirer's claim on the semiotic uniqueness of man. However, once again, the direct reading of Uexküll's texts tells us a different story. Even in the most elementary sense-perception, there is semiosis: "in the nervous system it is not the stimulus itself that sets forth" – writes Uexküll – "stimuli from the external world are globally translated as a nervous sign language [*in eine nervöse Zeichensprache*]" (Uexküll 1909, 192, my transl.). This position is reinforced by the idea that the species-specific, subjective animal environment is made of the interactions between perception signs [*Merkzeichen*] and effect signs [*Wirkzeichen*] (Uexküll 1928, 65, 119–121). Together with its correlative risk of idealism, this point has been highlighted by Thomas Sebeok in the following terms: "the phenomenal world [*Umwelt*] [is] the subjective world each animal models out of its 'true' environment [*Natur*, 'reality'], which reveals itself solely through signs" (Sebeok 2001, 33). In this respect, Cassirer's omission of any reference to the semiotic nature of the *Umwelt* formation bears testimony to his will to stick to the idea of an "animal immediacy" proving the uniqueness of man as *animal symbolicum*.

3.2 *"Things" or bundles of excitations?*

Among the many conceptual oppositions that underpin Cassirer's basic understanding of the qualitative novelty of man, we also find the idea that only man is endowed with the cognitive faculty of thing-like perception. Relying on Uexküll's studies on lower animals, Cassirer stresses that, in their *Umwelten*, "instead of 'objects' there are only 'series of excitations'" (Cassirer 1996, 61), which "do not condense into a firm, thing-like [*dinghaft*] 'nucleus'" (Cassirer 1996, 61). For example, in the perceptive environment of the pilgrim scallop, the enemy (i.e., the starfish) is given as a precise sequence of optical and chemical stimuli (Cassirer 1996, 61). The starfish does not appear as a unique and substantial object bearing properties and exerting effects. This description presupposes human categories and cognitive aptitudes and is, therefore, already a form of anthropomorphism. In outlining this theory of animal perception, Cassirer cannot fail to refer to the study of Hans Volkelt (1886–1964) on representations in animals, which not only supports the absence of thing-like perceptions but even suggests an alternative theory. Animals, Volkelt argues, leaning on Felix Krueger (1874–1948), would not control their action according to varying qualities of well-defined, thing-like nuclei but according to the appearing and disappearing of qualities generically referred to the global situation [*Komplexqualitäten*] (see Volkelt 1912, 90). Animal perception is holistic and not organized in substrata carrying the change of perceptive qualities; even slight modifications in the total situation would be enough to guide animal behavior, without reference to "substrates" or "causes."

Although the idea of the dissolution of objects in their properties is indeed key to Uexküll's arguments in the texts Cassirer takes into account, it is also true that

Cassirer unwarrantedly extends its scope to every animal species, and thus well beyond Uexküll. Moreover, according to Cassirer, the absence of a thing-like perception would make the formation of a subject impossible. Hinging on Uexküll's research, Cassirer says that "since the animals never receive back stimuli of their own movement," they cannot perform that "isolation of the body as a material substratum" from which any "closer definition in term of conscious awareness, or 'being-for-itself'" (Cassirer 1996, 65) necessarily depends. Consequently, "without the experience of the relative constancy of that empirical object that we call our body, no empirical feeling of self, no experience of one's own I can develop" (Cassirer 1996, 64f.).

Once again, Cassirer's interpretation of Uexküll's thought is one-sided. First, Cassirer extends Uexküll's claims about lower animals to all animal species. If it is true that, according to Uexküll, the pilgrim scallop does not re-perceive any stimulus coming from its effector organs, it is also true that, in *Theoretical Biology*, the biologist develops an articulated analysis of animal behavior in which many actions performed by higher animals, such as birds or mammals, are characterized by a constant increase in proprioceptivity (Uexküll 1928, 201–210; see also Brentari 2016b). Moreover, in Uexküll's theory of action – where categories such as plastic action, learned action, controlled action, and so on are introduced – there is no break between animal behavior and human actions. Under the label of controlled action, which Uexküll compiles based on the response to the organism's own previous movements, one finds both birdsongs and human motility. While this is certainly not enough to attribute a symbolic or representative self-consciousness to animals, the very fact that Uexküll establishes a scale of proprioceptivity where humans are also included casts more than a shadow on Cassirer's views of animals as incapable of distinguishing between themselves and the surrounding environment.

However, even this is not the central point. If one was to read Cassirer's arguments on the absence of thing-like perception in animals, with no direct knowledge of Uexküll's texts, one would most certainly fall under the impression that, for Uexküll too, the absence of self-consciousness coincides with the absence of subjectivity *tout court*. On the contrary, in Uexküll's work, the animal is a subject on an *a priori* level; its environment is a transcendental construction that arises following the provisions of the *Bauplan*. Whether the animal's body, as a phenomenon, is or is not included in the perceptive environment, does not affect the organism's "noumenal" spontaneity and operative capacity of self-determination. Here, the use of Kantian terminology is not accidental, since, for Uexküll, the construction of the environment is a species-specific transcendental process, which, to be understood, requires the extension to biology and physiology of the approach developed in the *Critique of Pure Reason* (Uexküll 1928, 3). In Uexküll's work, in short, the issue of animal subjectivity does not depend on the type of perception. The animal is always a transcendental subject, may it constitute its environment by using thing-like or not thing-like representations, substantial categories or fluid perceptions, and different degrees of re-perception of the body. Moreover, in this wide species-specific variety the experience world of man – relying on the Kantian intuition of space and time and on the categories of causality, substance ("thing"), and so forth – remains, for Uexküll, merely one particular *Umwelt* among others.[4]

4 Concluding remarks

It this concluding section, I try to sketch the overall image of Uexküll that emerges from Cassirer's pages. To this end, I test Cassirer's interpretation of Uexküll's *Umweltlehre* against a text from 1934, *A Foray into the Worlds of Animals and Humans* (Uexküll 2010), which Cassirer never mentioned – probably simply because he did not know it. This text presents to the reader, from a layperson's perspective, a series of increasingly complex environments, starting from elementary environments – such as those of jellyfish or *Paramecium* – to end with more elaborate and complex environments. The *Umwelten* of higher animals, in fact, appear as subjective constructions, where combinations of stimuli, diachronic and synchronic *Gestalten*, and even objects as unified entities phenomenologically appear. Furthermore, in order to illustrate some phenomena of the upper environments, Uexküll uses indifferently some cases taken from animal life and others coming from the human world, as to point out that what is growing in complexity and richness are not the faculties of the organisms (i.e. intelligence, memory, etc.) but the environments as configurations of meaning.

In this new context, it is hard to find any decisive support both for the idea of man as the unique truly semiotic being and for the supposition that all animal perception is based on flows of excitations. Moreover, whereas the texts known by Cassirer focus on the lower animals, and were therefore more suitable to him as described earlier, here phenomena taken from the human world are used to illustrate, without solution of continuity, behavioral processes taking place in the animal world. This happens both in the previously mentioned theory of action and in the case study of the oak (Uexküll 2010, 126–131), on which we shall now briefly focus.

The same environmental element, an oak tree, can have a high semantic variability in the environments of different animal species: for the fox, which has built its den among the oak's roots, the oak has "a protection tone"; for the squirrel, the tree has a "climbing tone"; and, for the songbirds, a "carrying tone" (Uexküll 2010, 130). With great ease, Uexküll then proceeds to show how the same *Umwelt* element has different meaning even for different human subjects: in the rational *Umwelt* of an old forester, who is interested in the monetary value of the oak, the tree has a "use tone"; while, for a little girl fascinated by the oak's bulging bark which resembles a human face, the tree has a "danger tone" (for a previous version of this case of study, which was very presumably accessible to Cassirer, see Uexküll 1928, 232). Based on Uexküll's theoretical framework, the perceptual and cognitive diversity between the forester's environment and that of the little girl, in connection with the same object, "oak," does not present any specific problem. Uexküll deals with this diversity as if it were entirely analogous to the difference in how a fox and a squirrel "interpret" the oak and give it a different meaning.

Cassirer chooses not to mention any of Uexküll's examples concerning the relationship between man and environment, since this would clearly reveal the intention of the Estonian biologist not to discriminate between human and animal environments. Regardless of Cassirer's strategy, however, here lies indeed a deeper problem. In Uexküll's account, but beyond his explicit intentions, the human species appears as the only one that shows such high variability in the individual elaboration of the *Umwelt*. This variability depends not only on the body's physiological state (i.e., hunger, reproductive impulse, etc.) but, above all, on individual characteristics, which are part of a wide repertoire of psychological and cultural possibilities. Moreover, humans are the only animals that show the inclination and the ability to distance themselves from their own species-specific environment. They break the innate perceptive limits of their own anatomical and physiological *Bauplan*, thus reaching something that is no longer a species-specific environment but the scientific world as an amalgam of recognized models and facts – which, incidentally, includes the theoretical reconstruction of the *Umwelten* of other species (Uexküll 2010, 133–135; Brentari 2015, 151–156). In this way, a critical analysis of Uexküll's gradualism brings us back to the qualitative novelty of the human world, characterized, as it is, by individual semantic variability of environmental elements and by a great inclusiveness toward other environments and sense constructions. Because these last problems only become really clear in *A Foray in the Worlds of Animals and Humans*, we cannot blame Cassirer for having missed on what ground exactly Uexküll could have helped him establish his philosophical anthropology.

Finally, the overall image of the living that emerges from the interaction between Uexküll and Cassirer can be summed up as follows. Uexküll's strong belief, taken up again by Cassirer, that mechanistic causality is not the only principle we can rely on in order to understand nature, can be interpreted in two different ways. For nonhuman living beings, teleology should be used as a heuristic principle that legitimizes, on a functional level, the use of a range of concepts such as "function" and "purpose," without which the very meaning of the observed phenomena could easily elude us. With the previously mentioned limits, which attest to the persistence in Uexküll of elements of substantial vitalism, the most convincing trait of Cassirer's interpretation of Uexküll is precisely the definition of a blind inner purposiveness that determines organic processes overlapping, without canceling it, mechanical causality. This form of determination is still largely unexplored from the philosophical point of view – except for the relevant discussion that Nicolai Hartmann devotes to it (Hartmann 1980) – but as a heuristic category, it continues to prove useful in the practice of biology.

As far as human beings are concerned, the confrontation between Uexküll and Cassirer actually creates the need for a reciprocal correction. On the one hand, with Uexküll, one can discover the semiotic dimension of all lived environments: every animal subject puts in place semiotic processes of mediation of the external reality. On the other hand, correcting Uexküll with Cassirer, a discontinuity between human and animal environments must be highlighted. The previously detected inclusiveness and high intraspecific variability of human experience require a specific analysis, with its own theoretical tools. The rise of the symbolic dimension, in fact, disrupts the spontaneous semiosis of the animal life; with the human animals, a categorical *novum* actually emerges: the conscious finality of human action based on the ability to anticipate future conditions and to imagine the world otherwise. This critical distinction, obtained through the intersection of some of the most convincing aspects of Uexküll's and Cassirer's perspectives, might well benefit many disciplines that deal with the living being from a philosophical perspective – from the philosophy of biology to the ontology of nature, from biophilosophy to philosophical anthropology.

Notes

1 The mediatedness of man is a central issue of the German philosophical anthropology of the 20th century (see Gehlen 1988; Plessner 2019). In general, there is a strong theoretical closeness between Cassirer's reflection on man and the German philosophical anthropology, as testified by many explicit or implicit references (see, for instance, Cassirer 1996). Moreover, all authors belonging to the German philosophical anthropology (and Heidegger) use the concept of *Umwelt* in opposition to the concept of world, as a way of emphasizing the qualitative difference of man. In this regard, see Chapters 2, 3, and 5 in this volume.
2 For a criticism of the traditional idea of the animality in man as a lower, but efficient, functional system, see Brentari (2016a).
3 The following statement by Andreas Weber should be mentioned for completeness: "In a manuscript [...] Cassirer discusses the symbolic worlds of certain animal species. Contrary to his statements in the *Essay on Man*, at least some seem to be fitted with a symbolic system which has a physiognomic character" (Weber 2004, 303). For the manuscript in question, see Cassirer (1995, 66f.); for an overview on Cassirer's (marginal) claims on the presence in animals of "elementary expressive moments," see Schwemmer (1997, 50f.).
4 The different ways in which Cassirer and Uexküll rely on the legacy of Kant is summarized very effectively by Frederik Stjernfelt: "We may schematically say that while Cassirer is an epistemologizing neo-Kantian with strong objectivist tendencies, Uexküll is a naturalizing Kantian with strong subjectivist leanings – a sort of interesting chiasm to be explained in further details" (Stjernfelt 2011, 171). On the subject, see Chapter 2 in this volume.

References

Brentari, Carlo (2015) *Jakob von Uexküll: The Discovery of the Umwelt between Biosemiotics and Theoretical Biology*. Dordrecht/Heidelberg/New York/London: Springer.

Brentari, Carlo (2016a) 'Behaving like an animal? Some implications of the philosophical debate on the animality in man'. In: Morten Tønnessen, Kristin Armstrong Oma, and Silver Rattasepp (eds.) *Thinking about Animals in the Age of the Anthropocene*. Lanham: Lexington, 127–144.

Brentari, Carlo (2016b) 'Jakob von Uexkülls Theorie der tierlichen Handlung zwischen Neovitalismus und vergleichender Verhaltensforschung'. In: Martin Böhnert, Kristian Köchy, and Matthias Wunsch (eds.) *Philosophie der Tierforschung*. Vol. 1. *Methoden und Programme*. Freiburg: Verlag Karl Alber, 209–240.

Cassirer, Ernst (1944) *An Essay on Man: An Introduction to a Philosophy of Human Culture*. New Haven: Yale University Press.

Cassirer, Ernst (1969) *The Problem of Knowledge: Philosophy, Science, and History since Hegel (1935–40)*. New Haven/London: Yale University Press.

Cassirer, Ernst (1995) [1928] *Nachgelassene Manuskripte und Texte*. Vol. 1. *Zur Metaphysik der symbolischen Formen*. Hamburg: Felix Meiner Verlag.

Cassirer, Ernst (1996) [1923–29] *The Philosophy of Symbolic Forms*, Vol. 4. *The Metaphysics of Symbolic Form*. Translated by John Michael Krois. New Haven/London: Yale University Press.

Cheung, Tobias (2004) 'From protoplasm to Umwelt. Plans and the technique of nature in Jakob von Uexküll's theory of organismic order'. *Sign Systems Studies* 32 (1/2), 139–167.

Cummins, Robert C. (2002) 'Neo-teleology'. In: Andre Ariew, Robert E. Cummins, and Mark Perlman (eds.) *Functions: New Essays in the Philosophy of Psychology and Biology*. Oxford: University Press.

Cummins, Robert C. and Roth, Martin (2010) 'Traits have not evolved to function the way they do because of a past advantage'. In: Francisco J. Ayala and Robert Arp (eds.) *Contemporary Debates in Philosophy of Biology*. Chichester, UK: Wiley-Blackwell, 72–85.

Gehlen, Arnold (1988) [1940] *Man: His Nature and Place in the World*. New York: Columbia University Press.

Geison, Gerald L. (1969) 'The protoplasmic theory of life and the vitalist mechanist debate'. *Isis* 60 (3), 272–292.

Godfrey-Smith, Peter (2014) *Philosophy of Biology*. Princeton/Oxford: Princeton University Press.

Hartmann, Nicolai (1980) [1950] *Philosophie der Natur. Abriss der speziellen Kategorienlehre* (2nd ed.). Berlin: De Gruyter.

Krois, John M. (2004) 'Ernst Cassirer's philosophy of biology'. *Sign Systems Studies* 32 (1/2), 277–295.

Kull, Kalevi (2004) 'Uexküll and the post-modern evolutionism'. *Sign Systems Studies* 3 (1/2), 99–114.

Langer, Susanne K. (1948) *Philosophy in a New Key: A Study in the Symbolism of Reason, Rite, and Art*. New York: New American Library.

Mayr, Ernst (1992) 'The idea of teleology'. *Journal of the History of Ideas* 53 (1), 117–135.

Nagel, Ernest (1951) 'The problem of knowledge: Philosophy, science, and history since Hegel by Ernst Cassirer, William H. Woglom, and Charles W. Hendel' (book review). *The Journal of Philosophy* 48 (5), 147–151.

Perlman, Mark (2004) 'The modern philosophical resurrection of teleology'. *The Monist* 87 (1), 3–51.

Perlman, Mark (2010) 'Traits have evolved to function the way they do because of a past advantage'. In: Francisco J. Ayala and Robert Arp (eds.) *Contemporary Debates in Philosophy of Biology*. Chichester: Wiley-Blackwell, 53–71.

Plessner, Helmuth (2019) [1928] *The Levels of Organic Life and the Human: An Introduction to Philosophical Anthropology*. Translated by Millay Hyatt. New York: Fordham University Press.

Schwemmer, Oswald (1997) *Ernst Cassirer: Ein Philosoph der europäischen Moderne*. Berlin: Akademie-Verlag.

Sebeok, Thomas A. (2001) *Signs: An Introduction to Semiotics*. Toronto: University of Toronto Press.

Stjernfelt, Frederik (2011) 'Simple animals and complex biology: Von Uexküll's two-fold influence on Cassirer's philosophy'. *Synthese* 179, 169–186.

Uexküll, Jakob von (1909) *Umwelt und Innenwelt der Tiere* (1st ed.). Berlin: Springer.

Uexküll, Jakob von (1920) *Theoretische Biologie* (1st ed.). Berlin: Paetel.

Uexküll, Jakob von (1921) *Umwelt und Innenwelt der Tiere* (2nd ed.). Berlin: Springer.

Uexküll, Jakob von (1928) *Theoretische Biologie* (2nd ed.). Berlin: Springer.

Uexküll, Jakob von (1930) *Die Lebenslehre*. Potsdam/Zürich: Müller & Kiepenheuer-Orell Füssli.

Uexküll, Jakob von (2010) [1934, 1940] *A Foray into the Worlds of Animals and Humans with a Theory of Meaning*. Translated by Joseph D. O'Neil. Minneapolis/London: University of Minneapolis Press.

Volkelt, Hans (1912) *Über die Vorstellungen der Tiere*. Leipzig: Engelmann Verlag.

Weber, Andreas (2004) 'Mimesis and metaphor: The biosemiotic generation of meaning in Cassirer and Uexküll'. *Sign Systems Studies* 32 (1/2), 297–307.

Welch, Rickey G. and Clegg, James S. (2010) 'From protoplasmic theory to cellular systems biology: A 150-year reflection'. *American Journal of Physiology: Cell Physiology* 298, 1280–1290.

7 The philosopher's boredom and the lizard's sun

Martin Heidegger's interpretation of Jakob von Uexküll's *Umwelt* theory

Francesca Michelini

Heidegger's interest in biology and in Uexküll's theories stems from a complex cluster of philosophical questions, revolving around the difference between human beings and animals, and their respective relationship to the world. In order to provide an account of this context, I introduce Heidegger's philosophical setup and the main concerns raised by it. As a result, Uexküll's impact turns out to be rich in consequences, even when not entirely consequential.

1 Introduction: behavior and comportment

Heidegger's famous lecture course held in 1929/1930 on *The Fundamental Concepts of Metaphysics* is divided into two parts that, at first sight, have little to do with one another. While the first is devoted to deep boredom [*tiefe Langeweile*] as existential condition of human beings, the second deals with the complex question of what the world is. Here, while attempting to shed light on the ontological structure of the living, Heidegger introduces his renown distinction between the *modus essendi* of animals and that of human beings. Whereas what pertains generally to animals is "behavior" [*Benehmen*], exclusive prerogative of human beings is what one calls "comportment" [*Verhalten*]. This terminological difference expresses, according to Heidegger, an "essential" difference. In this respect, the whole passage here at stake is worth reading:

> The *specific manner* in which man *is* we shall call *comportment* and the *specific manner* in which the animal is we shall call *behavior*. They are fundamentally different from one another. In principle it is also possible to reverse this linguistic usage and refer to *animal comportment*. [...] The behavior of the animal is not a *doing and acting*, as in human comportment, but a *driven performing* [*Treiben*]. In saying this we mean to suggest that an instinctual driveness, as it were, characterizes all such animal performance.
>
> (Heidegger 1995, 237, emphasis in the text)

The "driven performing" has clearly nothing to do with a decision or a goal-directed comportment. "Instinctual activity is not a recognitive self-directing toward objectively present things," says Heidegger (1995, 243). As specific forms of self-determination of existence, "doing" and "acting" pertain only to humans – a topic, the latter, later widely developed by Hannah Arendt (2013, 175f.).

Such a distinction between humans and animals is clearly not new but, rather, calls on a long tradition in Western thought. It is a story developing along a clearly traced demarcation line between the animal – as the place of instincts, impulses, and "necessity" – and the human – as the domain of rationality, language, choice, and freedom. One could therefore be tempted to place Heidegger along this established anthropocentric path, possibly as its culmination. He would have then contributed to drop into oblivion not only the issue of animals but also that of the animality of humans. This latter topic is indeed openly thematized by Heidegger in the *Letter on Humanism*, where humans are ultimately placed outside of life and nature themselves (on this point, see Cimatti, Chapter 10, in this volume, and Cimatti 2013).

However, the issue is more complex and thornier than what the earlier-quoted passage let us think. First and foremost, as soon as we take Heidegger as the culmination of the Western tradition, we also contradict his overall philosophical project, that is to say, his intention to break free from said tradition and notably from the notion of subjectivity typical of Western metaphysics. Heidegger wishes to rescue metaphysical questions from the supremacy of reason, the dictatorship of *logos* and, at least with reference to his production predating the *Kehre*, root them in the actual life of human beings, in the "facticity of *Dasein*" (in this regard, see the arguments in Heidegger's favor in Kessel 2011). Moreover, although he contributed to the marginalization of animal issues in contemporary philosophy, Heidegger has nevertheless opened the way to an approach on animal being to which many contemporary contributions are indebted. This is the case, for instance, for Derrida's and Agamben's accounts, as pointed out by Matthew Calarco (2008), despite the fact that these philosophers do not share Heidegger's positions and argue against them. In this respect, one can maintain that, although Heidegger has ultimately fallen back into a form of metaphysical anthropocentrism, some of his intuitions about animals, somehow precluded to him, can be fruitfully followed and developed in non-anthropocentric directions (Calarco 2008, 15). Two hints in particular would be worth pressing forward: the fact that, concerning animals and humans, Heidegger has explicitly avoided any distinction entailing hierarchy or degree and the fact that he attempted a metaphysical and phenomenological inquiry on the specific mode of Being of animals "on their own terms" (Calarco 2008, 22). Despite his rightful rejection of any hierarchy between animals and humans, Heidegger's greatest limit would then be, nevertheless, his maintaining the validity of the distinction as a fundamental reference in scientific and philosophical research (Calarco 2008, 23).

This chapter agrees – to an extent – with what it has been outlined so far following Calarco's lead. One can indeed accept that, despite the fact that his contribution has played an undeniably important role which cannot be easily written off, Heidegger has never left the anthropocentric framework after all. However, unlike Calarco – and, as we will see, also unlike Derrida – I argue that it is precisely Heidegger's obsession to avoid all reference to degrees and orders of Being – in other words, his fight against any kind of level continuity in nature – that forces him to remain within not only an anthropocentric framework but also one advocating the abyssal distance between mankind and the rest of nature.

Similar remarks have been made by Hans Jonas (1903–1993), Heidegger's pupil, who, in *The Phenomenon of Life* (1966), criticizes Heidegger for considering as a "lowering" of human beings their collocation "in any scale, that is, in a context of nature as such" (Jonas 1966, 227). According to Jonas, on the contrary, there is no structural difference between humans and other living beings; in virtue of their living corporeality [*Leiblichkeit*], human beings are part of life continuity, as organisms within a range of organisms. Jonas advocated then a natural gradation based on which biological phenomena are articulated in different levels. These levels should not be mistaken, however, as the revival of long-dismissed ancient models (e.g., the *scala naturae*). They are rather intended by Jonas as ontological levels of progressive detachment of the living being from the environment. He notably refers to different levels of "increase of mediacy" (Jonas 1966, 107). According to Jonas, then, Heidegger failed precisely to acknowledge this intrinsic belonging of human beings to the rest of the living, especially because, despite his insisting on the "facticity of *Dasein*" and on existence as care [*Sorge*], he would have entirely neglected the topic of corporeality.

> Heidegger had talked about existence as care, but he did so from an exclusively intellectual perspective. There was no mention of the primary physical reason for having to care, which is our corporeality [*Leiblichkeit*], [...] in *Being and Time* the body had been omitted and nature shunted aside as something merely present.
>
> (Jonas 2002, 31)

By placing human beings, in virtue of their living bodies, back within their vital and natural context, Jonas is able to positively reassess also what Heidegger completely excluded from his inquiry on human nature, that is, anthropomorphism. As Jonas sees it, anthropomorphism, if understood "critically," is not necessarily bad or antiscientific but, rather, a fruitful phenomenological mode granting man access to the living based on organic and spiritual kinship. Following Jonas, then, one could argue – as stated again in the conclusions to this chapter – that it is precisely Heidegger's radical refusal of this kind of "critical" anthropomorphism, that is to say, an approach to the living naturally and strongly anchored to human corporeality, that pushes him back into anthropocentrism.

It is within this complex cluster of problems that one finds Heidegger's references to Jakob von Uexküll, by him defined as "one of the most perceptive of contemporary biologists" (Heidegger 1995, 215). Uexküll's impact on Heidegger's theories is way superior than that of the other biologists he read and got to know along the years and should not be taken as a mere testament to Heidegger's scientific and biological knowledge – against the widespread idea that Heidegger had little understanding or little information about science. Heidegger's references to Uexküll should instead be read as a theoretical turning point, which is key to the understanding of his position in relation to a series of crucial points in his investigations, first among them the ontological difference between animals and humans. As we shall see, Uexküll is Heidegger's best ally, because he, more than any other biologist, would have shed light on the distance between the *Umwelten* of the animals and the human world – and not for his famous claim that each animal species is the subject of an *Umwelt* (Heidegger 1995, 264).

Against the earlier-outlined backdrop, I provide a short outline of Heidegger's ideas on animals (Section 2), emphasizing the points where he most noticeably relies on Uexküll both for "concrete" examples and for conceptual apparatus, even when he introduces radical linguistic alterations. I then look in some detail at Heidegger's agenda in his references to Uexküll's claims (Section 3) and point to how he bends them, often forcibly, to his philosophical project. This will help us, in conclusion, better understand in what respect, despite some correct basic intuitions developed by engaging with Uexküll, Heidegger is still fundamentally tied to an anthropocentric perspective.

2 The animal in Heidegger

2.1 Heidegger meets Jakob von Uexküll and bees

An unprecedented approach, never again repeated in his later inquiries, in order to exemplify the difference between behavior and comportment, in the *Fundamental Concepts*, Heidegger calls on the biology of his time. Jakob von Uexküll is clearly his main source. Direct and indirect references are made, however, also to many other biologists, mainly to the embryologist Karl Ernst von Baer (1792–1876); the founder of neovitalism, Hans Driesch (1867–1941); and the zoologist and anthropologist Frederik Jacobus Johannes Buytendijk (1887–1974), in addition to Darwin and in general to the Darwinism of his time.

Heidegger's acquaintance with Uexküll starts in all likelihood already in 1925/26, since the term *Umwelt* in connection with animals appears already in one lecture from 1925 (see Buchanan 2008, 92f.), while Uexküll's name is first mentioned in a *Logik-Vorlesung* from the winter semester 1925–26 (see Kessel 2011, 199). It should probably be dated to this time also his reading of *Theoretical Biology* (1920; second edition 1928), of Umwelt *and Inner-World of Animals* [*Umwelt und Innenwelt der Tiere*] (second edition with significant alterations in 1921), and of Uexküll's other papers, progressively published especially on the *Zeitschrift für Biologie*. Nevertheless, in order to grasp the full range of Heidegger's interest in Uexküll, one should not neglect the fundamental mediating role of Ernst Cassirer, which comes clearly to the fore especially when it comes to the idea that the animal is shut within the circle of instinctual life (Brentari 2015, 198). Cassirer is, back then, Uexküll's friend, besides being a regular attendee at the Institut für Umweltforschung founded by Uexküll in Hamburg. In 1929 – the same year of Heidegger's famous lectures – Cassirer holds in Davos a conference speech, attended by Heidegger, under the title "The Fundamental Problems of Philosophical Anthropology" [*Grundprobleme der Philosophischen Anthropologie*], where he deals at length with the Estonian biologist.

In the *Fundamental Concepts*, Heidegger points notably to some "*concrete examples of animal behavior drawn from experimental research*" (Heidegger 1995, 241, emphasis in the text). These concern, in particular, the world of bees – a species bound to rise, for Heidegger, to the status of emblem of animality itself. The first example recalls a bucolic scene, also included in Uexküll's *Theoretical Biology*, where a worker bee goes from blossom to blossom to collect nectar. Following Uexküll, Heidegger remarks that the bee's flying, which might appear totally random to an external observer, is absolutely not such. The bee's flying is directed by the flower scent. Heidegger talks about an animal "behavior," inasmuch as it takes place within what Uexküll presents as a functional circle [*Funktionskreis*]. In Uexküll's language, the flower scent is a perception-mark carrier [*Merkmalträger*], which corresponds for the bee to the effect-mark carrier [*Wirkmalträger*] "suck nectar." When a bee selects a flower based on scent and color and finally sits on it, one might think that it establishes a relation with it to suck its nectar, but actually, according to Heidegger, it never relates essentially to the flower *as* flower. Since it belongs to the functional circle of nutrition, the flower enters the bee's experience only as a trigger to the action leading to the nectar. In this sense, the flower is never apprehended [*vernommen*] as flower, that is to say, as an autonomous entity with its own features. For the bee the flower is only food, nectar. What's more, according to Heidegger, the bee does not grasp the nectar *as* nectar. In fact, once it has finished sucking the flower's nectar and flies to another flower, it does not do so because it has remarked the absence of nectar. This is emphasized also in the second example Heidegger extracts from Uexküll's research (Uexküll 1926, 169). A cut is performed on a bee's abdomen, where the nectar collecting organ is placed, while the bee is eating honey from a little bowl placed in front of it (Heidegger 1995, 242). The bee keeps sucking with no feeling of satiety, until the honey starts leaking from the bee's body. Notwithstanding, the bee does not stop. In fact, it cannot remark on either the excess of nectar in relation to its holding capacity or the very fact that its abdomen is missing. According to Heidegger, "there is no *apprehending* of honey *as* something present, but rather a peculiar captivation" (Heidegger 1995, 243) since "the bee is simply taken [*hingenommen*] by nectar" (Heidegger 1995, 242).

Heidegger also recalls another experiment, performed by one of Uexküll's co-researchers, Albrecht Bethe (1872–1954).[1] The experiment investigates the bees' orientation ability based on light, which allows them to easily find their way back to their hive. Upon leaving the hive, in fact, bees record the angle of sunlight incidence on them and the length of their flight to the meadow so that if imprisoned for some time in a dark box, as soon as they are free again, they will be able to fly back to the hive by taking into account the parameters recorded during their journey out. However, although the bee relies on the sun for orientation, also in this case, as in the previous two, it does not perceive the object as such. Hence, Heidegger concludes,

[t]he bee is simply *given over* to the sun and to the period of its flight *without being able to grasp either of these as such*, without being able to reflect upon them as something thus grasped. The bee can only be given over to things in this way because it is driven by the fundamental drive of foraging. It is precisely because of this *drivenness*, and not on account of any recognition or reflection, that the bee can be *captivated by* what the sun occasions in its behavior.

(Heidegger 1995, 247, emphasis in the text)

2.2 The essence of animality: captivation and being poor in world

Heidegger's reference to "concrete" examples has a quite clear goal. They are meant to confirm the idea that what he calls "as-structure" does not pertain to the animal: bees do not grasp flowers as flowers, nectar as nectar, sun as sun. This is explained by the fact that the relation they entertain with world entities is that of being fully captivated [*benommen*] or, even, taken in some sort of grip [*hingenommen:* taken or absorbed]. It is important to remark that this being captivated is not for the animal, as Heidegger sees it, a temporary condition or a transitional connotation of its behavior, but it is rather its essential structure. This is not about the structure of a given species, but of the "animal" taken as a single category, or, more precisely, of animality itself. In short: "The captivation of the animal therefore signifies, in the first place, essentially having every apprehending of something as something withheld from it" (Heidegger 1995, 240).

The animal is captivated inasmuch as it is taken by something that cannot be identified as the animal itself but also not exclusively as something that is other from it and coming from the outside. This something gathers in unity both terms of the relation. The animal is thereby tied to a chain of impulses that bind it fully to a life form articulated in a circle [*Ring*] of inescapable drives. "The activity does not simply cease – Heidegger writes – rather the drivenness of the capability is redirected into another drive" (Heidegger 1995, 243; on this topic, see Franck 1991, 142).

In order to capture the essence of animality in general, Heidegger establishes a relation between *Benommenheit* [captivation] and other variations on the German verb *nehmen* [to take] – often used in the past participle form, *genommen*. Above them all, *Hingenommenheit* [the being taken], far from being a temporary feature of the animal, describes the moment of receptive acceptance of the organism in relation to something that, despite not being identified as such, fully dominates it (see Bassanese 2004, 268). The animal's "being taken" is based upon, according to Heidegger, the *Eingenommenheit* – a term usually translated in English as *absorption* – that qualifies the indissoluble connection between organism and environment, to be here understood as an experience of involvement, absolute immanence, plenitude.

Rather than being just idle distinctions, these linguistic variations help us understand the twofold connotation of *captivation*. The latter defines animality as a sphere in which acting has no spontaneity whatsoever and is somehow predetermined and blind, and it presents the cognitive relation between the animal and the world as mere reactivity (Brentari 2015, 201). But it also simultaneously points to the plenitude of animal life. Said fullness is experienced precisely because what is experienced is not apprehended, perceived, or thought of as a given entity. In this respect, animal life (and life in general) can be taken as a form of openness and richness that is unknown to mankind (Heidegger 1995, 255). A comparison can be here drawn with the experience of falling in love (Cimatti 2013) or, as pointed out by Giorgio Agamben, with the two fundamental and contradictory aspects of mystical ecstasy:

> Heidegger seems [...] to oscillate between two opposite poles, which in some ways recall the paradoxes of mystical knowledge – or, rather, nonknowledge. On the one hand, captivation is a more spell binding and intense openness than any kind of human knowledge; on the other, insofar as it is not capable of disconcealing its own disinhibitor, it is closed in a total opacity.
>
> (Agamben 2004, 58)

Both falling in love and mystical knowledge share one distinctive feature: a form of forgetting, or loss, of the world. This is also what, on more than one occasion, Hannah Arendt states in *The Human Condition*, "[l]ove, by its very nature, is unworldly" (Arendt 2013, 242).

Being deprived of as-structure, being *benommen*, means then, so Heidegger, to have no access to the world, or better, according to the renowned, although fully ambiguous, language of the *Fundamental Concepts*, being poor in world [*weltarm*] (Heidegger 1995, 176f.). That the animal is poor in world is indeed Heidegger's pivotal claim in the comparative argument based on which the animal works as a middle term between the inorganic (e.g., the stone), which is wordless, and human beings, who are world-forming. By means of his comparative inquiry, Heidegger has clearly no intention to suggest a biological taxonomy though. This is confirmed, for instance, by the fact that, despite being mentioned several times, especially at the beginning of the account, (Heidegger 1995, 62, 179, 207), plants play no role in the rest of the argument. This disappearance seems to suggest, incidentally, that what is, for Heidegger, really decisive in the understanding of the essence of life is actually the determination of the essence of animality. As he writes, "what remains as a permanent and recurrent task of philosophy is precisely [...] to grasp the question concerning the essence of animality and thus the essence of life in general in all its questionableness" (Heidegger 1995, 207). "Life is nothing but the animal's encircling [*Ringen*] itself and struggling with its encircling ring" (Heidegger 1995, 257).

By saying "poor in world," Heidegger expresses the idea that the animal, unlike the inorganic, is not entirely deprived of access to other beings, although its mode is radically different compared to that of humans. The difference among inorganic, organic, and human is clarified by Heidegger by means of an often-quoted example portraying a lizard sunbathing on a rock. Unlike the rock, writes Heidegger, "[t]he lizard basking in the sun on its warm stone does not merely crop up in the world" (Heidegger 1995, 197) but, rather, entertains a given relation with the rock, the sun, and the other intra-mundane entities, or better said, with what we take as sun and rock, since for the lizard they are only "lizard-things." Sunbathing for the lizard is then radically different from sunbathing for the stone, but also from that of human beings:

It is true that the rock on which the lizard lies is not given for the lizard as rock, in such a way that it could inquire into its mineralogical constitution for example. It is true that the sun in which it is basking is not given for the lizard as sun, in such a way that it could ask questions of astrophysics about it.

(Heidegger 1995, 197f.)

It should be remarked here – what has been missed by the interpreters – that this example is again borrowed from the *Theoretical Biology*, where Uexküll – concerning a beetle, however, not a lizard – says, "The stone, that a beetle climbs over is merely a beetle-path, and does not in any way belong to the science of mineralogy" (Uexküll 1926, 130).

The "captivation" is then the deciding feature for the poverty in world that makes up for Heidegger's leading claim on animality in the *Fundamental Concepts*. One could certainly object that the status of being captivated is not a prerogative of animal beings, but also belongs to the human condition. The examples of falling in love and mystical ecstasy have already been mentioned. Heidegger himself concedes that in several existential situations – that in *Being and Time* he calls "inauthentic" – human beings, like animals, are captivated. This applies mainly to the condition of "deep boredom," the account of which makes up for the first part of Heidegger's inquiry in the *Fundamental Concepts*. In deep boredom, in fact, "we are [...] taken [*hingenommen*] by things, if not altogether lost in them, and often even captivated [*benommen*] by them. Our activities and exploits become immersed [*aufgehen*] in something" (Heidegger 1995, 101, see Mazzeo, Chapter 12, in this volume). One should here take into account the proximity, also terminological, with the condition of lizards and bees. However, the absolute proximity of the *Dasein* and the animal is after all only an illusion. Whereas the animal is irremediably absorbed in captivation, when an entity captures human beings and keeps them stuck denying their sense, they undergo only a temporary suspension. In deep boredom is always included the possibility of an unveiling [*a-lètheia*, in ancient Greek], hence the opening to the world, which is instead precluded in the "captivation" of the animal (on this topic, see Bassanese 2004, 295–296). Based on this perspective, it is then clear how the two parts of Heidegger's text, on deep boredom and on the world, although, at first sight, very distant from one another, have deep connecting roots (on this, see Agamben 2004).

3 Heidegger and Uexküll: proximity and differences

3.1 *Disinhibiting ring and* Umwelt

The animal's limited access to the world, conveyed by the expression "poor in world," is further elaborated upon by Heidegger by means of one more neologism:

"disinhibiting ring" [*Enthemmungsring*]. With it he notably identifies the environmental circle the animal carries its whole life as a key feature of its being (Heidegger 1995, 258). Its function is to prescribe "what can affect or occasion [the animal's] behavior" (Heidegger 1995, 255). Said encirclement should not be taken simply in spatial terms. It is not an encapsulation or a shell, as, for instance, the shell of turtles. While the "disinhibiting ring" pertains to the innermost organization of the animal and to the morphological structure of the organism, it mainly embodies its relations to the external world (Heidegger 1995, 255). In other words, it is the relational structure pertaining essentially to the captivation of the animal. As previously mentioned, unlike the inorganic, the mode of being of the animal refers to what is other than itself; it establishes some given relations, and in this respect, it is then open to the other. Simultaneously, however, unlike humans, animals have no access to beings in their "manifestedness". The other can only be apprehended by animals based on its function, which corresponds to a disinhibition: "This other is taken up into this openness of the animal in a manner that we shall describe as disinhibition [*Enthemmung*]" (Heidegger 1995, 254). This amounts to saying that the animal meets only what can strike it, what triggers, that is, instinctive motor sequences (see Brentari 2015, 200). This disinhibiting encirclement is what allows the survival of the animal and the preservation of its species. That an animal's instinct needs to be "disinhibited" might sound surprising at first, since the absence of inhibition or "uninhibitedness" seems to be a defining trait of instincts. According to Heidegger, however, this is true only for the results of instinctive actions but not for the instinctual drive intrinsically as such. As he puts it,

> if [...] we [...] consider the instinctual structure itself, then we can see that the instinctual drive precisely possesses an inner tension and charge, a containment and inhibitedness that essentially must be disinhibited before it can pass over into driven activity and thus be "uninhibited" in the usual, ordinary sense of the word.
>
> (Heidegger 1995, 254)

The indebtedness of the notion of "disinhibiting ring" to Uexküll's functional circle is quite clear and acknowledged explicitly by Heidegger himself, as the following passage shows:

> Even the fact that Uexküll talks of an environing world [*Umwelt*], and indeed of the "inner world" of the animal, should not initially prevent us from simply pursuing what he means here. For in fact he means nothing other than what we have characterized as the *disinhibiting ring*.
>
> (Heidegger 1995, 263, emphasis in the text)

Such an "alliance" with Uexküll and his notion of *Umwelt* allows Heidegger to investigate the living not in isolation but based on its relation with the other than itself. He ultimately relies on the biologist to reject once and for all the idea that the living exists on its own, independently of its environment, that the living adapts to an independently existing environment, behaving in conformity to it and adapting passively to it. This basic attainment allows Heidegger to take distance from the mere adaptationism he credits to the Darwinism of his time:

> In Darwinism such investigations were based upon the fundamentally misconceived idea that the animal is present at hand, and then subsequently adapts itself to a world that is present at hand, that it then comports itself accordingly and that the fittest individual gets selected. Yet the task is not simply to identify the specific conditions of life materially speaking, but rather to acquire insight into the *relational structure between the animal and its environment*.
>
> (Heidegger 1995, 263, emphasis in the text)

Uexküll's reference is then crucial to the aim of restating the original connection between the animal and the environment. In this respect, Heidegger's goal does not fall far from the theoretical intentions of other philosophers among his contemporaries, such as, for instance, Scheler and Plessner, although their respective perspectives entail fundamental differences (see Michelini 2017).

One could wonder why Heidegger avoids the direct reference to the notion of *Umwelt* but, rather, prefers the more complex (and maybe clunkier) notion of *disinhibiting ring*. The terminological replacement is no coincidence but actually brings attention to a substantial difference responding to divergent strategies. One will remark, first and foremost, that the word *world* is reserved for human beings, and carefully avoided for all other living beings, even when it comes to the expression *Um-welt*. In this respect, it is no surprise that Heidegger prefers the term *Um-gebung* (surroundings), without granting much importance to the differences between the latter and the *Umwelt*, as clearly exemplified by the following passage: "Captivation is the condition of possibility for the fact that, in accordance with its essence, the animal behaves within an environment [*Umgebung*] but never within a world [*Welt*]" (Heidegger 1995, 29f., 239). For Uexküll, on the contrary, such a difference is the very foundation of his biology. While the *Umgebung* only defines the surrounding space of the animal, in other words, what humans perceive as their generic environmental field, the *Umwelt* is the specific world of the animal, in which it acts as sensorial and operational center.

By preferring *Umgebung*, Heidegger also betrays another strategy of his, one which avoids, also from a linguistic point of view, any connotation or implication of perceptual nature. In so doing, he takes distance from the neo-Kantian constructivism of the zoologist (on Uexküll's Kantian intellectual background, see Brentari 2015 and Esposito, Chapter 2, in this volume). The kind of research Heidegger pursues is indeed an ontological one. He believes, for instance, that the *Dasein* has access to the manifestness of Being as such; Being reveals itself to *Dasein*, and it can be penetrated, so to speak, leaving nothing out. This is clearly a very distant position from Uexküll, according to whom not only animal environments but also nature, in general, remain at any rate for humankind a topic "[f]orever unknowable" (Uexküll 2010, 135). Brett Buchanan has clearly pin-pointed the difference between the two positions:

> Uexküll harbored no intention of popping the bubbles that surround ani-mals, for he was more content with observing their lives through the opaque transparency that separates each animal's experiential life. At best there is always a metaphorical film or residue that separates our access to their envi-ronments. With Heidegger, on the other hand, we encounter no disillusion in his wish to penetrate through the ring to get to the essential core of animality even if it also contends that such a breakthrough may never be achieved.
>
> (Buchanan 2008, 90)

Furthermore, as another strategic point, one should notice how Heidegger never misses the opportunity to emphasize the closure, almost the "imprisonment" of the animal in its encircling ring, in a way that is extraneous to Uexküll. One could read, for instance, Heidegger's statement, according to which the animal "holds and drives the animal within a *ring* which it cannot *escape* [emphasis mine] and within which something is open for the animals" (Heidegger 1995, 249). Or even its retrieval of a famous image of the *Theoretical Biology*, the *Umwelttunnel*, to which he attaches a negative connotation, inasmuch as the animal would be "locked up" in the environ-mental encircling, something Uexküll never suggested: "Throughout the course of its life" – Heidegger writes – "the animal is confined [in German: *eingesperrt*, locked up] to its environmental world, immured as it were within a fixed sphere [in German: *Rohr*, tube] that is incapable of further expansion or contraction" (Heidegger 1995, 198).[2] And this would be the widest opening granted to the animal.

Although the element of closure is fundamental also in Uexküll's understand-ing of *Umwelt*, this is differently interpreted by the two authors. Heidegger, in fact, despite all proclamation to the contrary, establishes an almost mechanical closure of the animal within the circle of its disinhibitors. On the one hand, he aims to outline a non-mechanistic theory of living organisms, rejecting, also linguistically, all terminology that hints back to that perspective (e.g., the word *Reiz*, stimulus), and laying emphasis instead on the anti-mechanistic elements of Uexküll's theory; on the other hand, in relation to environmental stimuli, the animal is described as merely receptive. The key element that Heidegger appears to neglect is the fact that, based on Uexküll's functional circle, the environmental stimuli are turned into actions; they trigger movement in effect organs [*Wirk-organe*] going through the subjective structure of the animal. The main goal of Uexküll's biology is indeed the outline of the range of influence of the environ-ment on the animal and the range of answers of the animal to the environment. Moreover, the animal never simply reacts; it rather responds:

> Every animal is a subject, which, in virtue of the structure peculiar to it, selects stimuli from the general influences of the outer world, and to it it responds in a certain way.
>
> (Uexküll 1926, 126)

It should also be added here that a key intuition for Uexküll is that animals do not respond to given effects surrounding them, but rather to the biological meaning that the environment has for them. In this respect, Ernst Cassirer is right when he claims that the biological research framework shifts, with Uexküll, from "what" questions applied to life to "how and why" questions (Cassirer 1944, 41f.). There is no doubt that, for Heidegger, the animal is not a subject in this sense.

3.2 The degrees of the organic and the "privative way"

As Uexküll understands it, the closure of the *Umwelt* has no "imprisoning" function but plays a key role under at least two other respects. First of all, undeniably, it ensures the certainty of the animal within its *Umwelt*, something the biologist deems way more important than "richness" (see, for instance, Uexküll 1934, 51). Second, closure is the flip side of the harmony or conformity to a plan [*Planmäßigkeit*] between the animal and the *Umwelt*:

> Every organism – Uexküll writes – can only be itself. But within itself is perfect. [...] Within it all resources are exploited to the full. And so we may make the following statement: – *every living creature is, in principle, absolutely perfect.*
>
> (Uexküll 1926, 164, emphasis in the text)

The here-mentioned perfection should not be misunderstood. It does not imply that living beings possess unlimited abilities or that it resembles any of the divine perfections, but it *"merely means the correct and complete exercise of all the means available"* (Uexküll 1926, 164; on this topic, see Romano 2009, 271f.). The conformity to a plan expresses the fact that animals are marvelously adapted (in the sense of the German *eingepasst*) to the peculiar *Umwelt* and not adapted [*angepasst*] to a physical, undifferentiated milieu [*Umgebung*].

To what extent the belief in said harmony is one of the most controversial points in Uexküll's proposal, and why today it cannot be accepted, is a topic that unfortunately cannot be dealt with here (see Di Paolo's Afterword to this volume, as well as Brentari 2015). What is relevant to the relation with Heidegger is that said conformity to a plan entails notably the elimination of any given hierarchy among living beings. Each species conforms to its specific environment, and this applies, according to Uexküll, also to human beings – although their *Umwelt* seems to have cultural connotations that diversify it already at the individual level and not only at the level of the species (see, e.g., Uexküll 2010, 126ff.). Very importantly, among all these environments, no level of distinction can be established. The world of the tick is not narrower than that of the dog, a squirrel, or a fox. In this regard, the theory of different and specific *Umwelten* is an antidote not only to the model of the old-fashion, pre-Darwinian *scala naturae* but also to the forms of evolutionary gradations of intelligence popular in Uexküll's time. This is also what Helmuth Plessner understands (see Krüger, Chapter 5, in this volume), as he believes one of Uexküll's undeniable merits to be that of having contributed to the dismissal of the idea of a unidirectional natural progression within the animal series, a progress supposedly leading to human beings, whose intelligence would be the highest in the natural realm (Plessner 1983, 57).

On the rejection of gradation, Heidegger sides then with Uexküll. As previously mentioned, it would be counterproductive for Heidegger to venture into the comparative analysis of the mundane relations among humans and animals, on the basic assumption that some or all animals are simpler or inferior compared to one another or/and to human beings or that animals are more or less perfect (Heidegger 1995, 255). By retrieving almost word by word Uexküll's statement, he therefore claims that "[e]very animal and every species of animal as such is just as perfect and complete as any other" (Heidegger 1995, 194). He knows very well, as previously anticipated, that under many respects many animal species are able to entertain extremely complex relations and that sometimes are even more complex than those established by human beings, as for instance the sense of smell in dogs and the sight of birds.

However, Heidegger's rejection of hierarchies, as further investigated in the conclusions to this chapter, responds primarily to the idea of a structural and ontological differentiation between the world of humans and the animal *Umwelt*. It therefore never aims at emphasizing the specificity and multiplicity of animal worlds, as it was instead the case in Uexküll's original intentions. According to this viewpoint, the model targeted by Heidegger's critique is not, first and foremost, the naturalistic idea of a progression in intelligence but, rather, the one advocated by Max Scheler in the famous conference of 1928, *The Human Place in the Cosmos*. Scheler believes that in human beings are gathered all previous degrees of the organic and that it is therefore not possible to understand human beings independently of the whole psychophysical world. Only with the stepping into stage of spirit [*Geist*], as the one principle beyond life, an essential difference – no longer one in degree only – is first introduced between human beings and the rest of the living (Scheler 2009).

From Heidegger's point of view, this theory is guilty of understanding man still within the animal series as the living to which spirit is added, according to an old-fashion simply additive idea, which he rejects categorically: "*Dasein* is never to be defined ontologically by regarding it as life (in an ontologically indefinite manner) plus something else" (Heidegger 1962, 75). According to Derrida, precisely in his taking distance from this line of thought lies the most interesting and worth treasuring side of Heidegger's contribution to animality. His proposal concerning the structural difference between human beings and animals would, in fact, have the advantage of avoiding anthropocentrism (Derrida 1989, 49), although not entirely. According to Derrida, Heidegger's account would still be

> bound to reintroduce the measure of man by the very route it claimed to be withdrawing from that measure – this meaning of lack or privation. This latter is anthropocentric or at least referred to the questioning *we* of *Dasein*. It can appear as such and gain meaning only from a non-animal world, and from our point of view.
>
> (Derrida 1989, 49f., emphasis in the text)

It is undeniable, in fact, that the key notions Heidegger selects to define the essential mode of being of animals – expressions such as "poor in world" and "captivation" – entail a privative and negative mode of access. Regardless of Heidegger's proclamation to the contrary, the animal is after all interpreted based on what makes it different from world-forming human beings. Already in the ordinary usage of the German language, *Benommenheit* [captivation] indicates the failing of consciousness or being deprived of awareness. Heidegger describes it as "the withholding of any possibility of apprehending beings [*die Genommenheit der Möglichkeit des Vernehmens*]" (Heidegger 1995, 255). When it comes to being "poor in world," the privative dimension is particularly clear already in the terminological choice, and it does not make much difference that Heidegger specifies that it should not be understood in terms of lack but rather as *Entbehren* [being deprived] (Heidegger 1995, 195).

The issue surfacing here is, then, that, despite his rejection of gradation-based and additive approaches, Heidegger does not drop the idea – already outlined in *Being and Time* – according to which life is posterior to existence, and would be accessible only starting from the *Dasein* in a derivative, or better "privative," mode:

> Life, in its own right, is a kind of Being; but essentially it is accessible only in *Dasein*. The ontology of life is accomplished by way of a privative interpretation; it determines what must be the case if there can be anything like mere aliveness [*Nur-noch-leben*].
>
> (Heidegger 1962, 75)

Heidegger's ontology of life, that is to say, is based upon a "privative" method, which takes *Dasein* as original and prior to life. However, as already remarked by Karl Löwith in his famous criticism – along the same lines of Helmuth Plessner – since *Dasein* is seen by Heidegger as "bodiless and genderless," it can never be taken as "something original" (Löwith 1957, 75). As soon as it is taken as an original starting point, in fact, it leads to what Heidegger, ambiguously and dangerously, calls "*Nur-noch-leben*."

4 Conclusion: humans are not animals

Why then, among the biologists of his time, does Heidegger chose to engage precisely with Jakob von Uexküll as an interlocutor? For sure a generalized interest in Uexküll was "in the air." He was widely popular thanks to the mediation of Scheler, the first to draw attention to his philosophical relevance (Brentari 2015, 177), and, as previously mentioned, of Cassirer. In line with German philosophical anthropologists such as Scheler and Plessner, Heidegger takes the indissoluble link established by Uexküll between organisms and their environments as a suitable antidote against the passive adaptationism of the living to a generic and undifferentiated milieu. Furthermore, more reasons to find Uexküll attractive could be found in his opposition to any attempt to analyze and reduce life to the laws of physics and chemistry (Calarco 2008, 19), which nevertheless resists the pitfalls of vitalist theories of the living. As also Chien argues, Uexküll's position is in agreement with the antireductionism Heidegger aims to support:

> He [Heidegger] thinks mechanism and vitalism are both dangerous trends, but praises Driesch and Uexküll for their holistic view in observing the growth of an organism in its developmental stages, and in observing how animals are bound to their environment.
>
> (Chien 2006, 74)

Finally, Uexküll comes across as a theoretical biologist who needs to anchor new experimental acquisitions to fundamental metaphysical questions (Bassanese 2004, 84). This aspect of his profile clearly matches Heidegger's philosophical project.

However, as this chapter has tried to show, Heidegger's interest in Uexküll's biology is mainly motivated by his need for an antihierarchical model of the living in order to support his claim concerning the clear demarcation line between humans and animals. Heidegger's main criticism of Uexküll is, in fact, not to have pursued his line to the end, in other words, not to have drawn the ultimate consequences from what he believed could have already been glimpsed, the difference, that is, between animal world and human world, as clearly stated by the following passage:

> Uexküll is the one who has repeatedly pointed out with the greatest emphasis that what the animal stands in relation to is given for it in a different way than it is for the human being. Yet this is precisely the place where the decisive problem lies concealed and demands to be exposed. For it is *not* simply a question of a *qualitative otherness* of the animal world as compared with the human world, and especially not a question of quantitative distinctions in range, depth, and breadth – not a question of whether or how the animal takes what is given to it in a different way, but rather of whether the animal can apprehend something *as* something, something *as* a being, at all. If it cannot, then the animal is separated from man by an abyss.
>
> (Heidegger 1995, 263f., emphasis in the text)

In this respect, the fundamental step undertaken by Heidegger is that of pushing to the extreme the specificity of animal environments identified by Uexküll, as to turn it into a factor of radical (ontological) separation between animals and humans, in a sense entirely alien to the biologist. The antihierarchism of Uexküll is very different from Heidegger's. According to Uexküll, the realm of human beings has no higher degree compared to that of other species, and all *Umwelten* are after all equivalent. Uexküll's goal is never to point to the abyssal difference between two domains – the animal and the human – but, rather, besides pointing to the limits also of our world, to understand, through the multiple and varied worlds of animals and humans, the score of nature's melody (on this topic, see Romano 2009, 275f.).

Within Heidegger's perspective, the price to pay to save the specificity of the mode of Being of humans and animals – provided one could talk about specificity, even though Heidegger always refers to it as a wide singular category – is then that of total discontinuity, of a difference, that is, in nature. The animal as poor, lacking, and simultaneously fully adherent to life – two sides of the same coin – serves the function of establishing an abyssal distance from the human being, thereby rejecting all continuity and gradation claims and thus, ultimately, avoids also anthropomorphism.

After all, the main issue tackled by all theories of radical difference, such as Heidegger's, is not anthropocentrism, but the possibly even more insidious one of anthropomorphism. According to Heidegger, in fact, animals' "lack" is not in the least an anthropomorphic projection but, rather, emerges from a careful investigation of animality itself. In its intentions, at least, Heidegger's project foresees an account of the essence of animality in the terms of animals themselves – a merit, this latter, acknowledged by Calarco (2008, 22). This project, however, clashed against the only possible way he sees to access life, that is to say, the privative way.[3]

Continuity theories, as the one mentioned at the beginning and advocated by Hans Jonas, are not able to so easily avoid the issue of anthropomorphism – even granted Heidegger was successful in this respect. For Jonas, not only it is not possible to escape a given form of anthropomorphism but also a "critical" usage of it is taken as preferred access to the living. Useful insight comes from following his arguments in this respect. Jonas reminds us, first of all, that anthropomorphism (and zoomorphism) is one of the basic prohibitions providing the ground to modern science; the ban on anthropomorphism is, in other words, the metaphysics on which science is founded. More precisely it is the result of the ban on dualistic and postdualistic metaphysics, which has alienated human beings from themselves and from the rest of nature and which therefore "in this extreme form" cannot but be a prejudice (Jonas 1966, 23). Jonas believes, in fact, that anthropomorphism is ingrained to human beings and inevitable. It should nevertheless be correctly understood. In itself, it is deprived of anthropocentric implications and it does not necessarily entail the cognitive and practical dominion of man over nature. It is not even the result of unjustified analogies and projections, but it is rather connected to our mode of access to "life." To say it better, it is grounded upon the basic fact that we have some sort of pre-familiarity with the living. We are, in other words, able to access life because we are ourselves living: "Life can be known only by life" (Jonas 1966, 91).

This statement is provided by Jonas, meaning that an opening on the living and on its understanding is available to us in virtue of the fact that we ourselves are organic bodies [*Leiber*]. Since we are observers of life and simultaneously living, we are perspectively at the core of the phenomenon of "life." This latter is not simply placed in front of us as a "neutral" object, as if it was observed from "outside" life itself. It is experienced from "within," that is to say, "through the exemplar of our psychophysical totality" (Jonas 1966, 23) – which is "the archetype of the concrete" (Jonas 1966, 24), "the maximum of concrete ontological completeness"

(Jonas 1966, 23) that we know. It is starting from our living body – and not from our rationality, self-consciousness or language – that us humans, as organisms in a world of organisms, are able to "recognize" life. In this sense only, the human being can be "the measure of all things" (Jonas 1966, 23) – as Jonas writes, while making new use of Protagoras's ancient motto – and only in this sense anthropomorphism is intrinsic and inevitable.

However, while the understanding of the living unavoidably relies upon the experience that human beings have of themselves, it is nevertheless necessary to take distance from "ingenuous" anthropomorphism and all the dangers implicit in it and against which science warns us. The remedy against all dangers, however, is not radical elimination – something by the way impossible – but, rather, a "controlled and critical" usage of anthropomorphism itself (Dewitte 2002, 458). This "legitimate" form of anthropomorphism, firmly anchored to human corporeity and basic element of a "biohermeneutics" of the living (see Michelini 2020, forthcoming), is also in agreement with Maturana and Varela's position on observations in biology, when, for instance, they remark that "[t]he observer is a human being, that is, a living system and whatever applies to living systems applies also to him" (Maturana and Varela 1980, 8).

Denying this fact means to withdraw human beings from the evidence they have about themselves, definitively alienating them from their own animality and, more generally, from life, thus, ultimately, separating them once and for all from nature.

Notes

1 The experiment is reported according to the interpretation offered in his *Untersuchungen über den Phototropismus der Tiere* [*Investigations on Animal Phototropismus*] (1903) by Emanuel Rádl (1873–1942), a Czech biologist, quoted by Heidegger with admiration also in other occasions, with whose work he got acquainted through the mediation of Uexküll's texts (Bassanese 2004, 229).
2 Seemingly the English translation of this passage by William McNeill and Nicholas Walker tries to dilute/downplay the negative connotation of the German original.
3 Also in the fight against anthropomorphism, Heidegger can find in Uexküll an ally. Since his early writings, Uexküll carefully avoids making reference to anthropomorphic notions and clearly rejects the – often Darwin-inspired – animal psychologies of his time. These are taken as philosophically naive, inasmuch as they would have applied anthropomorphic methodologies to understand the animal mind. Uexküll's goal, moreover, is not that of understanding animal minds – on which he embraces a Kantian agnosticism – but, rather, their perceptual worlds (*Merkwelten*). On this topic, see Harrington (1996, 41), Tønnessen in Brentari (2015, 11).

References

Agamben, Giorgio (2004) [2002] *The Open: Man and Animal*. Translated by Kevin Attell. Stanford: Stanford University Press.

Arendt, Hannah (2013) [1958] *The Human Condition* (2nd ed.). Chicago: University of Chicago Press.

Bassanese, Monica (2004) *Heidegger e von Uexküll: filosofia e biologia a confronto*. Trento: Verifiche.

Brentari, Carlo (2015) *Jakob von Uexküll: The Discovery of the Umwelt between Biosemiotics and Theoretical Biology*. Dordrecht/Heidelberg/New York/London: Springer.

Buchanan, Brett (2008) *Onto-Ethologies: The Animal Environments of Uexküll, Heidegger, Merleau-Ponty, and Deleuze*. Albany/New York: SUNY Press.

Calarco, Matthew (2008) *Zoographies: The Question of the Animal from Heidegger to Derrida*. New York/ Chichester/West Sussex: Columbia University Press.

Cassirer, Ernst (1944) *An Essay on Man: An Introduction to a Philosophy of Human Culture*. New Haven: Yale University Press.

Chien, Jui-Pi (2006) 'Of animals and men: A study of Umwelt in Uexküll, Cassirer, and Heidegger'. *Concentric: Literary and Cultural Studies* 32 (1), 57–79.

Cimatti, Felice (2013) *Filosofia dell'animalità*. Bari: Laterza.

Derrida, Jacques (1989) [1987] *Of Spirit: Heidegger and the Question*. Translated by Geoffrey Bennington and Rachel Bowlby. London: The University of Chicago Press.

Dewitte, Jacques (2002) 'L'anthropomorphisme, voie d'accès privilégiée au vivant. L'apport de Hans Jonas'. *Revue Philosophique de Louvain* 100 (3), 437–465.

Franck, Dieter (1991) 'Being and the living'. In: Eduardo Cadava, Peter Connor, and Jacques-Luc Nancy (eds.) *Who Comes after the Subject*. New York: Routledge, 135–147.

Harrington, Anne (1996) *Reenchanted Science: Holism in German Culture from Wilhelm II to Hitler*. Princeton: Princeton University Press.

Heidegger, Martin (1962) [1927] *Being and Time*. Translated by John Macquarrie and Edward Robinson. Oxford: Basic Blackwell.

Heidegger, Martin (1995) [1929/1930] *The Fundamental Concepts of Metaphysics: World, Finitude, Solitude*. Translated by William McNeill and Nicholas Walker. Bloomington/Indianapolis: Indiana University Press.

Jonas, Hans (1966) *The Phenomenon of Life: Toward a Philosophical Biology*. New York: Harper & Row.

Jonas, Hans (2002) 'Wissenschaft as Personal Experience'. *The Hastings Center Report* 32 (4), 27–35.

Kessel, Thomas (2011) *Phänomenologie des Lebendigen: Heideggers Kritik an den Leitbegriffen der neuzeitlichen Biologie*. Freiburg/München: Verlag Karl Alber.

Löwith, Karl (1957) 'Natur und Humanität des Menschen'. In: Klaus Ziegler (ed.) *Wesen und Wirklichkeit des Menschen. Festschrift für Helmuth Plessner*. Göttingen: Vandenhoeck & Ruprecht, 58–87.

Maturana, Humberto and Varela, Francisco J. (1980) *Autopoiesis and Cognition: The Realization of the Living*. Dordrecht: Reidel Publishing.

Michelini, Francesca (2017) 'Umwelt der Tiere und Welt der Menschen: Dichotomie oder Relationalität?'. In: Forschungsschwerpunkt Tier-Mensch-Gesellschaft (ed.) *Vielfältig verflochten. Interdisziplinäre Beiträge zur Tier-Mensch-Relationalität*. Bielefeld: Transcript Verlag, 301–314.

Michelini, Francesca (2020, forthcoming) 'Ontology versus anthropology: Plessner and Jonas's readings of Heidegger's philosophy'. *Revue Philosophique de Louvain*.

Plessner, Helmuth (1983) [1946] 'Mensch und Tier'. In: Helmuth Plessner. *Gesammelte Schriften*. Vol. 8. *Conditio Humana*. Frankfurt a. M.: Suhrkamp, 52–65.

Rádl, Emanuel (1903) *Untersuchungen über den Phototropismus der Tiere*. Leipzig: W. Engelmann.

Romano, Claude (2009) 'Le monde animal: Heidegger et von Uexküll'. In: Servanne Jollivet and Claude Romano (eds.) *Heidegger en dialogue 1912–1930. Rencontres, affinités, confrontations*. Paris: Vrin, 255–298.

Scheler, Max (2009) [1928] *The Human Place in the Cosmos*. Translated by Manfred S. Frings. Evanston, IL: Northwestern University Press.

Uexküll, Jakob von (1926) [1920] *Theoretical Biology*. Translated by Doris L. Mackinnon. London: K. Paul, Trench, Trubner & Co.; New York: Harcourt, Brace & Company.

Uexküll, Jakob von (2010) [1934, 1940] *A Foray into the Worlds of Animals and Humans with a Theory of Meaning*. Translated by Joseph D. O'Neil. Minneapolis/London: University of Minneapolis Press.

8 Animal behavior and the passage to culture

Merleau-Ponty's remarks on Uexküll

Tristan Moyle

1 Merleau-Ponty's second course on nature (1957–1958)

1.1 Background

Merleau-Ponty's most sustained discussion of Uexküll occurs in the second of three lectures courses, which were delivered at the Collège de France in the years 1956–57, 1957–58, and 1959–60, on the theme of the "Concept of Nature." The lecture courses, as a whole, are intended to study nature, understood as a "leaf or 'layer' of Being," as the route to a more general, ontological explication of the *vinculum* or *nexus* "Nature"–"Man"–"God." The ultimate task, Merleau-Ponty remarks, is to see how this *nexus* holds together (Merleau-Ponty 2003, 204). In particular, for his lecture courses, he chooses to focus on the unfolding of the "leaf of Nature," first, in relation to the intertwining [*Ineinander*] of life and physico-chemistry and, second, in relation to the intertwining of the human with "animality and Nature." His intention is to "grasp humanity [...] as another manner of being a body" rather than merely as "animality ... + reason" (Merleau-Ponty 2003, 208). Hence, the importance of the second course on animality, in which we find his remarks on Uexküll.

Aside from the overall, philosophical intent of the lecture series, there is a more specific interest that orders its general content and structure, which Merleau-Ponty refers to as the findings of the "new school of biology" (Merleau-Ponty 2003, 151). This school is commonly known as the movement of organicist biology, which flourished between the early decades of the 20th century and the late 1950s (see Esposito, Chapter 2, in this volume). The organicist movement, taking the whole organism as its fundamental object of investigation, attempted to provide a third way between mechanism, on the one hand, and vitalism, on the other. Its intellectual forebears were Aristotle but particularly Kant, Schelling, Bergson, and Whitehead. The key scientists associated with or belonging to the organicist movement, aside from Uexküll, included, after von Baer and Driesch, Haldane, Russell, and Waddington. So, one way to understand Merleau-Ponty's lecture series is as offering an interrogation of the philosophical foundations of the organicist movement (1956–57) and then a detailed consideration of the scientific work being undertaken by its contemporary adherents and associates (1957–58). This orientation explains not only the content of the first course but also the remarks on Uexküll and those influenced by him, such as A. Portmann, K. Lorenz, and E. S. Russell. Merleau-Ponty approaches Uexküll very much in the context of this community of thought.

1.2 Philosophical interpretation of the whole organism

In fact, at one point in the lectures, Merleau-Ponty explicitly mentions the "organicist idea" that the totality of the organism is more than the sum of its physico-chemical parts. He proceeds to describe the question of the status of this emergent totality as the key philosophical question, which is "at the center of this course on the idea of nature and maybe the whole of philosophy" (Merleau-Ponty 2003, 145). The key to answering the question is to avoid two pitfalls: either conceiving the organism as a "transcendent totality" or as a "totality by summation" (Merleau-Ponty 2003, 153). The latter conception is the error of mechanism, according to which the entirety of an organism's nature is explained by the interactions of its physico-chemical, material parts. Along with the organicist movement, Merleau-Ponty believes that this way of thinking is unable to grasp adequately either embryological development or mature, whole individual behavior. The former conception is the error of vitalism. The mistake here, apart from being "contradicted by the facts," is the commitment to metaphysical dualism, which Merleau-Ponty takes to be unacceptable. The vitalist falls back on a "positive principle" – for example, idea, essence, entelechy – which they assert to lie behind the observable phenomena. This is to "platonize" or "aristotelianize" and to "double the reality under our eyes without resolving the problem" (Merleau-Ponty 2003, 155). The problem, for example, of embryological development is not resolved by vitalist principles because the postulation of a causally efficacious, immaterial, transcendent living force ("entelechy") mysteriously acting on matter is both empirically unverifiable and as philosophically suspect as any other form of dualism.

Merleau-Ponty outlines his own response to the question in the first part of the second lecture course (the part before the discussion of Uexküll). This occurs during a discussion of Coghill's study of development in the axolotl (the Mexican salamander) and Gesell and Amatruda's work on human development. Merleau-Ponty's provisional conclusions are that organismic totality is immanent and emergent. It is immanent in that the living being "works only with physio-chemical elements" (Merleau-Ponty 2003, 177). That is, there is no immaterial, active, substantial force present in the organism; it is constituted content-wise by spatially and temporally located, physico-chemical events. However, organismic totality is more than the sum or aggregate of these events. The totality of an animal, such as an axolotl, is to be conceived, he remarks, as an emergent adhesion between the spatial parts of the embryo and the temporal parts of its life. This adhesion between the elements of the multiple is a "dimension" of "meaning" (Merleau-Ponty 2003, 156), such that one event in the maturation process, such as the axolotl beginning to curl and buckle, cannot exist, as the event that it is, without a later event, such as the axolotl learning how to swim (by curling and twisting its body repeatedly).

Thus, "the axolotl" is not an essence (axolotl-ness) existing outside of time, hidden behind appearances, in which the individual animal participates, but a totality, a total pattern of meaningful embryological behavior emerging across the domain of events that constitutes an axolotl's life. Although emergent, however, this organismic totality has real causal powers: it is a "directing principle." But it directs the developmental trajectory through a kind of "operant non-being," in which the animal's future potentiality is "there," in the present stage of matura-tion, only as a negative "state of imbalance" or "factor of disorder," which out-lines a future reequilibration. In this sense, Merleau-Ponty describes organismic totality as a "hollowed-out design" of which the animal's material organs are the "trace" (Merleau-Ponty 2003, 156).

Not only does the emergent, organismic totality have real causal powers, but the overall design is also surprising. I take this to mean surprising or unexpected from within the framework of physical or resultant causality (based on experience the manner of maturation of a particular animal species is generally predictable). Merleau-Ponty illustrates the point by referring to the principle of "reciprocal intertwining," which he takes from the dynamic morphology of Gesell. Biologi-cal processes, like threads in weaving, are reciprocally intertwined in that "the [overall T.M.] design must appear with a certain surprise, to the extent that it is born of the meeting of threads [i.e., biological processes T.M.] which have the air of having nothing to do with it" (Merleau-Ponty 2003, 149).

Later, when considering models for his notion of organismic development, Merleau-Ponty compares the intertwining of biological processes to the way in which films of artists show them brushing in apparently random places, only for the overall pattern to emerge during the course of construction. It is not that the idea of the artwork preexists and guides the process of material construction, as a blueprint precedes a finished house, determining its final form. Rather, the work is realized in and through the constructive process in such a way that one can imagine an artist standing back, when the painting is finished, and saying, "So, *that's* what I wanted to do." Likewise, the guiding pattern of animal develop-ment emerges unexpectedly through the crisscrossing of independent, physical processes. Like the production of a work of art, the law of construction of an organism, he concludes, is of a different kind.

This is the point that Merleau-Ponty reaches in his reflections on organismic totality, before he turns to Uexküll.

2 Merleau-Ponty's interpretation of Uexküll

2.1 *The concept of environment* [Umwelt]

As the basis for his remarks, Merleau-Ponty uses the first edition of Umwelt *and Inner-World of Animals* [*Umwelt und Innenwelt der Tiere*] (1909) and *A Foray into the Worlds of Animals and Humans* (1934). He offers a preliminary description of Uexküll's research and then a philosophical interpretation. During the interpretation, Merleau-Ponty argues that Uexküll's notion of an animal's *Umwelt* is especially important. In particular, he remarks, Uexküll's idea of the unfurling of an *Umwelt* as a "melody that is singing itself" is a "comparison full of meaning" (Merleau-Ponty 2003, 174). Indeed, "the theme of the melody [...] best expresses the intuition of the animal according to Uexküll" (Merleau-Ponty 2003, 178).

Let us examine Merleau-Ponty's remarks on the Uexküllian *Umwelt* before we turn to the significance of the melody. Two themes are especially prominent. The first is the way in which Uexküll allegedly downgrades the importance of consciousness in his explanation of animal activity. Thus, Merleau-Ponty comments, Uexküll's *Umwelt* marks the difference between the world as it exists in itself and the world of a living being, with the latter understood not as a "subjective domain" but as the *Umwelt* of behavior. This "behavioral activity orientated toward an *Umwelt* begins well before the invention of consciousness" (Merleau-Ponty 2003, 167). Indeed, it begins when an organism is disposed to receive and treat causal, physical stimulations as signals and to respond accordingly. Merleau-Ponty applauds the way in which Uexküll interprets consciousness merely as "one of the varied forms of behavior," a move that allows Uexküll to avoid "psychological speculation." Nonetheless, although consciousness is described by Merleau-Ponty, following Uexküll, as a type of behavior, this thesis does not amount to philosophical behaviorism. Consciousness, he remarks, "must [...] be defined [...] such as we grasp it across the bodies of others." It is an "institution," a new layer of meaning, experientially (and epistemically) read off forms of bodily activity (Merleau-Ponty 2003, 167).

It is worth considering the significance of this first theme in more detail. A lot depends on what Merleau-Ponty means by "consciousness." He is correct that, for Uexküll, an ethologist should not engage in psychological speculation. As a trained comparative physiologist, Uexküll distances his approach from early ethologists, such as Romanes, who inferred the existence of psychological states in animals from behavioral observation (or anecdote), using naïve analogies with human behavior (Brentari 2015, 76). Rejecting anthropomorphism, Uexküll is a methodological agnostic regarding animal consciousness. He argues that the correct approach is to investigate an animal's physiological (and behavioral) capacities, on the one hand, and to conduct a transcendental inquiry into the kinds of objects or properties that would be available to the animal, given these capacities, on the other. The issue of the existence and nature of an animal's subjective consciousness of these marks is explicitly put to one side.

But Merleau-Ponty clearly means more than this. It is tempting to interpret his focus on the *Umwelt* of behavior as drawing attention to Uexküll's innovative distinction between a perception world [*Merkwelt*], containing perception marks [*Merkmale*], and an effect world [*Wirkwelt*], containing effect marks [*Wirkmale*]. Insofar as the *Umwelt* "exists for the behavior of the animal" (Merleau-Ponty 2003, 167), on this interpretation, Merleau-Ponty is focusing on the Uexküllian *Wirkwelt*. But we must be careful here. For Uexküll, the *Merkwelt* and *Wirkwelt*, although ontologically and conceptually distinct, tend to exist alongside each other, in a coordinated way, within a single *Umwelt*. This is true of even primitive animals. A creature like the *Paramecium*, for example, still undergoes the perception sign, "obstacle," in response to which it undertakes a backward movement, with subsequent lateral turning (Uexküll 2010, 73f.).[1] Although behavior may not always be perceptually guided, it is generally coordinated with states of perceptual awareness, according to Uexküll. So, "pre-conscious" ought not to be taken as "pre-perceptual."[2]

However, although with the lower animals the perception and effect worlds are often merely externally coordinated, Merleau-Ponty points out that, for Uexküll, in the *Umwelt* of some animals, "[t]he *Wirkwelt* displaces the *Merkwelt*" in that "behaviors [...] deposit a surplus of signification on the surface of objects" (Merleau-Ponty 2003, 173). Indeed, according to Uexküll, a perceptual sign provided by an animal's sense organs can itself be completed and altered by an effect image [*Wirkbild*]. Thus, for example, a branch in a dragonfly's perception world is overlaid with a distinctive effect image, so that the creature sees the object "as" having a "sitting-tone," in other words, as offering opportunities for a certain kind of behavioral interaction (Uexküll 2010, 96). It would appear that Uexküll takes these functional properties displayed in an animal's perception world to be an ingredient of the experiential content rather than, for example, possibilities that are inferred from that content. For example, if a person learns what a ladder is, then, Uexküll argues, this person does not merely acquire a new perceptual belief, inferred from the perceptible shape, form, and size, but rather, the given bars and holes themselves take on a surplus of signification, that is, a "climbing tone" (Uexküll 2010, 94).

It seems likely that when he states that the Uexküllian *Umwelt* exists for the animal's behavior, Merleau-Ponty has this displacing of the perception world by the effect world in mind. Furthermore, we have reason to believe, given the references to Uexküll as far back as *The Structure of Behavior* (1942), that his own conception of motor intentionality was influenced by Uexküll's account of this particular phenomenon. For example, Merleau-Ponty describes the movements of players in a football match as perceptually guided by action-soliciting aspects of their surroundings. Certain features of the practical situation, that is, are grasped as affording opportunities for movement. These actions, being automatic and spontaneous, lack explicit (self-)consciousness, the presence of which would disrupt the flow. In that sense, the pitch exists for their behavior rather than for their consciousness. But the player's movements are nonetheless deeply integrated with states of sensory awareness, as are the motor intentions of much more primitive animals (see Merleau-Ponty 1963, 168f.). Indeed, in an Uexküllian passage from the *Phenomenology of Perception*, Merleau-Ponty argues that an animal's bodily recognition of features of its open situation, which call for its movements, does not entail any "objective consciousness" of those features (Merleau-Ponty 2012, 80f.). The modifier *objective* is significant: it leaves open the possibility of a pre-objective, sensory consciousness of environmental affordances.

The second theme that emerges from Merleau-Ponty's reflections on Uexküll's *Umwelt* theory, is the importance he places on the distinction between the *Umwelt* of the lower and the higher animals. Amoebae, paramecia, sea urchins, starfish, marine worms, and jellyfish, among others, comprise the lower animals. Higher animals, conversely, are organisms that have a developed nervous system and a sophisticated level of physical development, with sense organs, such as antennae, organized to give them "delicate information" about their surroundings. Many of them also possess proprioceptive capacities. In particular, the central nervous system of the higher animals establishes a counter-world [*Gegenwelt*], an alleged inner physiological landscape that formally "mirrors" external spatial relationships, albeit by "translating" these relationships into a unique, species-specific "nervous sign language."

The details of the *Gegenwelt* need not concern us (and Uexküll largely discards it in his mature work). What is philosophically salient are the different ways in which Uexküll and Merleau-Ponty treat the higher–lower distinction. Uexküll's approach is descriptive, for the most part. For him, the simple *Umwelt* of an urchin is as good, or perfect, as any other; there is a radically horizontal, relativist approach to the diversity of the *Umwelten*. Indeed, Uexküll argues that the human *Umwelt* is just another species-specific environment, produced by our nature (Uexküll 1985, 244). Merleau-Ponty, conversely, provides an evaluative gloss to the Uexküllian higher–lower distinction. Although he insists in his *Résumés de cours* that "one cannot conceive of the relations between species or between the species and man in terms of a hierarchy" (Merleau-Ponty 1970, 97), nonetheless, one does have the impression of a progression, of sorts, from lower to higher.

Merleau-Ponty makes a great deal of the contrast between the closed environment of the lower animals and the open environment of the higher animals, especially regarding the relationship each type has to its respective building-plan [*Bauplan*]. The movements of a sea urchin or a jellyfish, for example, are externally dictated by their building-plan. They do not form a "motor project," which would give to the living organism a kind of practical unity and which would constitute a "reply to the exterior world." For such creatures, their *Umwelt* is a closure that separates them from (most) external stimuli (Merleau-Ponty 2003, 170). The lower animal "*is moved*, does not move itself, does not support its *Umwelt*." By contrast, a higher, self-moving, animate animal

> dominate[s] its *Umwelt* itself [...] [it has T.M.] a regulation, that is [there is T.M.] an interaction with the outside and the centralized nervous system [...] that is, circularity exterior-organism [...] [The higher animals T.M.] are their *Bauplan*, they recreate it.
>
> (Merleau-Ponty 2003, 221)

Merleau-Ponty's thought seems to be that whereas a simple animal body (and its correlative *Umwelt*) receives its unity from the activity of the natural factor [*Naturfaktor*], ultimately responsible for its building-plan, the activity that unifies the being of a higher, self-moving animal body is to a greater degree *sui generis*, coming, as it does, in a sense, from the animal itself. We might say that with the higher animals, there is a necessary, participatory activity on the part of the organism, which mediates the exterior activity of the *Naturfaktor*. The higher animals realize the unity of their own functioning – they are their *Bauplan* – and do not simply receive it. Unlike the urchin, the bodily behavior of the higher animals, Merleau-Ponty comments, is more than the mere "effect" of their *Bauplan* (Merleau-Ponty 2003, 216). This is especially true of the human body that is both the body of a higher animal and something "different" and that is uniquely "open, transformable." Indeed, such is the plasticity of our body, it is the occasion of the "projection of a *Welt*" (Merleau-Ponty 2003, 222). This evaluative contrast between lower and higher animals is important in understanding Merleau-Ponty's ontology of nature, as we shall see.

2.2 *Nature's melody*

In Merleau-Ponty's philosophical interpretation of the notion of *Umwelt*, when he is again pondering the central issue of "organismic totality," it is Uexküll's idea of the unfurling of an *Umwelt* as a melody that is singing itself that takes center stage. There are a number of reasons why he takes the idea of the melody to be importantly illustrative for understanding animality. The first is that it foregrounds the receptive aspect of animal activity. Just as the melody "sings in us much more than we sing it" and the singer's body is a "type of servant" for the melody, so, too, does the organism receive its *Umwelt* (via its *Bauplan*) from nature. This is important because it distances Uexküll's *Umwelt* theory from the naïve idea that an *Umwelt* is established by the animal itself, as some kind of clear, distinct and explicitly self-conscious goal. A second, related reason is that

> we [cannot T.M.] distinguish the meaning [of the melody T.M.] apart from the meaning [of the notes T.M.] where it is expressed. As Proust says, melody is a Platonic idea that we cannot see separately. It is impossible to distinguish the means and the end, the essence and the existence in it.
>
> (Merleau-Ponty 2003, 174)

Whereas the first reason focuses on the passive, receptive element of animality, the second reason brings out the expressive, participatory element. Just as the meaning of a melody is constituted by its mode of expression, by the manner and order in which the notes are played, so, too, is nature – Uexküll's natural factor [*Naturfaktor*] together with its building-plans – constituted by the immanent unfolding of the biological processes and threads, and the behavioral repertoire, of those animal subjects who realize the plan with-which-they-conform. Nature's idea (or essence) cannot be separated from concrete existence, as this unfolds within the natural history of life. Likewise, again drawing on an analogy with artistic creation, the aesthetic whole (melody, painting, poem, etc.) that inspires and strikes the artist, rather than being a transcendent totality, is realized in and through his or her expressive activity. Thus,

> [t]he theme of the animal melody is not outside its manifest realization; it is a variable thematizatism that the animal does not seek to realize by a copy

of the model, but that haunts its particular realizations, without these themes being the goal of the organism.

(Merleau-Ponty 2003, 178)

Merleau-Ponty's comments indicate that he is significantly reinterpreting Uexküll's original melody metaphor. In particular, Uexküll believes that we are able to conceive the melody of nature as a Platonic idea (essence) separate from its instantiation. Thus, in *The Omnipotent Life* [*Das allmächtige Leben*], he uses (Neo) Platonic vocabulary to distinguish a world of appearance from the world in itself. Within the latter, he argues, there exists the natural factor [*Naturfaktor*], that is to say, the all-powerful nature-subject, as well as the ideas of distinct animal species and the archetypes of four fundamental functional circles, which predetermine the basic activities of all animals: movement, escape from enemies, search for food, and reproduction. These, he writes, "in tune with Plato" are the "four basic ideas of life" (Brentari 2015, 168).[3] But this is precisely to "platonize" reality in the way that Merleau-Ponty criticizes. Indeed, Merleau-Ponty recognizes in Uexküll a tendency toward dogmatic metaphysics, which he sees as "tak[ing] up [...] the intuitions of Schelling."[4] This tendency is manifest in Uexküll's mature conceptualization of the natural factor as a "nature-subject" [*Natursubjekt*], "hidden" behind the worlds it produces.[5] Alternatively, even if we consider the constraining influence of Kant on Uexküll's early work, in which he postulates the existence of an unknowable natural factor, for purposes of scientific inquiry, and likewise deduces the content of an animal's *Bauplan* strictly on the basis of detailed, empirical research, this still results in an unacceptable doubling of reality, which, as we have seen, Merleau-Ponty believes the melody metaphor, when thought through, helps us avoid.[6]

The third reason for the significance of the melody, for Merleau-Ponty, is that it offers us a different way of thinking about the causal and temporal structure of an animal's life, a point he has already broached in the discussion of embryology. In particular, the idea that the present state of the physical universe at t_1, together with the laws of nature, exhaustively determines the future state at t_2 – such that the "past secretes the future ahead of it" – is "refuted by the melody." Rather, in a melody, there is a "reciprocal influence" between the first and the last note, such that we have to say that "the first note is possible only because of the last, and vice versa" (Merleau-Ponty 2003, 174). This is not, however, to replace the framework of resultant causation with that of backward, teleological causation. Rather, it is the overall dimension of meaning that determines the "before" and the "after" and their interrelations. Merleau-Ponty illustrates the point with Uexküll's example of the tick's response to its *Umwelt*. Just as there is a reciprocal influence between the first and the last note in a melody, so, too, do the tick's movements (e.g., letting go of the branch) resolve stimulations from the outside (e.g., smell of the mammal), the execution of which in turn prepare for further tactile stimulations, and so on. Each component of the functional circle acts only as a part of the whole situation (Merleau-Ponty 2003, 175).

2.3 Methodological commitments

Now, although Merleau-Ponty remarks that behavior cannot be understood moment to moment, he admits, in his discussion of the tick's activity that "[c]ertainly we still find sufficient conditions from moment to moment" (Merleau-Ponty 2003, 175). Likewise, in his later discussion of the mating behavior of the stickleback, he admits that there is something that corresponds to the facts in Tinbergen's description of this as a "series of events chained together" (Merleau-Ponty 2003, 197). The problem is that, if we use the framework of efficient causation to understand these phenomena, we "do not grasp the relation of meaning," and it is this relation of meaning that is what "the expression *Umwelt* conveys" (Merleau-Ponty 2003, 175).

Naturally, we would like to know whether this dimension of meaning is taken by Merleau-Ponty to be a part of mind-independent nature or whether these meaningful patterns are simply there for human consciousness, as, for example, he argued in *The Structure of Behavior* (Merleau-Ponty 1963, 49). It is, after all, our experience of the melody that he uses to illustrate his Uexküllian ontology of nature. Merleau-Ponty remarks that for Uexküll, in his early work, "[a]ll that happens in animals is produced by chemical and physical forces" and that it is "we who have the right to coordinate them in the unity of a constitutional plan," with "we" referring to biologists who "must discern vectors in physiochemical phenomena" (Merleau-Ponty 2003, 169). These remarks are accurate. For early Uexküll, at least, the *Bauplan* is a model constructed by the biologist, who, for complete comprehension of the organism, must grasp it "as if" it has been constructed. This model has heuristic and orientating value, for the purposes of scientific inquiry (Brentari 2015, 54, 60).

Although Merleau-Ponty does not take the epistemic requirements of the scientific method to constitute the relations of meaning, nonetheless he does canvass a series of "models" of the idea of totality, which he takes from the "world of perception" (Merleau-Ponty 2003, 153). Bearing in mind his later remark that "[b]ehavior can be defined only by a perceptual relation" and that "Being cannot be defined outside of perceived being," these models are significant (Merleau-Ponty 2003, 189). We have already noted one, that is to say, the creation of artwork; another concerns the perception of animate movement, for which Merleau-Ponty draws on the experimental work of Albert Michotte. There is a "profound relation," he comments, between the "schema of the living thing" and the "perception of the living thing"; in other words, we see something as alive whether these are figures on a screen in one of Michotte's experiments or a caterpillar moving across a leaf:

> [We] wince [...] when we find a caterpillar where we weren't expecting it: we see [...] a living matter that moves; to the right, the animal's head, to the left, its tail. From this moment on, the future comes before the present. A field of space-time has been opened [...] the space in question is inhabited, animated. The perceived crawling is, in sum, the total meaning of the partial

> movements figured in the three phases, which make action as words make a
> sentence.
>
> (Merleau-Ponty 2003, 155)

In the perception of animate movement, we encounter a kind of corporeal sign language. We see a series of movements as crawling or swimming and so on. Subsuming these events under practical concepts, we give meaning to, categorize, the individual moments. Although we cannot specify the exact, spatiotemporal movements that constitute the perceptual schema "living thing," when we do see rhythmic, animate movement, it is unmistakable. Elsewhere, Merleau-Ponty mentions the "physiognomic perception of silhouettes, gestures, faces, signatures" (Merleau-Ponty 2003, 225). The point, to which I will return, is that this corporeal sign language – the perceptual schema living, animate thing (and its associated concepts) – is distributed throughout a significant portion of the animal kingdom. Animals recognize the movements of other animals, conspecifics, as well as those of other species, and respond accordingly. In other words, meaning relations are constituted within nature long before the human observer comes on the scene. As Merleau-Ponty remarks,

> [t]here is nature wherever there is a life that has meaning, but where, however, there is not thought. [...] Nature is what has a meaning, without this meaning being posited by thought: it is the autoproduction of a meaning.
>
> (Merleau-Ponty 2003, 3)

More precisely, nature produces meaning as perceived nature:

> The study of the appearance of animals takes on interest when we understand this appearance as a language. We must grasp the mystery of life in the way that animals show themselves to each other.
>
> (Merleau-Ponty 2003, 188)

So one might say that the point of view for which relations of meaning appear in the world is that of an animal that sees and is seen, that acts and responds to the actions of another. This intertwining [*Ineinander*], this mystery of life, is what Merleau-Ponty calls "inter-animality" (Merleau-Ponty 2003, 189).

The question is, For whom, or for what, is the emergent totality of meaning apparent? Not for the scientist, whose "myopia" and "attentive perception" (to molecular rather than molar being) Merleau-Ponty criticizes (Merleau-Ponty 2003, 155, 187). Not simply for the embodied, human subject, which, Merleau-Ponty suggests, is the methodological error of *Phenomenology of Perception*.[7] Rather, it seems that, although the later ontology of nature still works on the assumption that "it is from ourselves that living beings [...] speak to us," nonetheless, they speak to us insofar as we find ourselves thrown into a shared, inter-animal life-world [*Lebenswelt*]. Merleau-Ponty's remarks in his *Résumés de cours* support this view. He summarizes the final part of the second course as an epistemological inquiry into the conditions under which we "get at the nature of vital being," for example, by attributing to an animal "an associated milieu" or a "symbolic life," and concludes that

> [i]t emerged that all zoology assumes from our side a methodological *Ein-fühlung* into animal behavior, with the participation of the animal in our perceptive life and the participation of our perceptive life in animality.
>
> (Merleau-Ponty 1970, 97)

Merleau-Ponty's distance from Uexküll's scientific epistemology is striking. It is on the methodological basis of "empathy" [*Einfühlung*] that we are able to see relations of meaning in nature, including the perception of a series of physical events as an action. But this is not a projection from our consciousness into nature, conceived as a domain of mechanically-related, spatiotemporal objects. Rather, we are able empathically to understand the corporeal sign language of other animals because meaning is already there, in the life-world [*Lebenswelt*] in which we and other animals jointly participate. In that sense, methodologically speaking, meaning is real, and we discover it as we learn better to interpret nature's signs. Thus, Merleau-Ponty's ontology is still phenomenological. But the experiential prism through which we grasp the Being of nature is that of a participant immersed in an intersubjective, intercorporeal life-world that was there long before we came on the scene.

3 Symbolic behavior and the passage to culture

3.1 Symbolic behavior

Merleau-Ponty's understanding of natural history as progressive comes through in his discussion of another feature distinguishing the environments of lower and higher animals: symbolic behavior. Merleau-Ponty sees the emergence of symbolic behavior even within the *Umwelt* of the hermit crab, which he takes to represent the "beginning of culture" (Merleau-Ponty 2003, 176). He also argues for the existence of symbolic behavior in a species of Cyclades fish, herons, ducks, bees, and starlings, among others. When this sort of behavior has a social function, especially, then, Merleau-Ponty adds, "we can speak in a valid way of an animal culture" (Merleau-Ponty 2003, 198).

The example of the hermit crab comes from Uexküll. According to Uexküll, within the *Umwelt* of this kind of higher animal, there are additional features that he calls the "coloring" [*Färbung*] or "tone" [*Ton*] of stimuli. Depending on the mood [*Stimmung*] of the crab these can take on a feeding or defensive or protective tone. In this case, the animal's perceptual image is completed by a mood-dependent, operative image; the same anemone appears as for-feeding or for-dwelling-in and so on (Uexküll 2010, 93). Merleau-Ponty sees this "architecture of symbols" as a "species of preculture" because the crab's interpretative activity mediates its relation to the environmental stimuli. The contrast, of course, is with the lower animals, which either, as with some species of jellyfish, have no interaction with their surroundings or, like the sea urchin, are merely the effects of their *Bauplan*. They are, as Merleau-Ponty calls them, "planned animals." Conversely, higher animals, capable of symbolic behavior, are "animals that plan" (Merleau-Ponty 2003, 176). That is, although locked into "rails of behavior," higher animals enjoy the "freedom" that comes with the performance of actions that emerge from their own *sui generis*, coordinated, bodily and psychological capacities.

Whereas the hermit crab responds instinctively, albeit interpretatively, to features of its environment, at the next stage of symbolic behavior, animals are capable of undertaking a "characteristic gesticulation of the species," which has the form of an objectless ceremonial or stereotypical activity (Merleau-Ponty 2003, 191). Merleau-Ponty refers at this point to Lorenz's work on starlings, on which Uexküll also draws (Uexküll 2010, 120f.). Although actions, such as snapping gestures or diving movements are produced "most of the time by reference to an object," they are "something different from reference to an object" and are the "manifestation of a certain style." The motivation for these acts is to "resolve an endogenous tension"; indeed, they are not accomplished in view of an end but are an "activity for pleasure." Real or irreal ("imagined") features of the environment may "evoke" the expression of this style, this *a priori* of the species, but they do not "cause" the behavior (Merleau-Ponty 2003, 192). By "activity for pleasure," we ought not to take Merleau-Ponty to mean hedonistic activities, as if animals aimed solely at their own pleasure. Rather, these characteristic gestures are "appropriate" or "fitting" in an Aristotelian-Stoic sense of the term. I take Merleau-Ponty to mean that insofar as they undertake actions that are typical in their outline and progression for members of their species, for their own sake, then these animals undergo hedonic states.

This sort of instinctive activity passes to proper symbolic, cultural activity when the "empty" or "outlined" activities become "means of communicating for the animals." The outlined act is transformed into "seeming to do" and thus become a "signification." Thus, in the duck, the typical behavior of taking off in flight can become a sign for training the young; in a fish, the lateral movement of the head is a sign for moving off; suddenly stopped, in one species it becomes an appeal; in another, an alarm (Merleau-Ponty 2003, 195). In particular, Merleau-Ponty refers to the ritual of sexual display, taking as his examples behavior of different species of crab in the Barnave Islands and the mating dance of the stickleback. His overall point is that sexual behavior is not the "simple ornamentation of an essential fact," that is, the "reconciliation of male and female cells," because this would ignore the "richness of these manifestations" (Merleau-Ponty 2003, 188). Rather, it is an "action of presence [...] a ceremony in which the animals give themselves to each other" (Merleau-Ponty 2003, 196f.). He argues that symbolic behavior of this kind, cannot be exhaustively interpreted mechanically, for a number of reasons. The same behavior can take on different significations: for one species of Cyclades fish certain movements signal inferiority, for another a threat. Sometimes there is an exchange of roles in the ritualized display, such that there is a "double variation on the same theme," where the "effect would be the cause of what is normally its cause." This could not be exhaustively explained by a mechanical, "gradual causality" (Merleau-Ponty 2003, 196f.). There is not a "spirit of the species," he remarks, but a "dialog," in which roles and instincts (such as that of aggression) can be reversed (Merleau-Ponty 2003, 198). In summary, it is because symbolic behaviors have a "new value as social evocation" and thus have no direct physiological goal, being rather "indispensable conditions for the biological act of copulation," that we can speak of animal culture (Merleau-Ponty 2003, 197).

This line goes to the heart of the criticisms of Darwinism that one finds in Merleau-Ponty's second lecture course and that appears, particularly, during his discussion of Uexküll. Darwinism gives ontological priority to the actual world, conceived as existing *partes extra partes*, and thus views the animal as a collection of fortuitous elements, welded together for the purposes of survival and reproduction, as the "least bad arrangement" (Merleau-Ponty 2003, 175). Indeed, Merleau-Ponty's key criticism of Darwinian ideology targets its belief in utility as a criterion for life (Merleau-Ponty 2003, 186). The focus on survival and reproduction (along with the framework of efficient causation) for understanding nature, again, fails to understand that "sexuality, if it aims only at utility," could "manifest itself by more economic paths" and thus that what "the animal shows is not utility [...] [i]n a certain sense, the sexual ceremony is probably useful but it is useful only because the animal is what it is" (Merleau-Ponty 2003, 188). Merleau-Ponty's alternative picture is of an exuberant "prodigality of forms realized by life" and a "tide of natural production," a picture to which Uexküll also subscribes, among others, including Nietzsche (Merleau-Ponty 2003, 184).

Uexküll, of course, criticized Darwinism throughout his career. But Merleau-Ponty's approach, despite superficial similarities, is quite different. Above all, for Uexküll, the artistry of nature occurs at the transcendental level. The activity of the natural factor is responsible for the immaterial, timeless primal score [*Urpartitur*] or primal image [*Urbild*] of each animal species, as well as for any interspecific meaning relations that go into the construction of each building-plan. The natural factor, as Uexküll remarks, plays its "gigantic clavier" with an "invisible hand" (Uexküll 2010, 208). But no matter how richly expressive the symphonic production at the transcendental level, at the level of natural history the behavior of animals conforms to strictly utilitarian principles, that is, to the four archetypal functional circles. As far as I can see, there is nothing in Uexküll that matches Merleau-Ponty's emphasis on a distinctively nonfunctional, expressive core of behavior, at least in some higher animals. As such, in Uexküll's Platonic vision, individual animals, and their interrelations, being mere signifiers, do not constitute the unfolding symphony of life, in any significant sense.

By contrast, we find in Merleau-Ponty the foregrounding of the contingencies of actual, material, historical evolution.[8] A capacity for expression is understood to develop immanently, mediated by symbolic behavior and the emergence of culture. This perspective is particularly evident in the section of the second course in which Merleau-Ponty discusses the work of Adolf Portmann, and especially his *Animal Forms and Patterns* (1952). The lower animals, he remarks, possess merely utilitarian, functional capacities, which are transformed in the higher animals and take on

> an expressive value, a 'value of form'. If life consisted in forming coherent bands of animals, simple triggers would suffice. And so the same muscles of the face [...] have a utilitarian function in lower vertebrates [...] and in higher mammals, an expressive function.
>
> (Merleau-Ponty 2003, 188)

Portmann, like Lorenz, is closely associated with Uexküll's work; Portmann describes Uexküll as a "pioneer of the new biology" in a foreword to *A Foray into the Worlds of Animals and Humans* (1956). Indeed, it is during the discussion of Portmann, to which we now turn, that we have a clue as to Merleau-Ponty's thoughts on the overall direction of natural history of life.

3.2 Inter-animality

Whereas the remarks on Lorenz's research focus on symbolic behavior, the discussion of Portmann centers on animal form, in particular, on Portmann's study of patterns in nature. For example, the way in which the marks of a frog form a figure when it assumes a certain posture (folded legs) or how the features of a bird are joined to form a whole design. This convergence between the elements of the "design" is an "observable fact"; it is an "ensemble of marks" that contains a reference to "a possible eye," to a "semantic ensemble," that "allows the animal to be recognized by its fellow creatures." Again, we have a progressive account of this phenomenon. The richness of the exterior form of a lower animal, such as a spiraled mollusk, which "gives the impression of a product of art," is in "extension"; in the higher animals, it is "intensive." Thus, the form of a mollusk is "mechanically engendered," whereas

> [i]n higher animals, on the other hand, the appearance is more sober, but the expressive capacity is greater: the body is entirely a manner of expression.
>
> (Merleau-Ponty 2003, 187)

Using Uexküllian vocabulary, we might say that whereas the appearance of a mollusk is a mechanical effect of its building-plan, a plan to which it is subject, in the bodily gestures of which a higher animal is capable, the organism makes itself into a work of art and thereby dictates the arrangement of its expressive form. It is, of course, the "open, transformable" human body, which Merleau-Ponty will see as possessing the greatest capacity for expression.

Indeed, throughout the natural history of life, the "operation of Nature," Merleau-Ponty remarks, is "resemblance" (Merleau-Ponty 2003, 185). Whether this involves animals resembling their surroundings, or other animals, the principle of morphogenesis is not utility but the "design of expression." Nature is a vast web of interlinked "cryptogrammatical marks" (Merleau-Ponty 2003, 184). This thought underpins his suggestion that we understand the appearances of animals as language and his claim that the mystery of life is the way animals show themselves to each other. But the key point, which also pertains to his methodological stance discussed earlier, is that "to make of resemblance an operating factor in Nature is not to see that resemblance [as having T.M.] meaning only for a human eye" (Merleau-Ponty 2003, 189).

Nature, I take Merleau-Ponty to be saying, is geared in its unfolding to the intercorporeal communication that the emergence of symbolic, expressive behavior facilitates. Thus, what "exists are not separated animals, but an inter-animality," "inter-animality" involving a "perceptual relation" that "gives ontological value back to the notion of species" (Merleau-Ponty 2003, 189). This concept provides Merleau-Ponty's final answer to the question of organismic totality. As he puts it in the *Résumés de cours*, summarizing his reflections on Portmann, the "notion of an interanimality" is as "necessary to the complete definition of the organism as its hormones and its 'internal' processes" (Merleau-Ponty 1970, 94). The whole, therefore, is not simply the sum of an animal's physical elements, nor even the relation "animal-*Umwelt*" but, rather, "animal-animal-...," with the latter denoting intraspecific and interspecific relations. Another way of putting this point is that, for Merleau-Ponty, the ultimate end of nature is culture. But unlike Kant, who restricts culture to the second nature of human beings, for Merleau-Ponty there is no strict, qualitative division that separates us from the animals. Thus, for example, in the seventh sketch, he writes of the "human body [...] in a relation of intercorporeality in the biosphere with all animality" (Merleau-Ponty 2003, 268). Merleau-Ponty's vision is of an inter-organismic understanding spreading across species-specific horizons, unified in a widely ramifying culture of life. In short, not "organism-*Umwelt*" but "$organism_1$-$organism_2$-(...)-*Lebenswelt*," with the latter's holistic structure throwing its reflected light back on the meaning of embryological development.

This vision has Uexküllian roots. In a famous example, Uexküll describes how a spider's web is "fly-like," in that it represents the primal image of the fly, woven into the spider's original score (Uexküll 2010, 159). Indeed, generally, Uexküll's theory of point and counterpoint, comprising his compositional theory of nature, seems to echo Merleau-Ponty's concept of inter-animality. But, as we have seen, the harmonious *inter-* of Uexküll's inter-animality is timelessly preestablished by the all-powerful natural factor working invisibly behind the scenes. For Uexküll, an animal receives undifferentiated stimuli, which are processed physiologically, resulting in the outward, unconscious transcendental transposition of a series of distinctive perceptions and effect signs, constituting its *Umwelt*. These signs are species-specific; each animal, metaphorically, is trapped within its own "soap bubble." Another metaphor Uexküll uses, bringing out his Leibnizian commitments, is that of drops of dew in a meadow, with each drop mirroring the world through the prism of a species-specific *Bauplan* (Brentari 2015, 167). A species is, as it were, a monad without windows, an idea that generates the problem of "environmental solipsism," "solved" through the assertion of transcendental harmony.

But Merleau-Ponty's ontology of nature, although influenced by Uexküll, develops his thoughts in a different direction. He sees inter-animality not as dogmatically preestablished but as constituted by expressive, symbolic forms of animal behavior, especially as this behavior emerges in the higher forms of life. Inter-animality is conceived not as imposed from the *outside in* but as achieved from the *inside out*. Rather than being harmonized by a metaphysical nature subject, it is in the concrete modes of animal existence, as they unfold in natural history, that we are able to witness an increasingly intertwined symphony of nature. This position leaves behind Uexküll's vision of living organisms, including ourselves, serving merely as keys on the natural factor's clavier of life.

Notes

1 Likewise, a mosquito may not sense the blood to which its effect organ is adapted but it still has sophisticated sensory abilities, which it uses to find the host.
2 Could there be an *Umwelt* that was constituted only by a *Wirkwelt?* I argue for this in the case of plants, albeit without the Uexküllian framework. See Moyle (2017).
3 Uexküll refers positively to Plato throughout his career, from an early review of Chamberlain to his final "Platonic dialog." See Brentari (2015), 56, 105, ft. 8.
4 There is no suggestion Schelling influenced Uexküll directly, merely that "Schelling had already developed analogous ideas" (Merleau-Ponty 2003, 177). Merleau-Ponty is referring to Schelling's postulate of an "original productivity of nature" in his early *Naturphilosophie.*
5 Uexküll also refers to the nature subject as "the One" (Uexküll 2010, 135), recalling his Neoplatonist leanings.
6 Given his tendency to collapse Kant's distinction between constitutive and regulative judgments, together with his repeated affirmation of the real, causal efficacy of the natural factor, abiding within the strict constraints of transcendental idealism was never going to be a viable option, for Uexküll.
7 Merleau-Ponty (1968, 200): "The problems posed in *Ph.P* are insoluble because I start there from the 'consciousness'-'object' distinction."
8 Merleau-Ponty conceives the natural history of life as similar in its unfolding to the way various styles (archaic, Roman, etc.) emerge within the history of art (Merleau-Ponty 2003, 255, 259). The contingency involves "aesthetic" or "subjective" necessity rather than sheer randomness.

References

Brentari, Carlo (2015) *Jakob von Uexküll: The Discovery of the Umwelt between Biosemiotics and Theoretical Biology.* New York: Springer.
Merleau-Ponty, Maurice (1963) [1942] *The Structure of Behavior.* Translated by Alden Fisher. Pittsburgh: Duquesne University Press.
Merleau-Ponty, Maurice (1968) [1964] *The Visible and the Invisible: Followed by Working Notes.* Translated by Alphonso Lingis. Evanston, IL: Northwestern University Press.
Merleau-Ponty, Maurice (1970) [1968] *Themes from the Lectures at the Collège de France 1952–1960.* Translated by John O'Neill. Evanston, IL: Northwestern University Press.
Merleau-Ponty, Maurice (2003) [1956–1960] *Nature: Course Notes from the College de France.* Translated by Robert Vallier. Evanston, IL: Northwestern University Press.
Merleau-Ponty, Maurice (2012) [1945] *The Phenomenology of Perception.* New translated by Donald A. Landes. New York: Routledge.

Moyle, Tristan (2017) 'Heidegger's philosophical botany'. *Continental Philosophy Review* 50 (3), 377–394.

Portmann, Adolf (1952) [1948] *Animal Forms and Patterns: A Study of the Appearance of Animals*. Translated by Hella Czech. New York: Schocken Books.

Uexküll, Jakob von (1985) [1909, 1921] 'Environment and inner world of animals'. Translated excerpts by Chancey J. Mellor and Doris Gove. In: Gordon M. Burghard (ed.) *Foundations of Comparative Ethology*. New York: Van Nostrand Reinhold, 222–245.

Uexküll, Jakob von (2010) [1934, 1940] *A Foray into the Worlds of Animals and Humans with a Theory of Meaning*. Translated by Joseph D. O'Neil. Minneapolis/London: University of Minneapolis Press.

9   The organism and its *Umwelt*

A counterpoint between the theories of Uexküll, Goldstein, and Canguilhem

Agustín Ostachuk

1 The living and its milieu

The most explicit and detailed account of Uexküll's concepts in Canguilhem's work can be found in "The Living and its Milieu," one of the philosophical chapters of his book *Knowledge of Life*, published in 1952. This book collects various conferences delivered by Canguilhem, with only one previously unpublished text added as Introduction under the title "Thought and the living." The chapter "The Living and its Milieu" was originally a conference presented at the Collège Philosophique in Paris in 1946–47 (Canguilhem 2001). This was only three years after the publication of his first book, *The Normal and the Pathological* (1943), in which he studied and interpreted Kurt Goldstein's main work, *The Organism: A Holistic Approach to Biology Derived from Pathological Data in Man* (1934). In this text, Goldstein quotes Uexküll many times. It is then probable that Canguilhem got first acquainted with Uexküll's theories through Goldstein. In fact, in "The Living and its Milieu," Canguilhem deals with both their theories together. According to Canguilhem, Uexküll and Goldstein reverse the problem of the organism–milieu relationship, as they claim that characteristic of the living is to make a milieu for itself whereas the study of a living being under experimental conditions is to impose a milieu on it.

Canguilhem explains and distinguishes first Uexküll's concepts of *Umwelt*, *Umgebung*, and *Welt*. For Canguilhem, the *Umwelt* designates "the milieu of behavior proper to a certain organism"; the *Umgebung*, "the banal geographical environment"; and the *Welt*, "the universe of science" (Canguilhem 2008, 111). The *Umwelt* is "an ensemble of excitations, which have the value and signification of signals" (Canguilhem 2008, 111). The living does not react to all the physical excitations of the environment but only to those of which it is notified and which presuppose a previous interest. In this manner, the excitation comes ultimately from the subject and is anticipated by their attitude. In this regard, Canguilhem states that "a living being is not a machine, which responds to excitations with movements, it is a machinist, who responds to signals with operations" (Canguilhem 2008, 111). Consequently, among the almost unlimited number of excitations from the environment, the organism only detects some signals [*Merkmale*]. One could then say that the *Umwelt* is an "elective extraction from the *Umgebung*, the geographical environment" (Canguilhem 2008, 112). As the core of this *Umwelt*, then, one finds a subjectivity that organizes the milieu, centered according to the vital values that constitute the subject itself. To exemplify this, Canguilhem cites Uexküll's account on the *Umwelt* of the tick (Uexküll 2010, 44; Ostachuk 2013).

Of all the stimuli that could exist in the environment of the tick, only three of them have relevance for it and make up its world. After mating, the adult female climbs, guided by the photo-receptiveness of her skin, to the branch of a tree and waits. This is the first stimulus. She can wait, immobile and inactive, without feeding or taking refuge, up to eighteen years: only the right stimulus can take her out of that state of quiescence and start her up again in her life cycle. When a mammal passes under the tree chosen by the tick as a hunting post, she lets herself drop, guided by the smell of butyric acid secreted by the perspiration of the animal. This is the second stimulus. When she has fallen on the animal, she fixes onto it in response to the temperature of its blood. This is the third stimulus. Once fixed to the animal, she goes to the source of the heat, arriving at areas of the animal free of hair, and finally sucks its blood. Only when her stomach is full of blood does a biological response begin, which consists of releasing the spermatozoa that were encapsulated in the female and fertilizing the eggs that await in the ovary. Consequently, the tick can fulfill her life cycle in a few hours, after which she dies, having been able to wait up to eighteen years. During that long period of waiting and inactivity, nothing that may surround the tick has any meaning for her. Only the three stimuli mentioned earlier have meaning for her and constitute her *Umwelt* (see Uexküll 2010, 44).

Canguilhem then goes on to compare this theory with Goldstein's. Goldstein starts from the criticism of the mechanical theory of reflexes. The reflex is not an isolated reaction but is always a function of sense and orientation, which depends on the signification of a situation as an ensemble. In this respect, an animal in an experimental situation is in "an abnormal situation, a situation it does not need according to its own norms; it has not chosen this situation, which is imposed on it" (Canguilhem 2008, 113). An organism is never equal to the theoretical totality of its possibilities, but it has its own privileged behaviors that respond to its own vital norms. For Goldstein, the relationship between the organism and the environment is established as a debate [*Auseinandersetzung*, coming to terms] "to which the living brings its own proper norms of appreciating situations, both dominating the milieu and accommodating itself to it" (Canguilhem 2008, 113). The relationship does not consist then in a struggle or opposition, the latter concerning rather a pathological state: "The situation of a living being commanded from the outside by the milieu is what Goldstein considers the archetype of a catastrophic situation" (Canguilhem 2008, 113). To live is to organize the environment from a center of reference, and a healthy life is a life relying on its existence and its values. For Canguilhem, the organism is a being with sense and its individuality is a character in the order of values.

2 Uexküll: *Umwelt* and conformity to a plan

Uexküll proposes "a walk into unknown worlds" (Uexküll 2010, 41), worlds strange to us but known to other creatures, "as diverse as the animals themselves"

(Uexküll 2010, 42). In order to do this, he suggests, we must create an imaginary soap bubble around each creature. Each of these bubbles contains only the perceptions to which the creature has access and then forms its own true world. Each of these bubbles represents the world as it appears to the organisms themselves. As each organism perceives differently, there are as many of these worlds as there are organisms in nature. Uexküll does not consider organisms as mere objects but as subjects whose essential activity consists in perceiving and acting. Everything an organism perceives is part of its perceptual world [*Merkwelt*], while everything an organism makes is part of its operational world [*Wirkwelt*]. The perceptual and operational world together form a closed unit called *Umwelt* (Uexküll 2010, 42; Ostachuk 2013). In this manner, even if the same objects are present in a certain environment, they will not be perceived in the same way by the different organisms, and they will not have the same meaning for them.

The second important Uexküllian concept to consider in this context – that is to say, when it comes to the relation between an organism and its environment – is that of conformity to a plan [*Planmäßigkeit*]. Uexküll calls conformity to a plan the force of nature "that combines the manifold details into one whole by means of rules. Higher rules, which unite things separated even by time, are in general called plans" (Uexküll 1926, 175). Elsewhere, Uexküll defines conformity to a plan as "a rule stretching across time and space," "a rule in living Nature, which reveals itself even in the mechanical processes of the organism" (Uexküll 1926, 270), and as "a super-mechanical law" (Uexküll 1926, 271). Conformity to a plan is responsible for the creation of all organisms and their *Umwelten*, it is like the score laying out the "melody," which accounts for the whole of nature. Ultimately, conformity to a plan ensures the perfect complementarity between the different organisms and their *Umwelten*.

By means of this notion, Uexküll expresses himself against the concept of adaptation. From the point of view of the adaptation theory, in fact, "each organism is the product of influences to which it has been exposed for thousands of years" (Uexküll 1926, 319). Through innumerable cycles of trial and error, organisms reach their appropriate form, a final product adapted and congruent to the conditions of the environment in which they are present. Uexküll also criticizes the Darwinian theory, which aims to explain adaptation through the mechanism of natural selection. According to Darwinism, the struggle for existence determines organisms to compete with each other, a struggle in which only the "most adapted" will be able to survive. Differently, according to Uexküll, it is impossible for an organism, even a machine, in which all parts fit together properly, to arise through such a mechanism: "It certainly requires a powerful imagination to assume that any machine capable of functioning could arise in this way. But the Darwinians provide the requisite imagination" (Uexküll 1926, 320). In other words, Uexküll argues, it is impossible for a cooperative structure to emerge from a competitive mechanism.

According to Uexküll, then, nature produces all its organisms following a plan. The adjustment between the different parts of an organism, and between the organism and its *Umwelt*, is not produced by external erosion or molding, but they are adjusted, they are "congruent" with each other, from the beginning. This congruity is guaranteed by the plan, which is what builds the organisms in harmony with their *Umwelten*. The plan for the construction of the *Umwelt* of each organism at any given moment, as well as the tunnel formed by the addition of its successive vital moments, what Uexküll calls "life-tunnel," is fixed and is not subject to change. However, Uexküll also recognizes that deviations may occur which are responsible for generating the illusion of the variability of organisms. These deviations are of secondary importance, and Darwin's error was to have made them the main feature, when the main characteristic is the plan itself. Needless to say, the notion of conformity to a plan is not without difficulties. One could raise for instance the objection that the reduction of the whole of nature to a super-mechanical rule, that includes all the functional circles established between organisms and their *Umwelten*, entails also the reduction of the concept of subject to the mere assembly of functional circles, unlikely to allow much room for the characteristic autonomy and creativity of organisms. This point will be, as we shall see, the target of Kurt Goldstein's criticisms of Uexküll's theory.

In short, to ensure harmony and the perfect adjustment between organisms and their environments, Uexküll advocates the existence of a predetermined world, a world in which life-tunnels are fixed and predestined: "Proceeding from these immutable factors that determine all life in the world, we come to see that life itself is based on fixed laws, which are in conformity to a plan" (Uexküll 1926, 84).

3 Goldstein: the debate between the organism and its environment

In the section "Criticism of the Purely Environmental Theory" of his work *The Organism*, Goldstein criticizes, in the last instance, the theory of Uexküll. Goldstein argues that an organism not only lives in its own environment, to which it is perfectly adjusted, but must also deal with all other stimuli of the environment, including potentially negative ones. The organism does not live isolated or segregated in its own environment as if the rest of the world did not exist and all the stimuli it receives were adequate for it. For Goldstein, the environment of an organism is neither definitive nor static, but it is formed with the development and activities of the organism. In this manner, he affirms: "One could say that the environment emerges from the world through the being or actualization of the organism" (Goldstein 1995, 85). In other words, the organism, in order to exist, must find an adequate environment for itself; it must create it taking advantage of the opportunities offered by the world. "An environment always presupposes a given organism" (Goldstein 1995, 85) and not the other way around. An organism does not acquire order at the expense of its environment; an organism rather acquires order at the same time that the environment obtains it. However, Goldstein subsequently seems to change his mind, as he states that the "environment first arises from the world only when there is an ordered organism" (Goldstein 1995, 85), thus suggesting that first there must be an ordered organism for an environment to be created. According to this second version, order comes ultimately from the organism itself. In this regard, Goldstein brings the case of a diseased organism. For this organism, the environment prior to its state of disease has become strange and disturbing so that the essential requirement to exist and to return to a state of new normality is to make for itself once more an adequate environment. If we consider in-depth the previous case, we see in reality that the order is restored with the creation of a new adequate environment and not by reverting to a previous or anterior order of the organism on its own. Consequently, the restoration of order obviously requires the activity of the organism, but it is achieved and reached when it finds the appropriate conditions for the generation of a new adequate environment.

For Goldstein, the fundamental relationship between an organism and its environment is a debate or coming to terms (Goldstein 1995, 42; Ostachuk 2015). There is a fundamental separation between the organism and its environment, which makes them strange and which requires a constant debate and coming to terms between the two so that the relationship can be maintained. The organism achieves this coming to terms through a behavioral act called performance [*Leistung*]. When a performance is effective, the organism develops an ordered behavior. This is characteristic of a normal or healthy state. On the contrary, when a performance is ineffective, the organism develops a disordered behavior, which manifests itself in the form of a "catastrophic" reaction. This is the characteristic state of a pathological or disease state. In this case, the organism will proceed to recover the normal situation. This amounts to saying that there is a tendency in the organism to live in ordered behavior. Therefore, for Goldstein, disease consists of a disarrangement or disequilibrium that breaks the productive relationship between the organism and its environment. On the other hand, "an organism that actualizes its essential peculiarities, or – what really means the same thing – meets its adequate milieu [*Umwelt*] and the tasks arising from it, is 'normal'" (Goldstein 1995, 325). Goldstein arrives, in this manner, at a definition of normality. Normality, or the normal state, is the state in which the organism develops norms that allow it to respond adequately to its environment. Moreover, recovering health, that is, rehabilitation, consists of reaching a new order, a new normality, a new individual norm, which implies making for itself a new environment in which it can respond appropriately again.

The organism, in its relation to the environment, not only seeks its preservation [*Erhaltung*] but also its prosperity. For Goldstein, survival is the typical lifestyle of the disease state. A healthy organism not only seeks self-preservation but also aspires to self-realization [*Selbstverwirklichung*]. Self-realization is the tendency of organisms to realize their own essence and their peculiar individuality. This can be interpreted as a critique of Darwin's theory, which postulates natural selection as an evolutionary mechanism, based on the "struggle for existence," the competition among individuals, and the "survival of the fittest." For Goldstein, this would be a model of the world and society in pathological and disease state.

4 Canguilhem: normativity and institution of norms and values

The element of novelty introduced by Canguilhem in the relationship between the organism and its environment is that the former is characterized by the faculty of creating and instituting norms; that is to say, it possesses normativity or normative activity (Ostachuk 2015). In this respect, the organism not only has norms and is able to fulfill them, that is, it possesses normality, but its most characteristic and genuine feature also is that of creating and instituting new norms, that is, it possesses normativity. This is already stated by Canguilhem in his first book, *The Normal and the Pathological*:

> life is polarity and thereby even an unconscious position of value; in short, life is in fact a normative activity [...] Normative, in the fullest sense of the word, is that which establishes norms. And it is in this sense that we plan to talk about biological normativity.
>
> (Canguilhem 1991, 126f.)

A healthy organism then does not limit itself to self-preservation, resisting any variation and adaptation to new situations, but embodies norms that drive it forward:

> Health is more than normality; in simple terms, it is normativity. Behind all apparent normality, one must look to see if it is capable of tolerating infractions of the norm, of overcoming contradictions, of dealing with conflicts. Any normality open to possible future correction is authentic normativity, or health. Any normality limited to maintaining itself, hostile to any variation in the themes that express it, and incapable of adapting to new situations is a normality devoid of normative intention.
>
> (Canguilhem 1994, 351)

A healthy organism does not seek so much to preserve itself, to maintain its state and its environment, but, rather, to realize its own nature, which implies seeking new challenges, overcoming new obstacles and, ultimately, exposing itself to new risks. Consequently, a healthy organism is constantly exposed to the risk of losing its order and entering into situations of catastrophic reaction, that is, of becoming ill. A measure of the health of an organism is its capacity to overcome these crises and establish a new order, restoring an adequate relationship with its new environment. On the other hand, the pathological state does not imply the total absence of norms. Disease is itself a norm of life. However, it is a norm that does not tolerate deviations in the conditions of the relationship with its environment and is incapable of transforming itself into another norm. In other words, "the sick living being is normalized in well-defined conditions of existence and has lost his normative capacity, the capacity to establish other norms in other conditions" (Canguilhem 1991, 183).

5 Teleology and polarity in the organism

Canguilhem makes it clear that organisms have a polarity that consists in actualizing their own norms and values; that is to say, organisms are normative. This echoes Goldstein's suggestion that the natural tendency of organisms is to actualize their own essence, that is, the tendency toward self-realization.

Despite these claims, Goldstein explicitly opposes the teleological approach to organisms. His explanation is however not free of ambiguities and inconsistencies. In the first place, Goldstein rejects the teleological approach but recognizes that an "inner purposiveness in the sense of Kant" (Goldstein 1995, 323) could be accepted. He then introduces Karl Ernst von Baer's distinction between purpose [*Zweck*] and goal [*Ziel*]: "According to him 'purpose' is an intended task, whereas 'end' is a given direction of activity, an intrinsically predetermined effect" (Goldstein 1995, 324). Following von Baer, Goldstein maintains that the concept of purpose is inadequate and should be abandoned, while the concept of goal, which he interprets as the actualization of an essence, is useful and adequate for the understanding of the organism. With this explanation, Goldstein seems to support rather than reject teleology in organisms. The definitions of goal as "an intrinsically predetermined effect" and "actualization of an essence" are even compatible with the Aristotelian conception of teleology (Ostachuk 2016).

Uexküll had already pointed to the importance of von Baer's distinction between purpose and goal, although he explains it in more detail, quoting an example from von Baer himself:

> When a bullet leaves the barrel of a gun and hits the target, the target is the factor that prescribes the path for the ball. If we imagine the act of shooting to be eliminated, we must ascribe to the ball itself the property of being influenced directly by the target in the direction its movement takes. In such a case the ball possesses what Baer calls "effort toward a goal."
>
> (Uexküll 1926, 316)

According to von Baer, an embryo possesses this "effort toward a goal." Uexküll does not agree with this argument. In the first place, Uexküll considers that the goal is not the adult organism but the congruity with its *Umwelt*. In the second place, he believes that this goal cannot be achieved through this "effort toward a goal." For Uexküll, there are no influences from the *Umwelt* that can affect or alter the course of development of an embryo, since it does not possess the necessary organs to know the properties of the external world. And yet, says Uexküll, we see that the embryo "unerringly produces definite counter-properties, which fit into a definite group of properties in the external world" (Uexküll 1926, 317). This happens thanks to the perfect congruity between the organism and its *Umwelt* ensured by the conformity to a plan.

In fact, von Baer's example itself seems inappropriate. Eliminating the act of shooting in order to ascribe to the bullet the property of being attracted by the target makes for an unjustified assumption that leads to a wrong conclusion. In this particular case, the one responsible for determining the direction of the bullet is the shooter, not the target. Without a shooter, the bullet would not be fired, and if it were fired accidentally, its direction would be completely random. In general terms, without an agent there is no action, the action being a movement with a specific purpose. And it is the agent who contributes to the end of a certain action. Furthermore, von Baer's distinction between purpose and goal seems to be compatible with Driesch's distinction between dynamic and static teleology (Ostachuk 2016). Every machine has an end. However, this end is given externally by the designer or constructor of the machine. This is the external, static teleology, in which what is sought is the fulfillment of a goal or an end [*Ziel*]. An organism, on the other hand, is not only capable of fulfilling a given end but also of creating and choosing new ends, that is to say, it possesses a purposive capacity that allows it to adopt autonomic actions. An organism possesses internal, dynamic teleology, in which what is sought is the fulfillment or achievement of a purpose [*Zweck*]. In other words, it possesses purposiveness.

6 The organism and the machine

There is only one case in which von Baer's concept of "effort toward a goal" would be compatible with Uexküll's theory. It is the case in which the goal is the plan itself. If we accept this, we must accept at the same time that the world and nature are machines and that they work only according to mechanical laws. This is what Uexküll seems to indicate when he talks about the plan as "a super-mechanical law." However, this mechanical view does not seem compatible with his view of a world populated by subjects. Despite believing that there is a distinct difference and a discontinuity between living beings and physico-chemical processes, Uexküll states that this difference is that the latter are mere mechanical processes, while the former have "super-mechanical powers." These super-mechanical powers of organisms consist of: the construction of the machine, the running of the machine, and the repair of the machine (Uexküll 1926, 121). These powers come, ultimately, from the rules of conformity to a plan.

Canguilhem hardly agrees with these reasonings of Uexküll. He, rather, strongly supports the irreducibility of the organism to the machine: "it is an illusion to think that purpose can be expelled from the organism by comparing it to a composite of automatisms, no matter how complex" (Canguilhem 2008, 91). Inspired by Bichat, he remarks that there is no mechanical pathology; that is, a machine does not get ill. There is no distinction between health and disease in a machine. A machine cannot be healthy or diseased because it does not establish a relationship with an environment, and as we have already seen, for Goldstein and Canguilhem, health consists of an adequate and productive relationship with one's own environment. Based on Hans Driesch's experiments, Canguilhem states that embryological development cannot be reduced to a mechanical model. Whereas Uexküll argues in favor of the perfect congruity between all organisms and their *Umwelten*, Canguilhem advocates the autonomy of the living. One could even wonder if the relationship itself between the organism and the environment is not annulled by the very idea of a perfect but fixed and invariant adjustment. Once the relational element is eliminated, the "possibility" of change and variation over time is also erased. However, the autonomy of the living seems to require this possibility of "relation" with its own environment. And unless one wishes to promote the existence of a world populated only by machines, ordered and adjusted to each other mechanically, one should not do without the idea of autonomy. In short, as soon as teleology disappears from nature, the organism falls back into the status of a machine.

For Canguilhem, the organism presents the properties of self-construction, self-conservation, self-regulation, and self-repair, while for a machine, "its construction is foreign" and "conservation demands the constant surveillance and vigilance of the machinist" (Canguilhem 2008, 88). And the plan is not a machinist, but the blueprint that the machinist uses to build and repair the machine.

7 The archaic relationship between the organism and its environment

In later years, in his work *The Theory of Meaning* [*Bedeutungslehre*], Uexküll develops his theory more explicitly and extensively in musical terms. The development of organisms and their *Umwelten* is part of a great symphony in which all organisms play melodies that are assembled with each other by point and counterpoint: "Every animal, like every instrument, harbors a certain number of tones that enter into contrapuntal relationships with the tones of other animals" (Uexküll 1982, 63). He no longer explains the congruity and perfect adjustment between organisms and their *Umwelten* as a fitting between "pegs and sockets" (Uexküll 1926, 317) but, rather, in terms of the existence of an interrelation and interpenetration between them. In this regard, Uexküll tells us that just as the flower is bee-like, the bee is flower-like so that the melodies played by both resonate in unison.

A much more elaborate example, within this perspective, is that of the spider's web. The spider builds its web according to the structural characteristics of the fly so that the latter cannot see it and gets caught when flying toward it. However, the spider does this without even having come into contact with a fly. So, how does this correspondence occur? Uexküll's explanation is the following:

> It weaves its web before it is ever confronted with an actual fly. The web, therefore, cannot represent the physical image of a fly, but rather it is a representation of the archetype of a fly, which does not exist in the physical world.
> (Uexküll 1982, 42)

This first explanation then holds that each organism develops thanks to the existence of an original program or archetype and that, in some way, the archetype of the fly influences the archetype of the spider (Uexküll 1982, 43). Few pages later, Uexküll provides a more detailed and accurate explanation of this biological phenomenon:

> The spider's web is certainly formed in a 'fly-like' manner, because the spider itself is 'fly-like.' To be 'fly-like' means that the body structure of the spider has taken on certain of the fly's characteristics – not from a specific fly, but rather from the fly's archetype. To express it more accurately, the

spider's 'fly-likeness' comes about when its body structure has adopted certain themes from the fly's melody.

(Uexküll 1982, 66)

One way of interpreting this is to think that what we see as individualized and interacting elements in the real physical world exist archaically all included in a great invisible world, in which an overlap of resonant melodies occurs, and in which a clear and sharp separation between them cannot be established. Accordingly, a "relation" is the expression of an original resonant overlap. The evident complementarity and reciprocity between organisms and their *Umwelten* are the expression of a great symphony made up by a multitude of melodies resonating in unison.

There is no such melodic language, or the proposal of a universal interconnectivity in nature, in the works of Goldstein and Canguilhem. For them, the autonomy of the organism and the living prevails. However, it could be ventured that norms and values play this melodic role in the relations between organisms and their environments and that with each norm and value that is actualized, the resonance in unison Uexküll talks about is produced.

8 Meaning and sense

In his work, *The Theory of Meaning*, Uexküll also argues that meaning is the fundamental and key property for the understanding of life. All the objects of the *Umwelt* of an organism are subject-related meaning-carriers. Meaning is the connector that unites the organism with each object of its *Umwelt*: "In every instance a very intimate meaning rule joins the animal and its medium" (Uexküll 1982, 54). In even more explicit terms, he also says that "[m]eaning in nature's score serves as a connecting link, or rather as a bridge" (Uexküll 1982, 64). On the other hand, the same object can have different meanings for different organisms. For example, a flower stem acts as a different meaning-carrier for different meaning-utilizers: while for an ant, it is a path, for a cicada larva, it is a supplier of material for the building of a house, and for a cow, it is food.

This worldview transforms nature into a huge network of interconnections established through meaning. This ecological view of nature, which can act as an antidote to the mercantilist and competitive views of today, leaves out, however, the consideration of sense. Uexküll provides a very interesting example with regard to this question. The pea-beetle larva, thanks to its tunnel-boring activity, builds a tunnel that allows the adult beetle to leave the pea. If it were not for this tunnel exit, the adult larva would die. This example allows Uexküll to cast doubt on von Baer's claims regarding the presence of a goal-directedness in the origin of living creatures. Unlike the example of the spider web, in which other organisms intervene, which allows Uexküll to assign the anticipation of the spider to the participation in the fly's melody, in this example, this does not happen. It is the same organism that anticipates a future event of its own development: the larva knows in some way that it has to build a tunnel so that the adult organism can then leave the pea. There is an intentionality on the part of the organism that is not contained in the concept of meaning. Meaning can only act as an extrinsic connector of phenomena. But here we are in the presence of a prediction of a future event, which speaks of the existence of a subjective interiority. This interiority of the living is what anticipates, it has intentions, in short, it actualizes its potentialities.

Canguilhem adopts this second version of the matter. For him, the consideration of sense in biology can never be omitted:

> A center does not resolve into its environment. A living being is not reducible to a crossroads of influences. From this stems the insufficiency of any biology that, in complete submission to the spirit of the physico-chemical sciences, would seek to eliminate all consideration of sense from its domain. From the biological and psychological point of view, a sense is an appreciation of values in relation to a need.
>
> (Canguilhem 2008, 120)

A living being is then an irreducible center of reference, which has needs and institutes norms and values to satisfy them. Consequently, whereas Uexküll emphasizes nature as a network of interconnections mediated by meaning, Canguilhem insists on the centrality and irreducibility of an organism that establishes relations of sense with its environment.

9 Life

Our three authors make for an interesting panel for a discussion of the concept of life. Even between Goldstein and Canguilhem, who otherwise tend toward concordance rather than dissonance, this topic brings out the greatest differences and nuances.

Perhaps the most difficult position to decipher on this subject, Uexküll's position, on the one hand, establishes a clear difference between biology and physics and chemistry and argues strenuously for the irreducibility of the living to mere physico-chemical processes. On the other hand, with his concept of conformity to a plan, he seems to make organisms and their *Umwelten*, and ultimately life-tunnels, depend on super-mechanical rules and laws, supporting unchangeable congruity and perfect adjustment between each other. This leaves, as we have already seen, the autonomy of the organism in a rather inconvenient situation.

According to Uexküll's distinction between a machine and a living being, a machine is constructed based on a building-plan [*Bauplan*] in which the spatial arrangement and function of its different parts are made explicit. And that is enough, since "machines originate namely by assembling ready-made parts into a whole" (Uexküll 1913, 155; my transl.). Living beings, instead, do not originate by the assembly of ready-made parts. Uexküll explains that the problem of the origin of living beings has divided researchers into two groups. On one corner, there are those who at the origin of machines recognize two factors, human representation, that is, building-plan, and mechanical forces, while at the origin of living beings only recognize mechanical forces. On the opposite corner, there are those who maintain that mechanical, physico-chemical forces cannot originate any building-plan. An essential factor must exist "that stands above the mechanical forces, to which it directs, so that from diverse parts originates a whole that works in conformity to a plan" (Uexküll 1913, 156; my transl.). This super-mechanical factor is for Uexküll what we call life. Life is then conformity to a plan, that is, a super-mechanical law or rule. This leaves life in a diminished, regulated and de-autonomized condition, since it associates it with a fixed and preestablished plan.

Goldstein is notoriously the supporter of an organicist theory of life, according to which the fundamental characteristic of the organism and therefore of life is the maintenance over time of its organization, understood as a relation existing between its parts. Goldstein also states that the organism seeks to actualize its own essence and aspires to self-realization; that is, it aspires not only to maintain itself but also to thrive. However, it is Canguilhem who makes greater efforts to defend the autonomy and originality of life. In the first place, he does so by advocating the autonomy and specificity of biology with respect to the physico-chemical sciences. This position leads in general, says Canguilhem, to the qualification and accusation of vitalism. It should be made clear, here, that Canguilhem understands vitalism as a form of confidence in the organism's own reaction and self-defense, that is to say, in its own curative properties against the causal agent of diseases and beyond the constrictor power of remedies. Vitalism thus expresses a distrust in the power of technique over life and approaches naturism in its own terms. Vitalism is ultimately "a permanent exigency of life in the living, the self-identity of life immanent to the living" (Canguilhem 2008, 62). In this context, whereas vitalism comes about as an exigency, mechanism imposes itself as a method but as a method that creates nothing if not by human skill and art. This is why the mechanistic interpretation of the living automatically nullifies the living. Life has a spontaneity and a creativity that mechanism, in its eagerness to reduce it and decompose it into a simple set of machines, cannot account for or explain. For Canguilhem, the constant rebirth of vitalism expresses the unwavering resistance and rebellion of life to be subjected to mechanization. This resistance and rebellion of life to mechanization is also a resistance to its dissolution in an impersonal geographical environment and an exigency to place itself as a center. It is only from this center that it is possible to generate one's own surrounding world, an *Umwelt*.

Conclusion

A tension and counterpoint resembling the one between the organism and its environment is at play between the theories of Uexküll, Goldstein, and Canguilhem. While for Uexküll there is congruity and perfect adjustment between the organism and its *Umwelt*, for Goldstein and Canguilhem, there is debate and coming to terms; there is an actualization of the relation that is produced by the institution of

new norms and values, in other words, biological normativity. Potential estrange-
ment and maladjustment between the organism and its environment are, for Goldstein and Canguilhem, the origin of disease. It would be therefore inappropriate
to consider this debate and coming to terms as an adaptation. In each search of
normality, the organism does not seek to adapt to an environment that has become
strange and hostile but seeks to create for itself a new environment according to
its current conditions by establishing new norms. In this coming to terms, the
objective is not to survive, that is, to live in a constant situation of "struggle for
existence." Such a situation defines instead the pathological state. The ultimate
goal of the organism is instead to thrive and aspire to self-realization. Also in
disagreement with the concepts of adaptation and struggle for existence, Uexküll
sees instead a world in which harmony and perfect correspondence reign between
all organisms and their *Umwelten*.

The question of teleology and sense in organisms is another point in which
the tension and counterpoint between the theories of Uexküll, Goldstein, and
Canguilhem become clear. Heir to the Kantian issue of teleology as a regulative
principle, Goldstein's organicist approach denies the existence of purposiveness
in organisms, although he admits it with reservations, provided it is considered in
almost metaphorical terms. Like Goldstein, Uexküll resorts to von Baer's theory
in order to deny the existence of teleology in organisms. However, the case of the
pea-beetle larva makes him doubt and even admit the possibility of the existence
of a goal-directedness in the development of organisms. Canguilhem, due to his
inclination toward the vitalist theory, has a more original position on this subject
and deems sense essential for the understanding of the living.

Regarding the topic of life, the three authors defend the autonomy and specificity of biology with respect to the physico-chemical sciences, and the irreducibility of the living being to a set of mere mechanisms. However, they differ quite
a bit as to their respective notions of life. For Uexküll, life is the expression of
conformity to a plan, that is, a super-mechanical law or rule. This associates life
with a fixed and preestablished plan. For Goldstein, life is the maintenance of the
organization of a totality, but also the actualization of an essence whose goal is
self-realization. Finally, the most affirmative and positive position regarding life
is that of Canguilhem. For Canguilhem, life is an exigency that resists and rebels
against mechanization and that seeks to position itself as the generating center of
its own environment, an *Umwelt*.

In conclusion, the three authors might converge on the idea that the relationship
between the organism and its environment is an archaic relation, unfolding in an
invisible and musical world, in which the melodies of all the organisms interpen-
etrate each other and create a symphony that embraces all. This interpenetration
accounts for the correspondence between organisms and their *Umwelten* in the
real world, as well as for the existence of meaningful relationships between them.
This original and common source may well be called a principle, or plan, in reference to the Uexküllian plan, but it must also possess all the creative characteristics
of what we call life.

References

Canguilhem, Georges (1991) [1943] *The Normal and the Pathological.* Translated by Carolyn R. Fawcett in collaboration with Robert S. Cohen. New York: Zone Books.

Canguilhem, Georges (1994) *A Vital Rationalist: Selected Writings from Georges Canguilhem.* Translated by Arthur Goldhammer. New York: Zone Books.

Canguilhem, Georges (2001) 'The living and its milieu'. Translated by John Savage. *Grey Room* 3, 7–31.

Canguilhem, Georges (2008) [1952] *Knowledge of Life.* New York: Fordham University Press.

Goldstein, Kurt (1995) [1934] *The Organism: A Holistic Approach to Biology Derived from Pathological Data in Man.* New York: Zone Books.

Ostachuk, Agustín (2013) 'El *Umwelt* de Uexküll y Merleau-Ponty'. *Ludus Vitalis* 21 (39), 45–65.

Ostachuk, Agustín (2015) 'La vida como actividad normativa y auto-realización: debate en torno al concepto de normatividad biológica en Goldstein y Canguilhem'. *História, Ciências, Saúde-Manguinhos* 22 (4), 1199–1214.

Ostachuk, Agustín (2016) 'El principio de vida: de la *psyché* aristotélica a la entelequia drieschiana'. *Ludus Vitalis* 24 (45), 37–60.

Uexküll, Jakob von (1913) *Bausteine zu einer biologischen Weltanschauung. Gesammelte Aufsätze.* München: Bruckmann.

Uexküll, Jakob von (1926) [1920] *Theoretical Biology.* Translated by Doris L. Mackinnon. London: K. Paul, Trench, Trubner & Co.; New York: Harcourt, Brace & Company.

Uexküll, Jakob von (1982) [1940] 'Theory of meaning'. *Semiotica* 42 (1), 25–82.

Uexküll, Jakob von (2010) [1934, 1940] *A Foray into the Worlds of Animals and Humans with a Theory of Meaning.* Translated by Joseph D. O'Neil. Minneapolis/London: University of Minneapolis Press.

10 From ontology to ethology

Uexküll and Deleuze & Guattari

Felice Cimatti

1 Becoming-animal

Gilles Deleuze and Felix Guattari can be credited as the only European philosophers to apply the concept of "animality" in a "productive" way, and not as the merely speculative counterpart of the concept of "humanity." I argue in what follows that Uexküll's contributions have played a game-changing role in the path to such an achievement.

At least until Derrida's contribution, *The Animal That Therefore I Am* (2008), the concept of animality has indeed only been taken contrastively as the direct opposite of that of humanity. Traditionally taken as that living being which is not human (de Fontenay 1998; Keekok 2006; Cimatti 2013; Nance 2015), the animal has been mostly given for granted and left undiscussed. A common refrain in philosophical literature was for instance, that, since the human is the animal endowed of language, the animal must be deprived of it. Or even, since the human is *cogito* – as Descartes teaches us – the animal must be a machine. This clearly exemplifies in what sense the animal has been long understood only as the opposite of what is human. From this standpoint, contemporary animalist claims (Bekoff, Allen, and Burghardt 2002), arguing that animals are indeed endowed of language and reasoning, end up simply embracing the flip side of previous assumptions, acritically accepting the human as the reference term for what an animal is. But why should language be so important for an animal? What is "animality" per se? Deleuze and Guattari's great philosophical novelty lies indeed in their attempt to provide a conceptual understanding of animality that is not only the direct opposite of the traditional concept of "humanity." Thanks to Deleuze and Guattari, "animality" comes then to the fore as a fundamental notion in philosophy, allowing a redefinition of the main questions in our theoretical tradition.

In this regard, one of their most groundbreaking references in biology is Jakob von Uexküll, who provided them with authoritative insights on three main issues: (a) how to bypass the subject–object divide, (b) how to think about what bodies can do, and (c) how to transform the abstract concept of the "world" into a biological *Umwelt*. In what follows, I try to account for the positively disruptive impact of Uexküll's position within the framework of Western traditional ontology, as well as for how its understanding by Deleuze and Guattari has contributed to the outlining of their own radical ethology. One might want to start by asking, What is the only animal Western philosophy has ever taken an interest in?

Descartes makes for a famous case in this regard. In the *Discourse on the Method* (1637), as is well known, he claims that animals are radically different from human beings for two interconnected reasons: "they are incapable of arranging various words together and of composing from them a discourse by means of which they might make their thoughts understood" (Descartes 2000, 72). It is worth remarking here that this is not a description of animality, but it might well be taken as a "definition" of what "animality" should be. All in all, the Cartesian philosophical animal only exists in order to specify what "humanity" is. Therefore, it would be naive to rebut that, as a matter of fact, animal communication systems exist (Hauser 1999). Descartes is not speaking about existing animals; what he only cares about is humanity. Besides lack of language, animals, according to Descartes, lack also general reasoning:

> they have no intelligence at all, and that it is nature that acts in them, according to the disposition of their organs – just as we see that a clock composed exclusively of wheels and springs can count the hours and measure time more accurately than we can with all our carefulness.

> (Descartes 2000, 73)

Clearly, a "realistic" description of animality has never been Descartes's goal; what is at stake is instead a definition of "animality" which emphasizes the uniqueness and extraordinariness of the *Homo sapiens*. Therefore, animals are like "machines" because they "have to be without" reason and language. Deep-rooted theological assumptions lie at the very core of such a description of animality.

Granted that animals are machines, they cannot be credited with any form of subjectivity. It is not surprising, then, that the question of subjectivity remains a key problem in subsequent philosophical (and scientific) descriptions of animality, up to present days (Griffin 1976; Wynne 2004; Crystal and Foote 2009; Cimatti and Vallortigara 2015). This point is somewhat implicitly shown, for instance, by Heidegger's input on the matter. Although seemingly much different a philosopher from Descartes, in *The Fundamental Concepts of Metaphysics* (1929/1930; Heidegger 1995), Heidegger still describes the bee as some sort of machine. Machines are typically pushed to action without asking themselves why or knowing why. In this sense, one might claim that also according to Heidegger the animal is somewhat similar to a machine. In fact, he would say that the animal lives in a condition of "captivation" [*Benommenheit*] (Heidegger 1995, 333). As such, the animal would be completely caught by the stimulus that triggers "its own" behavior. In this respect, the animal is said to be unaware of what it is actually doing, although "its" behavior is not entirely casual or mechanical. "Captivation" is thus a controversial form of subjectless subjectivity:

> [I]s there any evidence that the bee recognizes the presence, or the absence, of honey?[1] Clearly there is. For, attracted by the scent of the flower, the bee

stayed on the blossom, began to suck up the honey, and then stopped doing so at a certain point. But does this really prove that the bee recognized the honey *as present*? Not at all, especially if we can and indeed must interpret the bee's activity as a driven performing and as drivenness, as behavior – as behavior rather than comportment on the part of the bee toward the honey which is present or no longer present.

(Heidegger 1995, 241)

Strikingly enough, three centuries after Descartes, animality is still investigated within the same dualistic framework. In other words, the underlying motivation for such a description of animal subjectivity is only to sharply differentiate the animal from the human being, and appoint this latter – and this latter only – the "shepherd of being" (Heidegger 1998, 252). It should go without saying that human beings are the only living beings to entertain an intrinsic relationship with the Being. Heidegger notably ventures into attributing to animals a peculiar kind of subjectivity, which still makes them very different from human beings:

The bee is simply taken [*hingenommen*] by its food. This being taken is only possible where there is an *instinctual* "toward ...". Yet such a driven being taken also excludes the possibility of any recognition of presence. It is precisely being taken by its food that prevents the animal from taking up a position over and against this food.

(Heidegger 1995, 242)

It can be argued, then, that the Heideggerian bee is still more similar to a "biological" machine than to an effective vital entity. The reader might then be very surprised to learn that Heidegger's main scientific source is precisely Uexküll (see Michelini's contribution, Chapter 7, in this volume). This is all the more surprising as Uexküll is notoriously famous for explicitly rejecting Descartes's idea that animals are "mere machines" (Uexküll 1957, 5). Astonishingly, then, Heidegger relies on Uexküll but does not fall much further than Descartes. It should be clear, here, that biological machines are, under many respects, not that different from mere mechanical machines. To be fair, Heidegger would differentiate between machine and stones, but all in all, he fails to explain why a stone is different from a bee-machine. The concept of "instinct" [*Triebe*] for instance only makes sense as counterpart of that of human "behavior" [*Benehmen*] (Heidegger 1995, 249). Heidegger's animal, just as the machine, is blindly guided by an uncontrollable force.

Take the case of Heidegger's *Dasein* as "Being-in-the-world" (Heidegger 1962, 65). Clearly, the position of *Dasein* still implies a dualism: the dualism of the subject and the object, the human and the animal, the mind and the body. Although it is true that Heidegger has attempted precisely to overcome such a dualism, his efforts were not fully successful. One could argue, for instance, that the position assigned by Heidegger to the *Homo sapiens* is still too humanistic to really overcome the dualistic heritage of Western metaphysics. On the one hand, in the *Letter on Humanism*, Heidegger writes that

[t]o think the truth of being at the same time means to think the humanity of *homo humanus*. What counts is *humanitas* in the service of the truth of being, but without humanism in the metaphysical sense.

(Heidegger 1998, 268)

On the other hand, Heidegger still insists on tracing a radical separation between *humanitas* and *animalitas*: "Metaphysics thinks of the human being on the basis of *annihilates* and does not think in the direction of his *humanitas*" (Heidegger 1998, 246–248). And this ultimately amounts to saying that *Homo sapiens* is not an animal (see also Michelini, Chapter 7, in this volume). Despite blaming metaphysics for being dualistic, Heidegger overcoming of metaphysics is not completely beyond dualism.

Committed to dismantle such an ontological dualism, Deleuze and Guattari's concept of "becoming animal" aims to establish a radical *animal* monism. The point is then to imagine a neither humanistic nor human way of life, that is, a life that is no longer based on human subjectivity. One should ultimately stay in the world, for instance, like a rat stays in the world. Thanks to Deleuze and Guattari, then, "we become universes. Becoming animal, plant, molecular, becoming zero" (Deleuze and Guattari 1994, 169), the underlying claim here being that the becoming animal of human beings leads to the becoming-world of humanity. For Deleuze and Guattari the concept of "animality" is therefore not the simple functional contrary of the concept of "humanity"; it provides instead a means to invent a different way to stay into the world. While the *Dasein* places itself "outside" the world – even and especially when man becomes the "shepherd of being" (Heidegger 1998, 252) – the becoming animal represents the opposite direction movement toward the world. Whereas the *Dasein* is "into" the world, the becoming animal is immanent to the world, "coincides with" the world (DeLanda 2009).

2 Antidualism

In general terms, the disruptive transition from the *Dasein* to the becoming animal can be understood within the framework of a crucial transition from metaphysics and ethics (i.e., theology) to ethology. Ethology is intended to focus on the connections between animals and environments. Whereas ethics always implies a hierarchy – right or wrong, high or low, transcendence or immanence – ethology is completely "flat." It remains at ground level, where the rat quickly moves its own legs. In Deleuze's words, "such studies [...] which define bodies, animals, or humans by the affects that they are capable of, founded what is today called *ethology*" (Deleuze 1988, 125). When Deleuze speaks of ethology, then, he is in actual terms referring to ontology "as the study of affects" (Buchanan 2008, 155; Vignola 2013). Ontology is here an ethology inasmuch as it is "a practical science of the modes of being. The mode of being is notably the status of beings, existing things, from the point of view of pure ontology" (Deleuze 1980, my transl.).

The transition from metaphysics to ethology implies a radical conceptual shift; while the *Dasein* approach envisages a more or less "classical" subject facing a world made of things and animals, Deleuze and Guattari's ethological world is simply made of "bodies" (Widder 2011; Gardner and McCormack 2017). When the two authors speak of bodies, they do not take them as the physical counterpart of the subject (Esposito 2015). Quite to the contrary, a body is nothing but the "power" of being affected: "affections (*affectio*) are the modes themselves. The modes are the affections of substance or of its attributes. [...] These affections are necessarily active" (Deleuze 1988, 48). What is worth remarking here is that in the ethological world the metaphysical distinction between human and animal – or artificial and natural – does no longer apply. In such a world, only bodies exist, and each body is identified by "an ethological chart of affects" (Deleuze 1978, 9, my transl.). A body is nothing but "its" own affections. Since it stands for a change in the affected body, an affection is strictly speaking neither active nor passive. However, one could point out that the active/passive distinction is a metaphysical one, inasmuch as it implies a hierarchy, where the active is superior to the passive. Therefore, when Deleuze says that affections are active, one should take it simply as an attempt to deactivate such a metaphysical couple. If all affections are somewhat active, there is indeed no room to distinguish between active and passive. This is all the clearer as soon as we see this ontology based only on bodies capable of affections as an extreme generalization of Uexküll's approach to the biological world. Deleuze's ontology is ultimately a "biologization" of the entire world. Accordingly, bodies are not bare things, that is, individuated entities placed in some point of the space-time; on the contrary, each body is an "individuation intensive" (Bergen 2001, 587). As is well known, in *A Thousand Plateaus*, Deleuze and Guattari define such an "individuation intensive" in terms of "block(s) of becoming" (Deleuze and Guattari 1987, 238), where the focus shifts on "the becoming itself [...] not the supposedly fixed terms through which that which becomes passes" (Deleuze and Guattari 1987, 238).

Breaking free from the separation of subjects and objects, Deleuze and Guattari's ethological world is a stratified *agencement* (often translated in English as *assemblage*)[2] of bodies and affections. To be precise, since bodies are nothing but affections and affections are changes, the world is made of changes. And changes are understood by the two authors within the framework of a unitary "plane of consistency" (Deleuze and Guattari 1994, 24). Interestingly enough, one of the main sources for such an idea can be traced back to Uexküll's theory of the functional circle. According to Uexküll,

> [e]very animal is a subject, which, in virtue of the structure peculiar to it, selects stimuli from the general influences of the world, and to these it responds in a certain way. These responses, in their turn, consist of certain effects on the outer world, and these again influence the stimuli. In this way there arises a self-contained periodic cycle, [...] the *functional circle* of the animal.
>
> (Uexküll 1926, 126, transl. slightly modified)

Although Uexküll seems to rely on a Kantian conceptual apparatus (see Esposito's contribution, Chapter 2, in this volume) – when, for instance, he takes the animal as a "subject" which "selects stimuli from the world" – according to Deleuze and Guattari such a functional circle successfully deactivates the long-established metaphysical couple subject–object. It is clear, in fact, that the functional circle does not feature a subject on one side and an object on the other. The functional circle is a unitary cycle, and biology is nothing but a stratified *agencement* of functional circles. These latter are, as we have already seen above, "the becoming itself [...] not the supposedly fixed terms through which that which becomes passes" (Deleuze and Guattari 1987, 238). In any functional circle it is indeed impossible to separate what pertains to the organism and what pertains to the environment. Deleuze and Guattari thus radicalize Uexküll's approach to find a way out from the misrepresentation of the animal as a separate entity from the environment. The only real onto-ethological presence is therefore the functional circle. Within such an ontology, a body, whatever it may be (natural or artificial), is nothing but the connections it partakes in. A body does not exist outside such connections: "[W]hat's a body? You never know in advance what a body is capable of. You never know how someone's modes of existence are organized and incorporated" (Deleuze 1980, my transl.).

In this regard, the biologization of ontology means that any entity of the world is a body that is no longer taken into account as an "object" of a certain type nor as a "thing" that belongs to a specific abstract category. The body, each body, is nothing but its power to take part in connections with other bodies. In such an ontology, the ancient notion of "essence" no longer plays any role. The ontology in the form of ethology is not made of essences, but it is made of functional circles that, in turn, participate in other functional circles, and so on.

3 Bodies as passions

Based on what has been argued so far, following Deleuze and Guattari's reworking of Uexküll's ideas, a body is made of its power to take part in functional circles. Such power, in turn, implies the capacity of each body to be affected – whether actively or passively it does not make much difference, inasmuch as both forms belong to bodily movements – by other bodies. The capacity of being affected, finally, means that a body is nothing but its own "passions." What is at stake here is "not what the thing is, but what it is capable of withstanding and do" (Deleuze 1980, my transl.). Affections therefore are passions. It should be remarked that, according to Deleuze and Guattari, passions are neither mental nor corporeal entities. In the onto-ethological Deleuzian world, such a venerable theological distinction no longer applies. Passions are just the peculiar ways of being of a body. Further exemplification of this point comes once more from the animal world. As Deleuze writes,

[a] fish can only what the neighbor fish can. Therefore there will be an infinite differentiation of the quantity of power according to existing things. Things can be differentiated inasmuch as they are correlated to a scale of power.

(Deleuze 1980, my transl.)

What really exists is this particular fish, which is only individuated by its passions (Beaulieu 2011). It is tempting to claim that the transformation of ontology in ethology transforms things into passions. But what is really striking is to find an anticipation of this innovative approach to ontology in Uexküll's claims concerning "qualities" as the "ultimate biological elements" (Uexküll 1926, 87). Uexküll explains clearly that a "quality" is a property neither of the perceiving subject nor of the perceived object. In fact, in every living body, he claims, while "constructing the world, mental sensations become properties of things; or, in other words, the subjective qualities build up the objective world" (Uexküll 1926, 77).[3] What a "quality" is for Uexküll closely corresponds to Deleuze's topical use of the word "affections." On this very point rests, moreover, the transformation of ontology into ethology. The world is not made of objects "and" relations, but on the contrary, the world is "nothing but" relations. This could be the Deleuzian way to interpret Uexküll's concept of *Umwelt*: an *agencement* which is nothing but relations, passions, affections, qualities. This is how Uexküll presents his perspectival achievement:

> Effect world [*Wirkwelt*] and perception world [*Merkwelt*] together make a comprehensive whole, [...] the *Umwelt*. The entire functional circle formed from inner world and *Umwelt* (the latter divisible into effect world and perception world) constitutes a whole [...] for each part belongs to the others.
> (Uexküll 1926, 127, transl. slightly modified)

Further generalization of the concept of *Umwelt* is achieved by Deleuze and Guattari by reading it through the conceptual lenses provided in particular by Bergson and Spinoza (Grosz 2007). Clearly, Deleuze and Guattari do not simply read Uexküll as a biologist, but they fully grasp the wider ontological implications of his approach. According to them, biology is much more than the science of living organisms. What Deleuze and Guattari see in biology is a way to redefine ontology as a generalized ethology:

> Long after Spinoza, biologists and naturalists will try to describe animal worlds defined by affects and capacities for affecting and being affected. For example, J. von Uexküll will do this for the tick, an animal that sucks the blood of mammals. [...] Such studies as this, which define bodies, animals, or humans by the affects they are capable of, founded what is today called *ethology*. The approach is no less valid for us, for human beings, than for animals, because no one knows ahead of time the affects one is capable of.
> (Deleuze 1988, 124f.)

The famous case of the tick is particularly compelling (even if Heidegger in the *Fundamental Concepts of Metaphysics* explicitly uses the case of the bee), as it allows us also to clearly differentiate between Heidegger's interpretation of Uexküll's biology (Chien 2006; Elden 2006; Goetz 2007; Buchanan 2008; Romano 2009; Brentari 2015, 198–204) and Deleuze and Guattari's otherwise directed understanding of it (Brentari 2015, 215ff.). Heidegger, as previously anticipated, relies indeed on Uexküll's theory to neatly separate the "animal" from the human being. As is well known, according to Heidegger, the "animal" lives in a condition of "poverty in world" (Heidegger 1995, 268; see Michelini, Chapter 7, in this volume).

He is also adamant that "the whole of Uexküll's approach does become philosophically problematic if we proceed to talk about the human world in the same manner" (Heidegger 1995, 263) as we talk about the animal environment. The concept of *Umwelt*, according to Heidegger applies only to animals, and human ontology has nothing to do with ethology. He can therefore conclude that "the animal is separated from man by an abyss" (Heidegger 1995, 264). Such a persistent dualism marks the difference between Heidegger's and Deleuze and Guattari's interpretation of Uexküll. Whereas Deleuze and Guattari find in Uexküll a way to biologize ontology, Heidegger objectivizes and de-animalizes the human environment.

Deleuze and Guattari generalize the case of the tick. The difference between the tick and humans, they argue, is not a difference based on their respective essences. It is rather a difference based on the capacity of passions only. Uexküll's inquiries, in this regard, highlights a world made of bodies *qua* passions, a world made of intensities:

> Ethology is first of all the study of the relations of speed and slowness, of the capacities for affecting and being affected that characterize each thing. For each thing these relations and capacities have an amplitude, thresholds (maximum and minimum), and variations or transformations that are peculiar to them. And they select, in the world or in Nature, that which corresponds to the thing; that is, they select what affects or is affected by the thing, what moves it or is moved by it. For example, given an animal, what is this animal unaffected by in the infinite world? What does it react to positively or negatively? What are its nutriments and its poisons? What does it "take" in its world? Every point has its counterpoints: the plant and the rain, the spider and the fly. So an animal, a thing, is never separable from its relations with the world. The interior is only a selected exterior, and the exterior, a projected interior.
>
> (Deleuze 1988, 125)

4 Onto-ethology

A key aspect of any functional circle is that it does not exist in isolation. That is, any functional circle is nested into many other functional circles. Take again the case of the tick. According to Deleuze, one of the first passions of the tick is to reach a warm source of blood where to find the food it needs to survive and reproduce itself. This passion, in turn, releases another passion, because, as Deleuze continues, "the tick abundant blood is also her last meal. Now there is nothing left for her to do but to drop to earth, lay her eggs and die" (Uexküll

1957, 7). This is the (temporary) end of the tick's life; however, this is not the end of the functional circle it partakes in. The tick, as a body of passions, exists only as one passage in a wider functional circle. According to this perspective, what is onto-ethologically real is only the functional circle. The world itself is nothing but such an indefinite circle of circles and so on. It is important to stress that these circles do not form a unitary hierarchy. There is not such a thing like a transcendent super-circle where all the other circles are harmoniously included. Accordingly, Deleuze and Guattari's onto-ethology does not rely on any form of transcendence. The world is just an endless circle of circles. On the other hand, as the tick's case openly shows, any functional circle identifies itself in respect to many other circles. In this sense, the onto-ethology is flat, but it is not internally undifferentiated. Along these same lines, Uexküll can consistently maintain what at first seems an outdated metaphysical position: "every living creature, is, in principle, absolutely perfect" (Uexküll 1926, 164). Careful understanding of this sentence is required. Everything is "perfect" because any body, even though it is but a phase of a nested circle of circles, nevertheless "exists" as a distinctive absolutely unique power of being affected. A similar point is made by Deleuze when he maintains that "the essence is the power" (Deleuze 1980) of – so-called active and passive – affections:

> We know nothing about a body until we know what it can do, in other words, what its affects are, how they can or cannot enter into composition with other affects, with the affects of another body, either to destroy that body or to be destroyed by it, either to exchange actions and passions with it or to join with it in composing a more powerful body.
>
> (Deleuze and Guattari 1987, 257)

Further elucidatory effects ensue from identifying the Uexküllian elements of Deleuze's "Spinozist" monism. What is notably at stake is how to introduce differences into the full immanence of the world *qua* infinitely nested circle of circles:

> To begin with, a stratum does indeed have a unity of composition, which is what allows it to be called *a* stratum: molecular materials, substantial elements, and formal relations or traits. Materials are not the same as the unformed matter of the plane of consistency; they are already stratified, and come from 'substrata'. But of course substrata should not be thought of only as substrata: in particular, their organization is no less complex than, nor is it inferior to, that of the strata; we should be on our guard against any kind of ridiculous cosmic evolutionism.
>
> (Deleuze and Guattari 1987, 49)

As a result, in Deleuze and Guattari, the explanatory mechanism of the functional circle becomes much more than a simple biological concept; it becomes the main "onto(-etho)logical" concept. The point worth stressing is the very idea of a circle as "a single *abstract machine* that is enveloped by the stratum and constitutes its unity" (Deleuze and Guattari 1987, 50). However, such an "abstract machine" should not be mistaken for a super-circle, which encompasses all other circles. Such a machine does not exist separately from its own functioning; as there are no such entities as the subject separated from the object, or an animal separated from its own *Umwelt*, similarly the machine is nothing but the circles it is actually circling. No separation can be traced between the machine, on one side, and its "effects," on the other. Such an "abstract" machine is like the tick, it coincides with its passions, that is, with its own "actions." In order to avoid introducing any form of transcendence into their onto-ethology, Deleuze and Guattari even give up on the unitary concept of "world," just as Uexküll teaches them to. There is not such an entity as "the world," inasmuch as such a unitary concept would entail an external entity who can think of it as "the world." In other words, a "world" can exist, strictly speaking, only for a subject to set against it. Differently, what Deleuze and Guattari claim is that

[i]t would be a mistake to believe that it is possible to isolate this unitary, central layer of the stratum, or to grasp it in itself, by regression. In the first place, a stratum necessarily goes from layer to layer, and from the very beginning. It already has several layers. It goes from a center to a periphery, at the same time as the periphery reacts back upon the center to form a new center in relation to a new periphery. Flows constantly radiate outward, then turn back.

(Deleuze and Guattari 1987, 50)

Probable links to Uexküll's (and Bergson's; see Grosz 2007) work can also be found in Deleuze and Guattari's references to the idea that nature, like the circle of circles, is, in fact, the continuous production of new circles. In this regard, Uexküll points out that

[a]ccording to the common linguistic use, the concept 'plan' does not connect itself to any activity. In this sense, the plan [*Plan*] is something inactive and inefficacious. On the contrary, the planes [*Pläne*] which govern the living beings, are by their own essence active and efficacious.

(Uexküll 1928, 205, my transl.)

Uexküll then concludes that the "natural plan [*Naturplan*] emerges like a maker of forms [*Formbildner*]" (Uexküll 1928, 205, my transl.). Nature is but such a production; that is, there is not such an entity as "nature" apart from the actual stratified functional circles:

The development of the associated milieus culminates in the animal worlds described by von Uexküll, with all their active, perceptive, and energetic characteristics. [...] Active and perceptive characteristics are themselves something of a double pincer, a double articulation.

(Deleuze and Guattari 1987, 51)

5 Nature as music

One last reference to Uexküll's work can be detected in Deleuze and Guattari's onto-ethological use of the concept of "biological meaning." A thus understood "meaning" is not properly speaking a "semiotic" or "mental" meaning. In fact, the biological meaning is neither subjective nor objective, neither in the mind nor in the world. It cannot even be any of the two. Sensations are not internal signs of the objects of the world. However, functional circles are not simply subjective, as the tick tells us. The tick does not choose to wait or not to wait on the tree to feel the butyric acid secreted by the skin of a mammal. The tick is extremely fond of such a scent. It is not the case that the tick falls over the fur of the dog, as if such a falling would not intimately concern it. The tick falls onto the fur in a similar way as someone falls in love with somebody else. Therefore, it should be concluded that the biological "meaning" is not a property arbitrarily attributed to an entity. Nevertheless, its not being subjective does not imply being simply objective. This is but another corollary of the concept of functional circle. Uexküll elucidates such a point with the example of a dog fetching a stone: "the stone, which lies as a relationless object [...] becomes a carrier of meaning as soon as it enters into a relationship with a subject" (Uexküll 2010, 140). That the stone "means" a certain "meaning" for the dog does not imply that such a meaning has been subjectively and arbitrarily attached to the stone by the dog. The dog does not "interpret" the stone in order to find its "meaning." The stone is by itself its own meaning-for-the-dog. This is a very important point, which is frequently missed, especially in the semiotic interpretation of Uexküll's work (see, e.g., Augustyn 2009). Take again the case of the relationship between the stone and the dog. Assuming that the key concept is that of the functional circle, there is not a principled reason not to take into account the "point of view" of the stone alongside that of the dog. The stone takes part in the functional circle dog–stone as much as the animal does. This radical move achieved by Deleuze and Guattari place them a step further compared to Uexküll. Whereas Uexküll gives us a biological theory of life, Deleuze and Guattari develop it and transform it into an onto-ethological theory of reality. In other words, they extend Uexküll's theory of biological "meaning" to the entire reality, be it natural or artificial. In this sense, they chase Uexküll's original goal even more faithfully than him. It is true, in fact, that Uexküll still assigns a privileged position to the living being. He takes into account the functional circle dog–stone but he does not even consider the possibility of a functional circle stone–dog. Admittedly, Uexküll limits his own theory only to the "classical" biological world. Conversely, Deleuze and Guattari transform Uexküll's biology into a biology-based ontology, in the form of a general ethology. For them the extreme and basic metaphysical distinction between living and nonliving beings no longer applies. For this reason, Uexküll's theory of "meaning" should not be understood in a strict "semiotic" sense (Brentari 2013), inasmuch as any semiotics is intrinsically dualistic in nature. A sign can only exist if it does not coincide with what it refers to. Semiotics stands then for a separation between signifier and meaning, signal and referent, expression and content.

On the contrary, Uexküll's theory of "meaning" is all but dualistic (Kull 1999). Uexküll theorizes notably the coinciding of meaning and meaning-carrier in the following terms:

> Through every [animal F.C.] relationship the neutral object [which is neutral only from a human external point of view F.C.] is transformed into a meaning-carrier, the meaning of which is imprinted upon it by a subject.
>
> (Uexküll 1982, 27f.)

What's most important, moreover, is that such a transformation is not at all arbitrary; the *Umwelt* of an animal is only the external side of the *Innenwelt*, and vice versa (Uexküll 1909). The *Umwelt* is a sign of the *Innenwelt* as much as the latter is a sign of the former. And this exactly proves that such a sign is not semiotic at all, because it is neither arbitrary (i.e., symbolic) nor natural (i.e., iconic). In short, the *Umwelt* is an external transformation and continuation of the *Innenwelt*, as the *Innenwelt* is an internal transformation and continuation of the *Umwelt*. Whether a radical interpretation of Uexküll's theories, such as that promoted by Deleuze and Guattari, could effectively contribute to push Uexküll's theoretical contributions beyond their Kantian framework remains an open question; more certain is instead that implicit in Uexküll one can find a radically antidualistic and antisubjectivist line of thinking, of which Uexküll himself might not even have been fully aware. On these premises, Deleuze and Guattari's theory can be used, very radically, as a way to construe a nonhumanistic and nondualistic ontology, that is, an ontology that coincides with an ethology. Such an ontology is made of passions more than bare things. This is the case of a "circle of circles," where always new *agencements* between passions of all sorts are established: animal, rocks, humans – an etho-ontology made of relations, not of essences. The key concept in this respect is that of "transcoding," understood as the continuous passage of a body into another body:

> One case of transcoding is particularly important: when a code is not content to take or receive components that are coded differently, and instead takes or receives fragments of a different code as such. The first case pertains to the leaf-water relation, the second to the spider-fly relation. It has often been noted that the spider web implies that there are sequences of the fly's own code in the spider's code; it is as though the spider had a fly in its head, a fly "motif," a fly "refrain." The implication may be reciprocal, as with the wasp and the orchid, or the snapdragon and the bumblebee. Jakob von Uexküll has elaborated an admirable theory of transcodings. He sees the components as melodies in counterpoint, each of which serves as a motif for another: Nature as music. Whenever there is transcoding, we can be sure that there is not a simple addition, but the constitution of a new plane, as of a surplus value. A melodic or rhythmic plane, surplus value of passage or bridging.
>
> (Deleuze and Guattari 1987, 314)

One can finally consider the formula "nature as music" as the arrival point of Deleuze and Guattari's radical reading of Uexküll's biological theory. Everything starts with a functional circle: all other entities – subject and object, animal and environment, genotype and phenotype – are only dead metaphysical abstractions. The circle, according to Uexküll, is analogous to a "musical composition" (Uexküll 2010, 171). Pursuing this line of thought, Deleuze and Guattari achieve even more radical results than those allowed, for instance, by Merleau-Ponty's reading of Uexküll (see Buchanan 2008; Umbelino 2013; see also Moyle's contribution, Chapter 8, in this volume), which takes language, rather than music, as reference "model" for natural life (Westling 2014). The kind of "music" in "nature as music" has, however, nothing to do with language and semiosis. Music means nothing but a circle of circles. Such a circle, Uexküll says,

> [i]s on the same level as every musical composition. [For example F.C.] the behavior of meaning factor in plants and of carriers of meaning in animals toward their meaning utilizers shows this especially clearly. As, in the composition of a duet, the two voices have to be composed for each other note for note, point for point, the meaning factors in Nature stand in a contrapuntal relation to the meaning utilizers. We will be closer to understanding the form development of living beings when we succeed in deriving a *composition theory of Nature from it*.
>
> (Uexküll 2010, 171, emphasis in the text)

Accordingly, for Deleuze and Guattari – as true heirs of Uexküll – "music is traversed by [...] becomings-animal, above all becomings-bird" (Deleuze and Guattari 1987, 272). Thus, ontology, qua music, becomes ethology, that is, composition – *agencements* – of passions (Heredia 2011). All in all, Deleuze and Guattari find in Uexküll the right tools to radically subvert Western metaphysics, notably applying the model of the functional circle to the whole of human and nonhuman life. As a result, human life is no longer separated from the rest of the natural world. By way of conclusion, one could say that, as Hume awakened Kant from his dogmatic sleep, Uexküll showed Deleuze and Guattari what a life – it does not matter whether human or nonhuman – really is. Moreover, Uexküll gave to Deleuze and Guattari the conceptual tools to imagine a life that extends beyond the boundaries of traditional biology. As a result, according to Deleuze and Guattari life does not simply mean cells and metabolism, life means passions and *agencements*. Such a life is ultimately beyond the metaphysical dualism of the living and of the nonliving (Herzogenrath 2008; Bennett 2010; Erlmann 2012; Roffe and Stark 2015; Cimatti 2018). To conclude, if Uexküll brought Deleuze and Guattari beyond the dualism of the subject and the object, somehow, they brought him beyond the dualism of the living and nonliving.

Notes

1 It is worth remarking that Heidegger repeatedly and erroneously speaks of "honey." As we know the bee actually searches the flowers for "nectar," which will be later transformed in honey. This is yet another proof that the philosophical animal does not at all

coincide with the "true" animal, *whatever* that may be. Heidegger seems to apply what only holds in an experimental situation (Uexküll 1926, 169) to the normality of the bee's life.

2 Although unsatisfactory, this translation is nowadays widely in use (Nail 2017).

3 Such a position is very close to Gibson's theory of "ecological perception" (Gibson 1966, 1979; Cimatti 1995).

References

Augustyn, Prisca (2009) 'Uexküll, Peirce, and other affinities between biosemiotics and biolinguistics'. *Biosemiotics* 2 (1), 1–17.

Beaulieu, Alain (2011) 'The status of animality in Deleuze's thought'. *Journal for Critical Animal Studies* 9 (1/2), 69–88.

Bekoff, Marc, Allen, Colin, and Burghardt, Gordon (eds.) (2002) *The Cognitive Animal: Empirical and Theoretical Perspectives on Animal Cognition*. Boston: The MIT Press.

Bennett, Jane (2010) *Vibrant Matter: A Political Ecology of Things*. Durham: Duke University Press.

Bergen, Véronique (2001) *L'ontologie de Gilles Deleuze*. Paris: Harmattan.

Brentari, Carlo (2013) 'How to make worlds with signs: Some remarks on Jakob von Uexküll's Umwelt theory'. *Rivista Italiana di Filosofia del Linguaggio* 7 (2), 8–21.

Brentari, Carlo (2015) *Jakob von Uexküll: The Discovery of the Umwelt between Biosemiotics and Theoretical Biology*. Dordrecht/Heidelberg/New York/London: Springer.

Buchanan, Brett (2008) *Onto-Ethologies: The Animal Environments of Uexküll, Heidegger, Merleau-Ponty, and Deleuze*. Albany/New York: SUNY Press.

Chien, Jui-Pi (2006) 'Of animals and men: A study of Umwelt in Uexküll, Cassirer, and Heidegger'. *Concentric: Literary and Cultural Studies* 32 (1), 57–79.

Cimatti, Felice (1995) 'La mente come risuonatore'. *Sistemi intelligenti* 7 (2), 263–286.

Cimatti, Felice (2013) *Filosofia dell'animalità*. Roma/Bari: Laterza.

Cimatti, Felice (2018) *Cose. Per una filosofia del reale*. Torino: Bollati Boringhieri.

Cimatti, Felice and Vallortigara, Giorgio (2015) 'So little mind, so much brain: Intelligence and behavior in non human animals'. *Italian Journal of Cognitive Sciences* 1, 9–24.

Crystal, Jonathon and Foote, Allison (2009) 'Metacognition in animals'. *Comparative Cognition & Behavior Reviews* 4, 1–16.

Fontenay, Élisabeth de (1998) *Le silence des bêtes: La philosophie à l'épreuve de l'animalité*. Paris: Fayard.

DeLanda, Manuel (2009) 'Ecology and realist ontology'. In: Bernd Herzogenrath (ed.) *Deleuze/Guattari & Ecology*. London: Palgrave Macmillan, 23–41.

Deleuze, Gilles (1978) 'Spinoza'. *Les cours de Gilles Deleuze*, www.webdeleuze.com/textes/14.

Deleuze, Gilles (1980) 'Ontologie-Ethique'. *Les cours de Gilles Deleuze*, www.webdeleuze.com/textes/26.

Deleuze, Gilles (1988) [1970] *Spinoza: Practical Philosophy*. Translated by Robert Hurley. San Francisco: City Lights Books.

Deleuze, Gilles and Guattari, Félix (1987) [1980] *A Thousand Plateaus: On Capitalism and Schizophrenia*. Translated by Brian Massumi. Minneapolis: University of Minnesota Press.

Deleuze, Gilles and Guattari, Félix (1994) [1991] *What Is Philosophy?* Translated by Hugh Tomlinson and Graham Burche. New York: Columbia University Press.

Derrida, Jacques (2008) [2006] *The Animal That Therefore I Am*. Translated by David Wills. New York: Fordham University Press.

Descartes, René (2000) [1637] 'Discourse on Method'. In: René Descartes. *Philosophical Essays and Correspondence*. Translated by Roger Ariew. Indianapolis: Hackett Publishing Company, 46–82.

Elden, Stuart (2006) 'Heidegger's animals'. *Continental Philosophy Review* 39 (3), 273–291.

Erlmann, Veit (2012) 'Klang, Raum und Umwelt: Jakob von Uexkülls Musiktheorie des Lebens'. *Zeitschrift für Semiotik* 34 (1–2), 145–157.

Esposito, Roberto (2015) [2014] *Persons and Things: From the Body's Point of View*. Translated by Zakiya Hanafi. Cambridge: Polity Press.

Gardner, Colin and McCormack, Patricia (eds.) (2017) *Deleuze and the Animal*. Edinburgh: Edinburgh University Press.

Gibson, James (1966) *The Senses Considered as Perceptual Systems*. Boston: Houghton Mifflin.

Gibson, James (1979) *The Ecological Approach to Visual Perception*. Boston: Houghton Mifflin.

Goetz, Benoît (2007) 'L'araignée, le lézard et la tique: Deleuze et Heidegger lecteurs de Uexküll'. *Le Portique. Revue de philosophie et de sciences humaine* 20, 1–15.

Griffin, Donald (1976) *The Question of Animal Awareness*. New York: Rockefeller University Press.

Grosz, Elizabeth (2007) 'Deleuze, Bergson and the concept of life'. *Revue internationale de philosophie* 241 (3), 287–300.

Hauser, Marc (1999) *The Design of Animal Communication*. Boston: The MIT Press.

Heidegger, Martin (1962) [1927] *Being and Time*. Translated by John Macquarrie and Edward Robinson. Oxford: Basic Blackwell.

Heidegger, Martin (1995) [1929/1930] *The Fundamental Concepts of Metaphysics: World, Finitude, Solitude*. Translated by William McNeill and Nicholas Walker. Bloomington/Indianapolis: Indiana University Press.

Heidegger, Martin (1998) 'Letter on "Humanismus"'. In: Martin Heidegger. *Pathmarks*. Translated by William McNeill. Cambridge UK: Cambridge University Press.

Heredia, Juan Manuel (2011) 'Deleuze, von Uexküll y "la naturaleza como música"'. *A Parte Rei. Revista de Filosofia* 75, 1–8.

Herzogenrath, Bernd (ed.) (2008) *An (Un)Likely Alliance: Thinking Environment(s) with Deleuze/Guattari*. Cambridge: Cambridge Scholars Publishing.

Keekok, Lee (2006) *Zoos: A Philosophical Tour*. New York: Palgrave Macmillan.

Kull, Kalevi (1999) 'Biosemiotics in the twentieth century: A view from biology'. *Semiotica* 127 (1/4), 385–414.

Nail, Thomas (2017) 'What is an assemblage?'. *SubStance* 46 (1), 21–37.

Nance, Susan (ed.) (2015) *The Historical Animal*. Syracuse, NY: Syracuse University Press.

Roffe, Jon and Stark, Hannah (eds.) (2015) *Deleuze and the Non/Human*. London: Palgrave Macmillan.

Romano, Claude (2009) 'Le monde animal: Heidegger et von Uexküll'. In: Servanne Jollivet and Claude Romano (eds.) *Heidegger en dialogue 1912–1930. Rencontres, affinités, confrontations*. Paris: Vrin, 255–298.

Uexküll, Jakob von (1909) *Umwelt und Innenwelt der Tiere*. Berlin: Springer.

Uexküll, Jakob von (1926) [1920] *Theoretical Biology*. Translated by Doris L. Mackinnon. London: K. Paul, Trench, Trubner & Co.; New York: Harcourt, Brace & Company.

Uexküll, Jakob von (1928) *Theoretische Biologie* (2nd ed.). Berlin: Springer.

Uexküll, Jakob von (1957) [1934] 'A Stroll through the Worlds of Animals and Men'. In: Claire H. Schiller (ed.) *Instinctive Behavior: The Development of a Modern Concept*. Translated by Claire H. Schiller. New York: International Universities Press, 5–80.

Uexküll, Jakob von (1982) [1940] 'The theory of meaning'. *Semiotica* 42 (1), 25–82.

Uexküll, Jakob von (2010) [1934, 1940] *A Foray into the Worlds of Animals and Humans with a Theory of Meaning*. Translated by Joseph D. O'Neil. Minneapolis/London: University of Minneapolis Press.

Umbelino, Luís António (2013) 'The melody of life: Merleau-Ponty, reader of Uexküll'. *Investigaciones Fenomenológicas* 4 (1), 351–360.

Vignola, Paolo (2013) 'Divenire-animale. La teoria degli affetti di Gilles Deleuze fra etica ed etologia'. In: Matteo Andreozzi, Silvana Castignone, and Alma Massaro (eds.) *Emotività animali*. Milano: Edizioni Universitarie di Lettere Economia Diritto, 117–124.

Westling, Luis (2014) *The Logos of the Living World: Merleau-Ponty, Animals, and Language*. New York: Fordham University.

Widder, Nathan (2011) 'Matter as simulacrum: Thought as phantasm; Body as event'. In: Laura Guillaume and Joe Hughes (eds.) *Deleuze and the Body*. Edinburgh: Edinburgh University Press, 96–114.

Wynne, Clive (2004) *Do Animals Think?* Princeton, NJ: Princeton University Press.

11 Hans Blumenberg

The transformation of Uexküll's bioepistemology into phenomenology[1]

Cornelius Borck

Hans Blumenberg explicitly discusses Jakob von Uexküll's work only in his late book, *Lifetime and Worldtime* [*Lebenszeit und Weltzeit*, Blumenberg 1986], a phenomenological exploration of the existential tension between the short individual life span and the enormous temporal dimensions of the universe.[2] In the respective chapter, Blumenberg juxtaposes Uexküll's notion of *Umwelt* and Karl Ernst von Baer's much earlier, speculative elaborations on the dependency of time relations (and thereby of the world's objectivity) on sensory experience and the timing of perception. For Blumenberg, Uexküll delivers, based on biological observations, a significant notion, but fails to develop it properly, especially when compared with the term life-world [*Lebenswelt*], phenomenology's key concept, coined at about the same time by Edmund Husserl. Although Blumenberg ridicules the *Umwelt* thinker in this chapter, Uexküll must nonetheless be counted among Blumenberg's important intellectual sources. Uexküll influenced Blumenberg both directly through the impact of his concept of *Umwelt*, and indirectly via Uexküll's reception by Ernst Cassirer and the German strand of philosophy of anthropology, especially Helmuth Plessner, Max Scheler, and Arnold Gehlen, all studied closely and extensively by Blumenberg.[3] With the posthumous publication of Blumenberg's two abandoned book projects, *Description of Man* [*Beschreibung des Menschen*] and *Theory of the Life-World* [*Theorie der Lebenswelt*], this indirect line has become more visible.[4] In both of these texts, Blumenberg develops anthropology as a phenomenological project at variance with the philosophical anthropology so prominent in Germany. He thereby interacts with Uexküll's theory of *Umwelt* to the aim of transforming bioepistemology into phenomenology, ultimately providing an exemplary case for how theoretical biology and scientific theorizing can intervene in phenomenology and transcendental philosophy. Embracing empirical data as significant evidence but siding with philosophy and its transcendental questions, Blumenberg arrives at a form of radical skepticism that questions contemporary conditions of thinking.

1 Two great Baltic biologists

In *Lifetime and Worldtime* (1986), Blumenberg explores the existential tensions between the short individual life span and the enormous spatiotemporal dimensions of the universe, further and further expanded by science, as characteristic of the modern condition. Uexküll is discussed here in a chapter that delivers a carefully crafted and highly condensed discussion of insights from biological research about the timing of perception and their pertinence for clarifying consciousness and the anthropological condition. Perception is obviously coupled, by its own timing, to a specific, temporally ordered form of consciousness. Since Friedrich Bessel's discovery in 1820 of the "personal equation," the individual differences in the precise recording of simultaneous events, biologists, physiologists, and psychologists debated at length on how the internal organization of the sensory apparatus shapes the perception of events and objects in space and time.[5] Philosophically speaking, this was part of the project to naturalize Kant's critical philosophy – and Uexküll positioned his work still in this tradition.[6] For Blumenberg, such a project is ill founded and wrongheaded, as it takes scientific objectivity as the foundation for consciousness instead of recognizing its conceptual dependence on human consciousness and philosophical reflection – as proved by Husserl's fundamental insight with the transcendental reduction. At the same time, however, Blumenberg acknowledges the relevance of biological theorizing for dealing with such an issue adequately, as well as for pursuing the phenomenological project beyond Husserl's limitations.

According to Blumenberg, two "great Baltic biologists" elucidate the biological side of the philosophical problem, Karl Ernst von Baer in a series of thought-provoking speculations, and Jakob von Uexküll in distilling the concept of the perceptual "moment" out of von Baer's reflections and providing it with experimental determination. Nevertheless, Uexküll earns Blumenberg's sharp criticism for judging his measurements as a sound basis for natural philosophy. Combining condensed philosophical argumentation with rhetor craftsmanship and weaving together the latest research and eternal philosophical problems, the chapter showcases Blumenberg's magisterial style of writing. This should not be mistaken for poetic license but should be linked to Blumenberg's decisive move and rhetorical strategy for addressing the issues at stake in linguistically appropriate ways, mobilizing language's nuances and differentiality against the shortcomings of schematized thinking in black-boxed concepts (Blumenberg 1987).[7] The chapter is hence difficult to read and its main intervention easy to miss. In fact, Blumenberg mobilizes here insights from biological research against Husserl's phenomenology to the aim of saving phenomenology as a philosophical project. However, in order to arrive at the subtlety of Blumenberg's insights, one has to connect dispersed sentences and to dig beneath the surface of the text.

The reason why Uexküll is introduced in this chapter (and thereby in Blumenberg's publications) as "the other great Baltic biologist," is that the primary role in the chapter's discussion is reserved to Karl Ernst von Baer, "the first great Baltic biologist" from the village of Piep in Estonia, professor of zoology and anatomy in Königsberg (today Kaliningrad) and later in St. Petersburg. Writing in a period when comparative anatomy and zoology were still part of natural history rather than branches of experimental science, von Baer pushes biology forward, according to Blumenberg's introductory remarks in the chapter, by "forming the appropriate concepts necessary for starting empirical investigations" (Blumenberg 1986, 282). Referring to von Baer's famous thought experiments in his opening address of 1860 to the Russian Entomological Society, founded by von Baer himself, Blumenberg presents what is also Uexküll's main source for the analysis of time in his *Theoretical Biology*. According to Blumenberg, von Baer represents the theoretician who comes first in every respect, and he is contrasted with Uexküll, who comes second as only confirming experimentally the species-specific constancy of the perceptual moment postulated by von Baer but who falls short of unfolding the philosophical implications of his theory.

On the surface of the text, it seems that Blumenberg, instead of engaging properly and critically with Uexküll's core concepts, *Umwelt* and functional circle, chooses unfairly to comment on an odd metaphor fashioned in one of Uexküll's smaller publications. According to his *Theoretical Biology*, Uexküll's main work, organic life is bound to its milieu and interconnected with it in many ways, but each organism lives strictly in its own world-as-perceived and world-as-acted-upon, with the *Merkwelt* [perception world] and *Wirkwelt* [effect world] forming together the *Umwelt* of each organism – leaving only to human beings the ability to recognize this specificity and to construct an objective world. The *Naturwissenschaften* paper, from which Blumenberg retrieves a misguided metaphor to be exploited by his critique, was a summary of Uexküll's biological theory for a general audience. There Uexküll explained his concept of the "perception world" [*Merkwelt*] by describing how perception cuts out at each moment disk-like units of the world that for the individual organism connect to strings, existing separately and (largely) independently of other such strings in the world, because of the species-specific differences in the sensorial apparatus. Inasmuch as these disc-like segments of the perception world connect over time to strings, according to Uexküll, they were hence to be described as a "tunnel":

> The temporal sequence of segments in the perception world can be visualized as a tunnel [*Merkweltentunnel*], if one conceives of every segment as a two-dimensional disk and forms a chain out of them. The thickness of each disk represents the duration of a moment. [...] We can expand the tunnel of the world-as-perceived to the environment-tunnel [*Umwelttunnel*] by taking on the effect world, because we know how closely the perception world and the effect world are tied to the same objects. [...] In this way, we get for the earthworm a life-tunnel which is comprised exclusively of earthworm-things and for the dragonfly a life-tunnel which comprises only dragonfly things.
>
> (Uexküll 1922, 300f., my transl.)

It is precisely the image of the tunnel that kindles Blumenberg's criticism, as it contradicts his idea of intersubjectivity as the basis of objectivity and consciousness. According to Blumenberg, consciousness and subjectivity emerge together from intersubjectivity and on this basis lead to the formation of an objective world. Uexküll's concept of perception world captures precisely the problem of the constitution of an objective world, but his analogy with a tunnel takes the explanatory efforts in the wrong direction, Blumenberg maintains, inasmuch as it misses the very conceptual nature of the perception world. This is why Blumenberg treats von Baer, the "first great Baltic biologist," more generously, while "the other great Baltic biologist" receives his ridiculing criticism, which pushes Uexküll's analogy to the absurd, namely to a "bundle of macaroni" (Blumenberg 1986, 285). Blumenberg's ludicrous discussion of Uexküll here, however, contrasts sharply with the honorable tone by which he introduces him earlier in the chapter as the experimentalist confirming von Baer's speculative insights.

Based on Blumenberg's philosophy and in line with the chapter's argument, the motivation for a strongly uneven treatment of "the two great Baltic biologists" may be summarized as follows: a naturalist of von Baer's stature may speculate and indulge in thought experiments, especially if he thereby advances disciplinary studies, and biologists shall experiment and systematize their empirical evidence, but as soon as biologists start to philosophize with the help of misguided metaphors, they earn the strongest possible verdict, especially by a metaphorologist of Blumenberg's stature and especially if fundamental philosophical concerns are at stake (cf. Haverkamp 2017). It would be worth it, then, not necessarily to argue in Uexküll's defense but, rather, to unearth the reasons for Blumenberg's harsh criticism and to elaborate more on his multiple reliance on Uexküll's insights here as elsewhere. In order to do so, a brief reconstruction of the chapter's main argument is in order.

2 Biological time and the timing of perception

Blumenberg's chapter revolves around the question concerning how the timing of perception and the inner temporal organization of the perceiving apparatus relate to the constitution of consciousness and of a perceived reality. Consciousness is structured by time, but its internal organization – as Blumenberg's argument has it – cannot be determined merely by the timing of perception, since consciousness constitutes units of perception that are already structured by meaning and concepts, that is to say, forms of order of a different origin than the signals perceived. The recognition of a melody, for example, seems to indicate an internal time consciousness with a different timing for acoustic stimuli compared to visual perception; such a "differentiation may be impossible with regard to the physiological background of the specific moment but its regulation by sense and meaning may still require it" (Blumenberg 1986, 288f.). This brief sentence gives the gist of Blumenberg's way of arguing. If physiological possibilities contradict philosophical requirements, as here the ready availability of forms and concepts, their epistemic functionality, on which ultimately all scientific explanations have to rely, proves that physiology cannot found philosophy. The apparent physiological impossibility, however, must guide the philosophical investigation on the formation of concepts and the phenomenological analysis of consciousness. The shortest moment of the sensorial apparatus, differing from species to species, differentiates perception from consciousness, as the latter must always already extend beyond individual incidents in order to avoid fragmentation and atomism – similarly, Husserl describes consciousness as the synthesis of retention and protention. "The sensorial moment and the experiential presence cannot be identical – Blumenberg writes – if experience of the world and the relating to the world-time in the context of the lifetime shall be possible" (Blumenberg 1986, 286). It would have to be a time point inside of the perceived moment where there is no further differentiation possible as its lower and upper limit escape perception.

In light of the earlier-outlined core point in Blumenberg's phenomenological analysis, the biological basis of the moment and its relation to consciousness and subjectivity become philosophically problematic, precisely because of, and thanks to, Uexküll's scientific work. Uexküll offers new insights for phenomenological reflection in at least three respects. First, he defines the term "moment" as the perceived segment of the world. Second, he conceptualizes this segment in terms of perception world as an always already internally structured world, not just a single stimulus or set of impressions. Each moment has *Gestalt* quality and is hence more than the sum of the perceived signals. In biological terms, the moment may be a measurable slice in time, but it is always already structured from within as meaningful; it is not an atom of perceptual time but the kernel out of which the objects of the world get constituted conceptually. And third, Uexküll clarifies the theoretical construct "simultaneity" as transgressing empirical evidence, since it characterizes what escapes experiential differentiation. The philosophical solution, however, that Uexküll develops out of these insights with the misleading analogy to a tunnel would immediately run into absurdity, as Blumenberg demonstrates playing ironically with the metaphor and exemplifying it as a bundle of macaroni. A proper philosophical reflection, Blumenberg would claim, must deal with biological insights in different ways, for instance, in terms of a phenomenological analysis. In order to follow Blumenberg's argument to its conclusions, one should have an overview also of his account on von Baer's thought experiments.

In his lecture from 1860, von Baer started with the observation that some organisms respond faster than others to stimuli and that some organisms become irresponsive to brief intervals stimuli which others are instead still able to differentiate. This perceptual timing inspired him a series of spectacular thought experiments in shifting time perceptions, decoupling human perception from the rhythm of the heart and magnifying or shrinking it by factors of tens, hundreds and even thousands. He thus speculates how the apparently objectively perceived world may, in fact, depend for its objectivity on the internal structuring of the sensorial apparatus. Von Baer starts with examples from zoology, but his thought experiments quickly leave the known world and extrapolate to the unknown, thus addressing the fundamental questions of how the timing of human perception is responsible for the constitution of a stable reality and how this reality must therefore be conceived as formatted by the internal organization of the sensory organs. Based on his biological knowledge, von Baer wonders whether the old philosophical dictum, the human be the measure of things, may critically be true in the sense that all things depend in their essence on the timing of human perception because of an inescapable anthropomorphism.

A human life accelerated by a factor of a thousand would last only about a month, but with a much speedier sensorial apparatus, this human being would perceive many things imperceptible to ordinary men. And, possibly, the person would interpret the shrinking of the moon as an apocalyptic event, claims Blumenberg quoting von Baer. Similar connections to mythology are particularly welcomed by Blumenberg as they confirm the existential needs that structure consciousness's constitution of a life-world out of perception (Nicholls 2015). A life span of a single day would link the perceived time to the rise and decline of the sun; animals and plants could hardly be recognized as living entities, lacking any perceptible change: "Only fellow human beings would appear as truly living – and all the more would he contemplate their likely ruin together with the declining sun," Blumenberg quotes from von Baer's wondering. All the more striking is the lecture's shifting of gears at this moment, when von Baer starts to question whether the accelerated sensory organs might work too fast for sensing acoustic signals, just like science had to come up with new methods for registering mysterious forces such as gravity or electricity. Here, von Baer comes close to Blumenberg's last big topic: Science is not the breaking of the world's spell and intellectual history, nor the path from myth to reason, but the endlessly ongoing *Work on Myth* (Blumenberg 1985).

The thought experiment in the opposite direction of an extremely prolonged life with equally extended perceptual moments would result in a drastic shrinking of experience, each single effect lasting hours or days – very much complicating any comprehension of the sun's rhythm in rising and declining. Another extension of the perceptual moment to a single pulse for an entire year would result in the earth being perceived in dynamic motion without any stability, without any chance for the constitution of a world of stable forms. No Platonism is possible under such an assumption, as Blumenberg comments on von Baer's speculation, and nature would be merely process – in other words, one human life would conflate with the time of the universe. The very idea of stable objects and of an objective space must be conceived of, in light of this thought experiment, as the product of the specificities of human consciousness, of its internal temporal organization. Because it introduces the concept of the perceptual moment and paves the way to questioning the very concept of nature and objectivity, Blumenberg follows von Baer's lecture at great length. With von Baer's final conclusion that time instead of space is the fundamental issue, the lecture turns into an essential step in the genealogy of phenomenology. Von Baer's speculative thought experiments mobilize imagination for "questioning experience from the largest possible distance" (Blumenberg 1986, 269).

Sixty years later, Uexküll took up von Baer's speculative exploration of biological time as a scientific problem. In his *Theoretical Biology*, he introduced "moment" as a theoretical term, defining it as the subjective timing in the perception of an objective world, and he made its experimental determination the research topic of his newly opened Institut für Umweltforschung in Hamburg – historically exactly when the "moment" became the operative principle of the cinematographic illusion of movement, as Blumenberg remarks (Blumenberg 1986,

286).[8] This theoretical clarification of "moment" out of Uexküll's notion of the world-as-perceived represents for Blumenberg a remarkable conceptual achievement, regardless of the harsh critique to follow just a few pages later. Similar praise for Uexküll is hidden in the text when Blumenberg describes the perceptual *Gestalt* in relation to the perceived *Umwelt* as the organism's conceptual craft – an ordering of the environment by imprinting on it a simplifying *Gestalt* that already anticipates the replacement of complex and manifold signals by a proper "thing." By using Uexküll's vocabulary and theorizing, Blumenberg honors him here for pinpointing the philosophically crucial step from impressions to ideas, from stimulus to concept:

> The necessity for *Gestalt* things contradicts the biological drive for constant metabolic turnover and the assimilation of matter through the entire series of naturally given forms. All pausing is deception originating from the grid-like timing of the sensory apparatus. [...] [The formation of concepts] corresponds to the trick of the organism to provide its relation to an *Umwelt* with some simplification in the form of an orienting *Gestalt*; at the moment of the sensation of a "signal" as isolated from the sensorial manifold, the signal already represents – or at least announces – a "thing."
>
> (Blumenberg 1986, 272)

This brief passage already indicates how in his anthropology Blumenberg accounts for the emergence of symbols and a conceptual world out of the stimulus–response circle, that is, Uexküll's functional circle, as I argue in some detail further on. These few lines also show his (again mainly implicit) indebtedness to Ernst Cassirer, whose philosophical anthropology links the formation of concepts to insights from *Gestalt* psychology, ensuing from Cassirer's intensive discussions with his cousin, the neurologist Kurt Goldstein. Along the same lines of what we read in Blumenberg, Cassirer claims, in the decisive third volume of *The Philosophy of Symbolic Forms*, that already the human sensorial apparatus identifies and separates significant and meaningful units (Cassirer 1957, Vol. 2, Ch. 6; see also Brentari's contribution, Chapter 6, in this volume). Differently from Cassirer, however, the disparaging qualifications of "trick" and "deception" point here to the skepticism of Blumenberg's negative anthropology. In this respect, one may look at Uexküll's *Umweltlehre* for a crucial starting point in reviewing ill-placed anthropomorphism, but his conceptualization of reality is too indebted to the objectivity of biological experiments for providing a philosophically sound basis. Although Uexküll's concept of reality is therefore not well founded, according to Blumenberg, he is definitely in good company; most of his fellow philosophers working on a philosophical anthropology base indeed their conceptualization of rationality and human thinking on too safe a ground even where they opt for the negative anthropology of "*Organausschaltung*" (Paul Alsberg), "neoteny" (Arnold Gehlen following Louis Bolk), "excentric positionality" (Helmuth Plessner), "*Weltoffenheit*" (Max Scheler), or "symbolization" (Ernst Cassirer) (see Blumenberg 1981).

The support Blumenberg finds in Uexküll while reaching his skeptical perspective should be then taken into account when attempting to appropriately evaluate Blumenberg's critical remarks. These apply for instance to Uexküll's sharp definition of "moment." According to Blumenberg, a moment is not – and cannot be – the theoretical conclusion ensuing from experimental results; however, it provides a very valuable conceptual clarification. It defines simultaneity as a theoretical construct, not as an observable incident or empirical fact. The experimental determination of the moment rests on procedures of measurement that conceptually rely on complex intersubjective arrangements. The phenomenological exploration of the scientific construction of simultaneity thus arrives at intersubjectivity as its transcendental foundation:

> Simultaneity is not the concrete form for a concept that could then serve as [the] basis for all objectifications, because objectification operates under the condition of simultaneity not just and only for the measurement between unit and object, but already in the construction of that form of inter-subjectivity which is required for the concept of measurement. The reactions by others regarding a given situation can confirm an assessment only if I already rely on the simultaneity of the subjects involved. [...] They must have, and share, the concept of simultaneity. [...] Simultaneity does not exist [in nature]. This follows for the immanence of subjectivity from the paradox of fusion: Inside of a moment, the 'world stands still.' But it must be possible for the subject to overcome the distance of separate impressions.
>
> (Blumenberg 1986, 284)

According to Blumenberg, Uexküll contributed enormously to the phenomenology of consciousness with the introduction of simultaneity as a theoretical construct, but instead of exploring and embracing its true philosophical implications, he left his analysis at the level of experimental investigations and followed in his interpretation of their results a misguided analogy.

For Blumenberg, von Baer's and Uexküll's strength resides in their conceptual clarification of biological observations. Once these concepts have been properly formed, their phenomenological analysis provides the insight that simultaneity is a "theoretical construct, not a fact which could be described phenomenologically" (Blumenberg 1986, 283). With this statement, Blumenberg does not intend to support the idea that simultaneity is a mere construct; for him, simultaneity is the conceptual basis, reached by human beings collectively thanks to their ability to think and reflect, and that can be used for exploring and determining subjective and objective time. In this respect, Uexküll shares with von Baer the coining of meaningful concepts; *Umwelt*, perception world [*Merkwelt*] and effect world [*Wirkwelt*] mark decisive philosophical advances toward a biologically grounded epistemology. Instead of fully exploring their philosophical potential, however, Uexküll – according to Blumenberg – conflates their constitutional role with the assumed explanatory power of the "moment."

The problem rests on the one hand upon the ambiguity of the term "world," which continues to signify the whole, even though Uexküll limits it to the subjectively perceived world, and on the other hand on the all-too-human need to arrive at definitive answers by finding refuge in "absolute metaphors," according to Blumenberg's terminology, that is to say, seemingly perfect images substituting proper concepts.[9] The misleading, absolute metaphor of the tunnel of the perceived world is outrageous, in Blumenberg's view, inasmuch as it forecloses the analysis short of further conceptual advances and shows ignorance about contemporary philosophical concerns in the same direction. The philosophical alternative to Uexküll's *Umwelt* is, according to Blumenberg, Husserl's concept of the life-world [*Lebenswelt*], introduced at exactly the same time. This missed chance invites Blumenberg to outbid the imprudent metaphor by the ironic image of macaroni. And yet, with life-world, Husserl provides only an appropriate concept but does not deliver its proper exploration, as he ignores the relevant insights from biological research, as Blumenberg elucidates in this chapter in reference to von Baer and Uexküll. Whereas Uexküll takes easy recourse to absolute metaphors, convinced as he was of Nature's provisional ordering of life and world, Husserl relies too swiftly on a solipsistic exploration of consciousness. Taken together, the problematic concept of the life-world and the misguided metaphor of the tunnel point to the intellectual richness of the period – and to the missed chances to translate both into appropriately distanced concepts. Blumenberg ridicules Uexküll here, but he does so because of the strength and originality of his *Theoretical Biology*, which reaches out, and comes close, to phenomenological reflection. Blumenberg criticizes Uexküll so harshly precisely because his theory has what it takes to inform phenomenology, although his philosophizing was insufficient. From Blumenberg's perspective, the image of the tunnel isolates the perception world and separates it from the constitutive role of intersubjectivity for the formation of concepts in human thinking – and hence from the common ground from which also the notion of *Umwelttunnel* (environment-tunnel) emerged. The metaphor makes its own philosophical basis inconceivable.

This is the central argument in Blumenberg's critique of Uexküll. And it relates to the fundamental issue that Blumenberg is treating in this book, the anthropological threat of the limited and far-too-short human life span, especially its ever shorter extension with regard to the enormously enlarged "world-time," the extension and duration of the universe as determined by modern science with its dimensions far beyond human experience. The shortness of human life, and especially the limits of human beings in grasping and dealing with the enormity of reality, was one of Blumenberg's central topics. There was also a personal twist to the importance of this topic for him, as he had been prevented from using his personal time according to his own needs and preferences, because of him being classified as belonging to the Jewish population in Nazi Germany. For this reason, he could not enter university and hardly escaped persecution. After this existential experience of a deferred and risky start, the limited availability of time put constant pressure on his work. Here turned into the topic of a book, time pressure ultimately helped Blumenberg explore, clarify, and reconceptualize his relation to phenomenology as a transcendental, foundational project of philosophy.

3 Uexküll in the background of Blumenberg's phenomenological anthropology

On the surface of the polished and elaborate text of *Lifetime and Worldtime*, Uexküll might appear as a brilliant experimentalist who falls short of his own philosophical aspirations. By contextualizing Uexküll's concept of *Umwelt* in both the history of thought-provoking biological experiments and the trajectory of phenomenological analysis, however, Blumenberg hides his actual indebtedness to Uexküll. The chapter, so typical of Blumenberg's style of writing, bears only evasive testimony to Uexküll's eminent place in his philosophy at the crossroads of phenomenology and anthropology. This nexus has nevertheless come much more to the fore with the publication of *Description of Man* (2006) and *Theory of the Life-World* (2010). While Blumenberg was previously regarded primarily as a philosopher of history, mythos, and rationality, focusing on the history of ideas with his unique metaphorology method, now his philosophical anthropology (and also his philosophy of technology) receive increasing attention.[10] It has now become much clearer that anthropology has always been a core topic in Blumenberg's thinking. He actually gave lectures on philosophical anthropology already in the Winter term 1963/64 in Gießen, but also in Münster in the 1970s and again in the early 1980s; the manuscripts now published partly derive from these lectures.[11]

In *Description of Man*, Blumenberg discusses primarily Paul Alsberg's evolutionary theory of the human species and culture through *Organausschaltung* (literally, the turning off of organs), in other words, the increasing independence of the human body obtained by reducing the functional specificity of organs like the hand. Blumenberg combines this concept with the idea of a decisive moment in human evolution, the erection of the body, the switching to bipedalism, and the biotope change enabled thereby from the woods into the savanna. In Blumenberg's view, this transition changed the temporal regimes of survival dramatically from immediate signal–response chains to hitherto nonexistent periods of expectancy and planning. The migration into the open with the change of the environment forced human life to a new mode of prevention based on anticipating and evaluating risks. Thinking and reflection had to be there in order to be able to survive the new existential dangers of visibility – a fundamental combination in Blumenberg's thinking.

The entire argument is obviously built on Uexküll's notion of *Umwelt*, although Uexküll is not referenced in *Description of Man* (there are only very few references in the manuscript). Especially in the section describing the emergence of consciousness and intentionality, Blumenberg clearly relies on Uexküll's concept of functional circle of species-specific stimulus–response chains; however, he does not link its emergence causally to the change of the environment but regards

some primitive form of consciousness as a requirement for the successful management of the biotope change:

> Intentionality as the structure of an unspecific form of consciousness that
> interrupts the reflex arc cannot have arisen from the change of biotope. It
> must have been available for it and in it as the advantage of an endogenous,
> preparatory development of the brain. [...] That human beings had at least
> been beings prepared for theory can only be explained by a predisposition
> for a not at all self-evident and by its nature not at all ideal structure of con-
> sciousness. [...] Consciousness is in its mature structure an interruption of
> immediacy in the excitatory chain from the afferent to the efferent site of the
> black box of the organic system. But this is true only because consciousness
> has already had its origin in a disturbance of the preconditions of this imme-
> diacy, in a breaking apart of the specific relationality of signal and response,
> stimulus and reaction. Consciousness is not the mysterious intruder that has
> interrupted the closed system, but it is an arrangement with the disturbance;
> consciousness is the organized catching up with a situation of leveled-out
> reality by using still available sensorial data.

(Blumenberg 2006, 559f.)

In a rhetorical move very similar to chapter 12 of *Lifetime and Worldtime*, Blumenberg grants Uexküll the privilege of describing accurately sensorial physiology by removing all misplaced anthropomorphisms, but by doing so, theoretical
biology can only clarify the philosophical position of consciousness in the evolutionary trajectory and not explain its emergence.

This figure of thought is illustrated further by one of Blumenberg's famous
little cards of notes. Under the title "reality check" this card shows a schematic
drawing apparently depicting a functional circle, connecting the afferent signals
perceived by a "receptor" via a "black box" with an "effector" (see Fig. 11.1).[12]
So far, this would be the typical scheme of the functional circle linking the organism to its environment. Blumenberg, however, deviates in his drawing from the
standard of the physiological stimulus–response chain in at least two important
respects: The circle is no longer closed but operates against the elapsing time,
indicated by vertical arrows marked "t(ime)." The sketch thus illustrates the dealings of an organism in the situation of an already interrupted stimulus–response
chain. And the "black box = *Seele* (soul)" has been inscribed with an "I think" in
quotation marks that sends arrows in opposite directions, labeled "I want" and
"I experience," again in quotation marks. Blumenberg's drawing thus represents
consciousness in the form of a functional circle that has arrived at dealing with
time and processing objectivity ("reality check") by means of the capacity of
thinking, always already residing in the center.

Evolution has prepared this peculiar organism to survive under the different
temporal constraints of the new environment by a capacity of anticipation that
materializes experientially as an "I think," a willing and effectuating subjectivity.
Blumenberg's philosophical anthropology project thus aims to mediate between

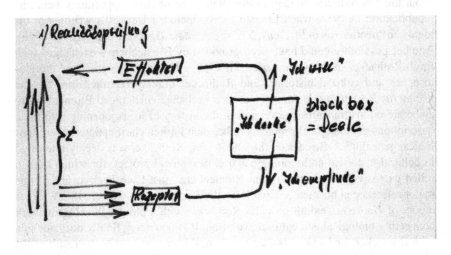

Figure 11.1 Reality check

the worlds of empirical knowledge, such as sensorial physiology and evolutionary theory, and the philosophical clarification of the transcendental requirements of reflection. In this regard, biological theories can help elucidate the precise position of consciousness in its evolutionary genealogy but fall short of justifying its emergence epistemologically, inasmuch as such a transcendental problem, according to Blumenberg, can only be addressed by philosophy, which, as reflection about the present condition, has to take into account scientific research without giving up its radical questions.

Blumenberg coined a specific formula for this peculiar position of philosophy vis-à-vis the sciences already in his inaugural address when taking up his professor position in Gießen in 1960. His address celebrated the (belated) reopening of the philosophical faculty that had to be discontinued in 1945 because of its severe alliance with National Socialist ideology. In this programmatic lecture, Blumenberg states, "Philosophy transcends the sciences not outwardly but inwardly" (Blumenberg 1961, 74).[13] The schematic drawing from his later anthropology lectures captures accurately this formula: The "I think" is a consequence of the opening of the functional circle to the elapsing time, its genealogy can be linked to changes of/in the black box, but the "I think" is an act of conceptual conclusion that transcends biological reasoning inwardly. The black box has not arrived at, or gained, subjectivity because of empirically specifiable biological characteristics, but these have become recognizable because of the black box's existence (by inward transcendence) as subjectivity.

There is no evident link to Uexküll in this early lecture and assuming one would predate Blumenberg's acquaintance with Uexküll's work even before his personal exchanges with Uexküll's son, the physician and fellow professor in Gießen, Thure von Uexküll. It is, however, rather likely that Blumenberg heard about the *Umweltlehre* already in the 1950s, one obvious opportunity being its republication in 1956, when Ernesto Grassi included Uexküll's writings in his highly influential *rowohlts deutsche Enzyklopädie* (Uexküll and Kriszat 1956). Another possibility could have been provided by Blumenberg's exchanges with Erich Rothacker in the same period. In 1957, Blumenberg approached the philosopher and cultural historian, and Rothacker quickly recruited him for the *Archiv für Begriffsgeschichte*. In fact, this exchange encouraged Blumenberg to elaborate on his transgressive version of the history of ideas, opening it to thick descriptions of conceptual problems rather than limited clarifications of terminological genealogies. Besides editing the *Archiv*, Rothacker was deeply interested in cultural-historical anthropology as a philosophical project, for which he had relied extensively on Uexküll, and Blumenberg must have been aware of this major reference in his mentor's work (Rothacker 1948, esp. 103–124).[14] With his theory of *Umwelt*, Uexküll provides Rothacker with a philosophical framework connecting biological and cultural evolution. If Blumenberg finally does not side with Rothacker for Uexküll but against him, this can be read as the result of his phenomenological approach to the integration of the history of problems into the history of ideas. Whereas with his philosophy of cultural anthropology Rothacker arrives at the history of culture, Blumenberg discovers the abyss of unsolvable problems, inasmuch as he envisions philosophical anthropology as a foundational project carrying on an epistemological critique by mediating between Kant and Husserl.

A fine example of Blumenberg's craft in navigating this course is finally provided by the brief manuscript on "self-evidence, self-erection, self-comparison" now included as chapter 3 in *Theory of the Life-World*, that starts with the provocative statement that the life-world is no *Umwelt*. The text weaves several by now familiar topics and themes into an extremely dense speculation on how the emergence of bipedalism and consciousness are linked to intersubjectivity, mythos, and religion with intersubjective comparison acting as cultural driving force:

> The preventive constitution of human beings is related to the inconsistency of the horizon of their life-world, especially their earliest life-world which must always be regarded as being in transition from a biological, automatically functioning *Umwelt* to a pre-modally organized life-world which was in many respects self-evident but not universally secured. In this preventive attitude, everything depends on the ability to project by attention and expectancy from the life-world into the unknown that lies in the periphery beyond the horizon. [...] Magic and mythos are such attempts, they reverse the direction of the projection at the periphery of the life-world, especially with the aim to get control over the powers that lie behind the visible. [...] *Angst* as the general correlate of the uncertainty of the horizon of the life-world gets de-potentiated, transformed into fear as concrete behavior in relation to specific powers. [...] The technology of this mythical border trafficking is the metaphor, integrating the worldly unfamiliar by means of the life-worldly familiar.
>
> (Blumenberg 2010, 136f.)

Here, everything comes together, Blumenberg's metaphorology, his intellectual history of technology, his interest in evolution theory and anthropology, his reflection on mythos and religion, though in a rather speculative way.

Blumenberg's indebtedness to Uexküll as well as his distancing from him become increasingly clear a few pages later, where he derives his specific conceptualization of life-world as lost security to which there is no possibility of returning. Clarifying the relation between *Umwelt* and life-world, Blumenberg argues that the precondition for the "I think" is a stepping out of the life-world which in itself already indicates a departure from mere *Umwelt*. Any return to its security would thus entail giving up the "I think":

> Life-world is always the topic of a process of stepping out, the term conceptualizes its origin. This can never by the idyll of a primordial harmony or final destination. [...] The human being, emerging in this process by the *Urakt* of erection, gains for the first time spatial as well as temporal distance: What affects the human being can be detected earlier and thereby processed timely. A widening horizon of possibilities gets discovered. Ever before the life-world was left, the *Umwelt* had been left, transcended; this was the first act of a fundamental transgression. Such transgressions always correspond to a widening of the perceptual horizon, finally to the transition from a perceptual world into an observed world, the intentional, question-guided selection of perceptual possibilities. Observation in contrast to perception requires concepts, the availability of hypotheses. [...] Self-erection into verticality increases not only the quantity of the perceptible and observable into the distance of no longer acutely imminent objects, but also indirectly as distance to the perception world by a self-comparison of the organism which has now turned into a human being with her fellow beings. [...] The achievements of species companions can be perceived, creating a competition in self-comparison that influences the horizon of the perception world, widening it further and transforming it into an eccentric position.
>
> (Blumenberg 2010, 139, 142f.)

The concept of *Umwelt* is then the starting point for any meaningful reflection about the life-world, pointing to its rootedness in the intrinsic organization of the organism. A careful investigation of the *Umwelt*, not just as term for biological theorizing but as springboard for philosophical reflections, finally yields three important results: (1) since biological theory is always already anchored in the life-world, one can conclude that biological theory requires concepts it cannot found on its empirical data; (2) the concept of the life-world, however, is a *terminus a quo*, capturing the transcendence of a thinking form of life that already comprises intersubjectivity and (3) cannot, hence, be elucidated by focusing on subjectivity.

This was also the reason why Husserl – and not von Baer or Uexküll – remained Blumenberg's central reference in the discussion of *Umwelt*. The introduction of von Baer's thought experiments and of Uexküll's concepts in chapter 12 of *Lifetime and Worldtime* functions as a strategic and programmatic move for mobilizing biology as philosophical questioning: if biology can demonstrate how the concept of reality (together with its contents) is dependent on the empirical conditions of the sensory apparatus, phenomenology can no longer remain ignorant regarding biology, and there is no way to secure the life-world against empirical evidence. Philosophical speculation cannot be done exclusively on biological data, but phenomenology cannot be a self-founding endeavor either; both have to enlighten and pass each other under review. And in the end, the self-reflecting animal, successfully turned into an observer, can become a spectator, living at a distance from many forms of *Umwelt*, conscious about having lost the life-world and yet fully aware of living.

Notes

1 I want to thank Marco Mauerer, Angus Nicholls, Tessa Marzotto Caotorta, and the editors for many helpful comments on earlier versions of this chapter. All translations from German sources are mine, if not otherwise stated.

2 Hans Blumenberg (1986). The book has not been translated into English; all translations are mine. References to this source are given in brackets directly in the text. For an English description of the book, see Harries (1987). For an overview of his work from the North American perspective, focusing on him as a philosopher of modernity, see Palti (1997).

3 There is a growing body of studies on Blumenberg's anthropology, the following two edited volumes provide a good starting point: Moxter (2011) and Klein (2009).

4 Blumenberg (2006, 2010).

5 Blumenberg mentions Bessel right at the chapter's beginning, although without including a reference (Friedrich Bessel 1820). For a contextualization of this episode, see Hoffmann (2007).

6 In his *Theoretical Biology*, Uexküll describes his project as follows: "The task of biology consists in expanding in two directions the results of Kant's investigations: (1) by considering the part played by our body, and especially by our sense-organs and the central nervous system, and (2) by studying the relations of other subjects (animals) to objects" (Uexküll 1926, xv). See Esposito, Chapter 2, in this volume.

7 On Blumenberg's rhetoric see Koerner (1993) and Zill (2015).

8 Blumenberg gets surprisingly close here to Henri Bergson's famous discussion of the cinematographic apparatus in *Creative Evolution* (Bergson 1911, 272ff.), where he criticizes that the dissection of time into a sequence of moments misses the essence of movement. On Uexküll's use of the new medium, cf. Kühnast (2010).

9 In his "Prospects for a Theory of Nonconceptuality," Blumenberg remarks that his metaphorology should be seen in terms of its direction back toward "the connections with the life-world as the constant motivating support (though one that cannot be constantly kept in view) of all theory" (Blumenberg 1997, 81). See also Adams (1991).

10 See note 5 and Borck (2013a) and Blumenberg (2015).

11 Ferdinand Fellmann took the class in Gießen; cf. Boden and Zill (2017, 121). Manfred Sommer described the later cycles in his editorial comment to *Beschreibung des Menschen*.

12 The card has been published in Bülow and Krusche (2013, 275).

13 On the figure of "inward transcendence," see Borck (2013b).

14 On Blumenberg's relation to him, cf. Kranz (2013).

References

Adams, David (1991) 'Metaphors for mankind: The development of Hans Blumenberg's anthropological metaphorology'. *Journal of the History of Ideas* 52, 152–166.

Bergson, Henri (1911) [1907] *Creative Evolution*. Translated by Arthur Mitchell. New York: Holt.

Blumenberg, Hans (1961) 'Weltbilder und Weltmodelle'. *Nachrichten der Gießener Hochschulgesellschaft* 30, 67–75.

Blumenberg, Hans (1981) 'Anthropologische Annäherung an die Aktualität der Rhetorik'. In: Hans Blumenberg. *Wirklichkeiten in denen wir leben*. Stuttgart: Reclam, 104–136.

Blumenberg, Hans (1985) *Work on Myth*. Translated by Robert M. Wallace. Cambridge, MA: MIT Press.

Blumenberg, Hans (1986) *Lebenszeit und Weltzeit*. Frankfurt a. M.: Suhrkamp.

Blumenberg, Hans (1987) 'An anthropological approach to the contemporary significance of rhetoric'. Translated by Robert M. Wallace. In: Kenneth Baynes, James Bohman, and Thomas McCarthy (eds.) *After Philosophy: End or Transformation*. Cambridge, MA: MIT Press, 429–458.

Blumenberg, Hans (1997) [1979] *Shipwreck with Spectator: Paradigms for a Metaphor of Existence*. Translated by Stephen Rendall. Cambridge, MA: MIT Press.

Blumenberg, Hans (2006) *Beschreibung des Menschen*. Frankfurt a. M.: Suhrkamp.

Blumenberg, Hans (2010) *Theorie der Lebenswelt*. Berlin: Suhrkamp.

Blumenberg, Hans (2015) *Schriften zur Technik*. Berlin: Suhrkamp.

Boden, Petra and Zill, Rüdiger (eds.) (2017) *Poetik und Hermeneutik im Rückblick: Interviews mit Beteiligten*. Paderborn: Wilhelm Fink.

Borck, Cornelius (ed.) (2013a) *Hans Blumenberg beobachtet: Wissenschaft, Technik und Philosophie*. Freiburg: Verlag Karl Alber.

Borck, Cornelius (2013b) 'Philosophie als "Transzendenz nach innen": Einleitende Bemerkungen zu Hans Blumenbergs Ortsbestimmung der Philosophie zwischen Wissenschaft und Technik'. In: Cornelius Borck (ed.) *Hans Blumenberg beobachtet: Wissenschaft, Technik und Philosophie*. Freiburg: Verlag Karl Alber, 9–22.

Bülow, Ulrich von and Krusche, Dorit (2013) 'Vorläufiges zum Nachlass von Hans Blumenberg'. In: Cornelius Borck (ed.) *Hans Blumenberg beobachtet: Wissenschaft, Technik und Philosophie*. Freiburg: Karl Alber Verlag, 273–288.

Cassirer, Ernst (1957) [1929] *The Philosophy of Symbolic Forms*. Vol. 3. *The Phenomenology of Knowledge*. Translated by Ralph Manheim. New Haven/London: Yale University Press.

Harries, Karsten (1987) 'Lebenszeit und Weltzeit by Hans Blumenberg' (book review). *Journal of Philosophy* 84, 516–519.

Haverkamp, Anselm (2017) 'The scandal of metaphorology: Blumenberg's challenge'. In: Anselm Haverkamp (ed.) *Productive Digression: Theorizing Practice*. Berlin: de Gruyter, 34–52 (originally 2012 in *Telos* 158, 37–58).

Hoffmann, Christoph (2007) 'Constant differences: Friedrich Wilhelm Bessel, the concept of the observer in early nineteenth-century practical astronomy and the history of the personal equation'. *British Journal for the History of Science* 40, 333–365.

Klein, Rebekka A. (ed.) (2009) *Auf Distanz zur Natur: Philosophische und theologische Perspektiven in Hans Blumenbergs Anthropologie*. Würzburg: Königshausen & Neumann.

Koerner, Joseph Leo (1993) 'Ideas about the thing, not the thing itself: Hans Blumenberg's style'. *History of the Human Sciences* 6 (4), 1–10.

Kranz, Margarita (2013) 'Blumenbergs Begriffsgeschichte: Vom Anfang und Ende aller Dienstbarkeiten'. In: Cornelius Borck (ed.) *Blumenberg beobachtet: Wissenschaft, Technik und Philosophie*. Freiburg: Verlag Karl Alber, 231–253.

Kühnast, Katja (2010) 'Kinematographie als Medium der Umweltforschung Jakob von Uexkülls'. *Bild – Wissen – Technik* 4, 1–14, www.kunsttexte.de.

Moxter, Michael (ed.) (2011) *Erinnerung an das Humane: Beiträge zur phänomenolo-gischen Anthropologie Hans Blumenbergs.* Tübingen: Mohr Siebeck.

Nicholls, Angus (2015) *Myth and the Human Sciences: Hans Blumenberg's Theory of Myth.* New York: Routledge.

Palti, Elias Jose (1997) 'In memoriam: Hans Blumenberg (1920–1996), an unended quest'. *Journal of the History of Ideas* 58, 503–524.

Rothacker, Erich (1948) *Probleme der Kulturanthropologie.* Bonn: Bouvier.

Uexküll, Jakob von (1922) 'Wie sehen wir die Natur und wie sieht sie sich selber?'. *Natur-wissenschaften* 10, 265–271, 296–301, 316–322.

Uexküll, Jakob von (1926) [1920] *Theoretical Biology.* Translated by Doris L. Mackinnon. London: K. Paul, Trench, Trubner & Co.; New York: Harcourt, Brace & Company.

Uexküll, Jakob von and Kriszat, Georg (1956) *Streifzüge durch die Umwelten von Tie-ren und Menschen: ein Bilderbuch unsichtbarer Welten. Bedeutungslehre* (rowohlts deutsche Enzyklopädie: Das Wissen des 20. Jahrhunderts im Taschenbuch mit enzyk-lopädischem Stichwort, Vol. 13). Hamburg: Rowohlt.

Zill, Rüdiger (2015) 'Auch eine Kritik der reinen Rationalität: Hans Blumenbergs Anti-Methodologie'. In: Michael Heidgen, Matthias Koch, and Christian Köhler (eds.) *Per-manentes Provisorium: Hans Blumenbergs Umwege.* Paderborn: Fink, 53–73.

12 Giorgio Agamben
The political meaning of Uexküll's "sleeping tick"

Marco Mazzeo

1 *The Open*: Uexküll between Heidegger and Agamben

In one of his most well-known texts, *A Foray into the Worlds of Animals and Humans*, Jakob von Uexküll illustrates "the fundamental aspects of the structure of the environments [*Umwelten*]" through the following example:

> In order to increase the probability that its prey will pass by, the tick must be capable of living a long time without nourishment. And the tick is capable of this to an unusual degree. At the Zoological Institute in Rostock, they kept ticks alive that had gone hungry for eighteen years.
>
> (Uexküll 2010, 51f.)

Apparently, the unassuming hero of a marginal and bizarre passage, this annoying parasite, the tick, displays an unusual skill discovered by experiments conducted in a small German town. In *The Open* (2002, English translation 2004) Giorgio Agamben has the merit of emphasizing the importance of this passage. According to him, in fact, the tick's behavior hints heavily to an unexpected proximity between humans and animals and the possibility to acknowledge it, in spite of all the traditional emphasis on the distance between the two realms. The Italian philosopher notably supports his proximity claim by means of two strategic moves. The first calls upon another philosopher, Martin Heidegger, who was also deeply impressed by Uexküll's work (see Michelini's contribution, Chapter 7, in this volume). In *The Fundamental Concepts of Metaphysics* (1929–30), Heidegger moves indeed from the work of the Estonian biologist to further investigate a distinction previously touched on by Max Scheler. According to Heidegger, animal environments [*Umwelten*] should not be distinguished from the human world [*Welt*] in terms of differences in degrees but, rather, in terms of internal structure. While the animal is "captivated" [*benommen*] by its environment (Heidegger 1995, 268ff.), human beings are open to the world they "form" (*weltbildend*; Heidegger 1995, 274ff.). It should also be noted that, before dwelling on the distinction between the human world and animal environment, Heidegger focuses at length on one emotion considered crucial to the understanding of the human *Dasein*, the feeling of boredom. Agamben's first step then lies in emphasizing the close relation between the two theoretical moments. A close connection is thereby established within the pair world/environment and boredom, as in the latter, Agamben believes, the intersection between *Umwelt* and *Welt* can be effectively explored. Following Heidegger's account, boredom comprises two structural traits. The German philosopher presents in this regard one clear and simple example. We are at a railway station, and "it is four hours until the next train arrives" (Heidegger 1995, 93). While waiting, we walk back and forth; we count the trees; we keep looking at the clock. The first structural trait is thus presented as "being held in limbo," in which we experience "a specific kind of oppressing, the dragging of time" (Heidegger 1995, 99): time does not pass; the train does not arrive. The second aspect is what Heidegger calls "being left empty" (Heidegger 1995, 106) in which, conversely, things "leave us completely in peace [...] and this is precisely the reason why they bore us" (Heidegger 1995, 102). The situation does not offer us what we expected, things "abandon us to ourselves" (Heidegger 1995, 103). Agamben borrows the same distinction in his analysis of Uexküll's example of the tick. And here comes his second argumentative step: boredom lives on "being in limbo," the tick can live in suspension even for eighteen years; hence, a close link can be established between boredom and the suspended life of the tick.

According to Agamben (2004, 68), boredom is a "metaphysical operator" as it displays the intersection between animality and humanity: "*Dasein* is simply an animal that has learned to become bored" (Agamben 2004, 70). As he describes the boredom of "being left empty," Heidegger (1995, 101) argues that we are usually "captivated" [*benommen*] by our daily commitments and concerns, using the same expression he would use shortly after to indicate the typical state of the animal in its environment. Thanks to boredom, then, men would temporarily suspend their usual state of captivation. Against the backdrop of this understanding of boredom, as what experienced by humans as the opposite of any form of captivation, Uexküll's tick represents, in Agamben's words, an awkward and unresolved case:

> Under particular circumstances, like those which man creates in laboratories, the animal can effectively suspend its immediate relationship with its environment, without, however, either ceasing to be an animal or becoming human. Perhaps the tick in the Rostock laboratory guards a mystery of the 'simply living being,' which neither Uexküll nor Heidegger was prepared to confront.
>
> (Agamben 2004, 70)

Through the suspension of daily activities, boredom may lead humans to evade a condition of captivation; similarly, through the suspension of food and organic functions, the tick suspends its captivated relation with the environment. It is precisely on this point that Agamben grafts his philosophical proposal. Boredom gives access to a crucial turning point of what today is called "biopolitics" (Agamben 2004, 15), since it enables both humans and animals to escape the typical framing apparatuses of Western history. In *The Open*, Agamben, in fact, argues that two paths lie ahead of us: one seeks to handle the gap through technique; the other supports an appropriation of the gap that avoids burying it or trying to dominate it (Agamben 2004, 80). In this respect, the "time of boredom" is no longer an empty time nor a waste of time. It is instead the time that cancels out any distinction between the animal world and the human world. It is instead the time of un-distinction, when the animal world and the human world are no longer distinct.

In Agamben's view, only the second path enables a radical rethinking of our contemporary world and its ethical-political crisis. Boredom thus allows us to understand two key concepts in his investigations: "inactivity" (Agamben 2004, 76) and "messianic" time (Agamben 2004, 61). The former "represents an Energy that has not been exhausted and that cannot be exhausted in the passing of the potential to the actual" (de la Durantaye 2009, 331), a paradigm of human activity that is alternative to that of work but that does not coincide with stasis or mere inertia. The latter indicates instead a nonchronological time mode, namely, a mode that is not associated with the linear sequence of moments described by classical science, but it is rather a time related to Paul of Tarsus's discussion of *kairòs* (Attell 2015, 212ff.). At variance with the theological paradigm of eschatology (i.e., the waiting for the Messiah), Agamben draws from the Gnostic idea based on which the Messiah has already returned. It should be remarked, moreover, that for him, boredom is the paradigmatic state of mind for both concepts: human activity free from any reproductive form, a time that is not crushed in the linear form of a geometric sequence.

Agamben's overall objective is, in other words, that of cutting out a notion of time that is radically different from the leading notion in our tradition. References to religion serve then the purpose of challenging the well-established equation time = measuring = money, typical of capitalist societies and justifying the whole concept of "salary." In this regard, then, boredom is not useless time that should be instead devoted to entrepreneurship and financial speculation but, rather, the fundamental drive allowing us to reframe differently how we live our lives today. Boredom is no longer a waste of time but rather the birthplace of a new time, free from all institutional forms at work today – in national states' neoliberal markets.

2 Uexküll's problem: human *Umwelten* and animal surroundings

Agamben has the merit of emphasizing the philosophical potential of one particular example, which, to be sure, appears to be especially treasured also by Uexküll. It is, in fact, the tick that opens *A Foray into the Worlds of Animals and Humans* to the aim no less of testing "the soundness of the biological point of view as opposed to the previously common physiological treatment of the subject" (Uexküll 2010, 45). Furthermore, as Brentari (2015, 161f.) points out, the parasite takes center stage also in the *Theory of Meaning*, as to illustrate the relation between the animal and its *Umwelt* by "points and counterpoints" (Uexküll 2010, 178). Although there is no need to wonder "what would Uexküll be without his tick" (Winthrop-Young 2010, 235), this animal is undoubtedly, still today, the poster child of the theoretical outreach of our Estonian biologist. It is clear, nevertheless, that, in this respect, more open questions than ready-made answers ensue from Uexküll's legacy (see Geroux 2012, 147).

One more element of this legacy is also duly emphasized by Agamben. As is well known, Uexküll does not counter animal *Umwelten* with the human world [*Welt*] – a theoretical strategy that characterizes instead the line of thought usually referred to as "anthropological philosophy" – but, rather, with what he calls "surroundings" [*Umgebung*]. While the *Umwelt* is subjective, that is to say, it comprises a selection of what animals can perceive and do, the surroundings require a different concept. In the various examples provided by Uexküll, the characterization of the surroundings is twofold. They mainly appear to include invariant aspects of the physical universe, such as the impenetrability of bodies or universal gravitation. Each animal's *Umwelt* then carves out of this wider common context: "the whole rich world surrounding the tick is constricted and transformed into an impoverished structure" (Uexküll 2010, 51). In other words, even though the mole is blind while many birds have a sharp vision, it is nonetheless true that every form of life has to come to terms with gravity and the density of matter. In Uexküll's work, however, the surroundings also play a further theoretical role, no less important than the former:

> The animal's *Umwelt*, which we want to investigate now, is only a piece cut out of its surroundings, which we see stretching out on all sides around the animal – and these surroundings are nothing else but our own, human *Umwelt*.
>
> (Uexküll 2010, 53)

To this interesting overlapping between animal surroundings and human *Umwelt*, Agamben, as previously anticipated, has the merit of drawing philosophical attention. However, he cannot be said to have gone much further than this. Stepping into the realm of the limits of Agamben's investigation of Uexküll, we might want to look at one specific passage of his interpretation of these concepts:

> In reality, the *Umgebung* is our own *Umwelt*, to which Uexküll does not attribute any particular privilege and which, as such, can also vary according to the *Umwelt* point of view from which we observe it.
>
> (Agamben 2004, 40f.)

As I will shortly explain, I believe this statement contains at least two problematic aspects. First, provided that the human *Umwelt* is understood as coinciding with the surroundings, which is to say, what the other animal *Umwelten* draw on, it is not without difficulties that one can claim that Uexküll does not attribute any privilege to the position of humans. Second, it is equally difficult to take the relation between the unity of the surroundings and the variety of human points of view entirely for granted. Especially at the end of *A Foray into the Worlds of Animals and Humans*, Uexküll (2010, 126ff.) insists that the human *Umwelten* are varied and are such not just in comparison to the *Umwelten* of other species (the tick, the owl, the fox, etc.). Human environmental diversity *is such in itself*. In *A Foray* the different ways in which the fox and the owl perceive a tree are compared to the different responses of a little girl and a lumberjack to the same tree (Uexküll 2010, 127f.). Different approaches to human life are therefore explicitly equated to the difference between very different animal species, such as a feline and a bird. Furthermore, in his most complex work, *Theoretical Biology*, the biologist discusses explicitly the two *Umwelten* that characterize every human being: one concerning the natural task of reproduction and belonging to the family, the main cell of a group of people [*Volk*]; the other corresponding to the role each of us assumes in the reproductive cycle, a historical role associated with the state (Uexküll 1973, 330–332).[1] What Uexküll, let alone Agamben, misses is the fact that such diversity, as it characterizes the *Umwelt* of every individual (the little girl, the lumberjack, etc.) can be said to be intrinsic to the species (the *Homo sapiens*). Finally, as Uexküll claims that the *Umwelten* are in contradiction (Uexküll 2010, 54, 135), in other words, largely incommensurable with each other, the fact that to every human being corresponds a different *Umwelt* produces a major theoretical and ethical-political problem.

How can the *Homo sapiens* get on well with and communicate with each other? It is not unlikely that Uexküll's conservative and authoritarian political ideas[2] were aimed precisely to solve this problem. A strong and centralized government could solve the internal conflicts of a paradoxical species that extends its *Umwelt* into the *Umgebung* (surroundings) and, at the same time, fragments it in small individual *Umwelten* (Mazzeo 2010). This is probably the key to understanding Uexküll as one of the precursors of biopolitics in the 20th century. In his texts, in fact, biological and political notions reciprocally integrate each other to the point of indistinguishably overlapping. As remarked by Esposito (2008, 17), in *State Biology* [*Staatsbiologie*], "the discourse revolves around the biological configuration of a state body that it is unified by harmonic relations of its own organs, representative of different professions and competences." Similarly, in *A Foray*, the example of the tick is, once more, at the core of such a problematic constellation. On the one hand, the parasite appears to offer an interesting solution. Uexküll (2010, 76f.) defines the sea urchins as "reflex republic" because they lack a central structure. This life-form appears indeed to be lacking a nervous system coordinating perceptions and movements. He then continues the political analogy by suggesting that the tick "represents a higher type of animal" since it "possess a common perception organ" (Uexküll 2010, 77) that is independent from what he calls the "reflex person" (Uexküll 2010, 76) in the sea urchin, that is to say an "external organ [that] harbors a complete reflex [...] without central direction" (Uexküll 2010, 76). The sea urchin's spines retract independently of one another. In the same text, the case of the tick exacerbates the problem of human *Umwelten* fragmentation precisely when it shows its otherwise outstanding capacity of resisting for eighteen years without doing anything. In this respect, Agamben fails to say that, when he illustrates the Rostock tick experiment, Uexküll feels the need to add a note to the text in which he suggests a clear opposition between *Umwelt* and surroundings [*Umgebung*]. The tick reveals indeed the existing conflict between an *Umwelt* that has to be always "optimal" for the species and the surroundings that, on the contrary, can be "pessimal":

> Bodenheimer is entirely right when he speaks of a "pessimal" world, i.e., one as unfavorable as possible, in which most animals live. But this world is not their environment, only their surroundings. An *optimal environment*, i.e., one as favorable as possible, and *pessimal surroundings* will obtain as a general rule. For the point is that the species be preserved, no matter how many individuals perish. If the surroundings of a certain species were not pessimal, it would quickly predominate over all other species thanks to its optimal environment.
>
> (Uexküll 2010, 250, fn. 5, emphasis in the text)

It is probably no coincidence that Uexküll refers here to Shimon Fritz Bodenheimer (1897–1959),[3] a two-faced intellectual who is both an obsessive biologist and a political activist. If Bodenheimer can be credited with reviving entomology, "the 'Cinderella' of the sciences" (Harpaz 1984, 3), "in post-WWI Germany," he is also a dedicated Zionist who moved to what will become Israel as early as 1922 (Harpaz 1984, 3f.) and who in 1923 married the daughter of M. Ussishkin, one of the leaders of Russian Zionism (Harpaz 1984, 6). In Bodenheimer, insect biology and the centrality of the state go hand in hand to the point of overlapping, for instance, in his cataloguing of the existing species in Palestine (Harpaz 1984, 18) or in his study of the ecology of entomophagy, the human practice of eating insects (Harpaz 1984, 10). Although he is standing on the opposite side of the political scenario (his *Materials for the History of the Entomology until Linné* is banned by the Nazis for being a "Jewish book"; Harpaz 1984, 11), Bodenheimer works on a double track that Uexküll cannot but share some affinities with. The tick, in fact, shows the extent of a biological as well as of an *anthropological* problem. Since *Umwelt* and surroundings tend to coincide in human beings, the opposition between optimal *Umwelt* and pessimal surroundings collapses into an ambivalence. On the one hand, since human surroundings are our *Umwelt*, they can become optimal. Humans possibly realize what Uexküll passed off as a mere hypothesis in his note; they end up predominating on all the other species on planet Earth. On the other hand, the human *Umwelt* appears to reveal a pessimal side as it usually happens to the surroundings – and not the *Umwelt* – of plants and animals. The fragmentation into the individual micro-*Umwelten* of the little girl and the lumberjack is then the most blatant evidence of a species-intrinsic difficulty when it comes to ethical-political conciliation. We are a form of life that tends to dominate other animals while, at the same time, being torn apart by an ongoing internal conflict. Based on what Uexküll says, the lumberjack can indeed take no notice of the little girl, exactly as the fox could not care less about what a flying animal is doing.

3 Does the tick get bored?

It has been argued that, by referring to the example of the tick, "Agamben presents [the tick] as a sort of third term between or beyond [...] the system of the anthropological machine" (Attell 2015, 175), namely with regard to the structures of separation and domination within the animal sphere. According to Agamben, the anthropological machine is the set of social and institutional mechanisms, based on which one establishes "the articulation between human and animal, man and nonman, speaking being and living being" (Agamben 2004, 38). The tick lingering boredom is then the symptom of the possible collapse of the whole mechanism leading to conceptual artifacts such as "the man-ape, the *enfant sauvage* or *Homo ferus*, but also and above all the slave, the barbarian, and the foreigner" (Agamben 2004, 37). Current scholarly literature also maintains that Agamben's reading of Uexküll lends itself to a twofold criticism. As very clearly summed up by Oliver (2009, 230): "first, he does not consider the function of women in that binary, and second, he does not consider violence to animals" (see also Wadiwel 2004; Calarco 2008). In *The Open*, Agamben discusses indeed the relationship between animal condition and human condition, as outlined by Uexküll, without taking into any account all the problems opened by gender differences, nor does he make any remark on human violence on other forms of life. Without going into details, I believe, however, that this criticism points more to a much-needed thematic addition than to an actual theoretical lacuna. Referring to women with regard to the established operational modes of the "anthropological machine" and its devices supporting the animalization of men or the humanization of animals, could introduce an important specification to *The Open*, without nevertheless shaking its foundations. Agamben's "andrological machine" (Stone cit. in Oliver 2009, 231) could be welcomely replaced by the description of a more complex anthropological machine, but the issues connected to the machine itself and its possible stalling would remain nonetheless central. The same is true when it comes to violence on animals. The idle "letting be" that closes Agamben's discourse actually leaves the door open to an animal-friendly perspective, since it may also mean "letting the animal be" (Prozorov 2014, 169). In these core pages, I wish therefore to focus on a different critical remark which concerns the analogy between the tick and boredom, as to shed further light on the limits of Agamben's overall account on Uexküll's contributions. There are at least two main points that deserve thorough discussion but that I can only hint at here:

a Is boredom a suitable middle term bridging the human and the animal condition?

b Is the anthropological phenomenon of "boredom" fairly accounted for by Heidegger's and Agamben's descriptions of it?

As will become increasingly clear, I am inclined to argue that drawing an analogy between the eighteen years of suspended life of the tick and boredom is a very suggestive yet far-fetched move. With regard to one of boredom's two structural moments, as described by Heidegger, Agamben argues:

> In being left empty by profound boredom, something vibrates like an echo of that "essential disruption" that arises in the animal from its being exposed and taken in an "other" that is, however, never revealed to it as such.
>
> (Agamben 2004, 65)

Despite the resoundingly suggestive wording of this passage, one might still wonder what exactly is this environmental "echo" that supposedly vibrates in boredom. It is true that, as Agamben reminds us, Heidegger refers to a possible captivation of human life in relation to boredom. But it is also true that Heidegger refers to boredom as something that shakes humans out of the captivation of their daily activities. Boredom is thus the "antidote" to captivation and not one of its – even if just approximate – human forms.

Clearly, under more than one respect, Agamben's interpretation stretches far beyond its original Heideggerian framework. He first claims that one structural aspect of boredom, the "being in limbo" aspect, provides humans with a possibility that is completely alien to the animal ("the step which boredom takes beyond it [animal captivation]"; Agamben 2004, 64). Later on, however, he argues for the Rostock tick as the new touchstone for the second structural aspect of boredom. The experiment would, in fact, demonstrate that "under particular circumstances, like those which man creates in laboratories, the animal can effectively suspend its immediate relationship with its environment" (Agamben 2004, 70). This amounts to saying that the condition of the tick usually resembles one of the two structural traits of boredom, the "being left empty," but under particular circumstances, it can grasp something that resembles the other trait, the "being in limbo." Through this double comparison, Agamben sets up a questionable framework at least in light of the categories he claims to be using. Agamben suggests implicitly a gradual approach that is not at all what Heidegger argued for. The latter is indeed adamant that either we have both structural elements or there is no boredom. The German philosopher points out that these are two essential aspects and he lists them in the following order: "being in limbo" and "being left empty." Agamben subverts the sequence (defining "being left empty" as the first step and "being in limbo" as the second) to argue more easily that they are rather two *moments or stages* of boredom rather than two inseparable sides of the same coin. If Heidegger's understanding of boredom is considered to be correct, its two structural aspects cannot be read as stages; conversely, if one considers them as such, then it is not clear why one should want to use Heidegger's approach as the basis of one's argument.

In this regard, namely, the analogy between boredom and the tick, Uexküll himself offers the reader very different suggestions from the ones put forward by Agamben. The biologist claims explicitly, for instance, that the Rostock tick emphasizes the *distance* between human and nonhuman *Umwelten*: "The tick can wait eighteen years; we humans cannot" (Uexküll 2010, 52). The comparison between the two species, in fact, prompts a significant inversion: while "moments, i.e., the shortest segments of time in which the world exhibits no changes" (Uexküll 2010, 52) can last eighteen years for a tick, they only last "one-eighteenth of a second" (Uexküll 2010, 52) for a human. The recurring number, eighteen, indicates the distance rather than the proximity, between the two *Umwelten* that are so far away that they have an inverse relationship. Only sleep ("a state similar to sleep"; Uexküll 2010, 52) seems to provide a common comparison.[4] However, even the analogy with sleep proves equally misleading for one specific reason, as emphasized by the footnote I quoted earlier. The reason is as simple as fundamental: that of the Rostock tick is a strategy of survival. Although he was not interested in Darwin's "struggle of life," as he was rather a supporter of the harmonious idea of the biological kingdom, Uexküll here clarifies a recurring problem: every animal looks for a "security" (Uexküll 2010, 96) of behavior in its own *Umwelt* against adverse, hostile, dangerous surroundings. Contemporary biology defines as "diapause" (Apanaskevich and Oliver 2014, 72) the possibility the tick has, along with other invertebrates, to suspend its own activities. According to one of the most important scholars of the physiology of these parasites, this state is distinct from "quiescence" for the following reasons:

> *Quiescence* is an arrest of development or activity arisen under direct impact of either adverse environmental conditions, or deficit of vitally essential factors, that is due to some exogenous constraints, and recovered after cessation of their action. By contrast, *diapause* is an anticipated arrest of development or activity arisen according to internal program, either genetically fixed, or induced through effect of token factors signaling in time an approach of unfavorable conditions, and eliminated by special mechanism of reactivation.
> (Belozerov 2008, 80, emphasis in the text)

Back to the issue mentioned by Uexküll, when there are signals anticipating such adverse conditions of the surroundings that they can also become environmentally dangerous (e.g., freezing temperature, absolute scarcity of forms of life), the animal reacts with a state similar to sleep that is thus not at all "a mystery" (pace Agamben 2004, 70). The diapause is a conflicting biological response to adverse conditions. This aspect is finally heavily smoothened by Uexküll, whose vision of nature is harmoniously organized by what he calls *Naturplan* (natural plan; Uexküll 2010, 86). Differently, by relying on Heidegger's notion of boredom to understand the tick's behavior, Agamben ends up completely downplaying the problem of the intrinsically conflicting characters of both human and nonhuman nature.

On this matter, an inversion might be worth a try. Rather than reading Uexküll's tick as something akin to Heidegger's boredom in the animal sphere, it could be interesting to reconsider the merely anthropological phenomenon of boredom following the cue offered by the tick itself and its strategies of survival. Far be it from me to suggest further reckless analogies, but Uexküll's tick can perhaps provide the opportunity to read in conflicting terms a series of very different phenomena that all share the feature of looking calm and quiet. Both the invertebrate's diapause and *Homo sapiens'* boredom seem to be, each in its own way, harmonious forms of suspension. In both cases, looks can be deceiving – although in very different ways. Under apparent stasis, conflict is brewing.

4 The anthropology of boredom: endless regression and conflict with the limit

Agamben seems to fully rely on Heidegger on one more point. Heidegger explicitly conflates a state of mind he considers crucial for the *Dasein* – thus for any human being – with the description of boredom offered by his own language:

> This *profound boredom is the fundamental attunement*. [...] Boredom, long time: especially in the Allemannic usage it is not an accident that 'to have a long time' means the same as to 'homesick'. [...] Boredom, *Langeweile* – whatever its ultimate essence may be – shows, particularly in our German world, an obvious *relation to time*, a way in which we stand with respect to time, a feeling of time. Boredom and the question of boredom thus lead us to the problem of time.
>
> (Heidegger 1995, 80, § 20, emphasis in the text)

Agamben adopts this line of reasoning without discussing it, yet it conveys the untenable idea that the German language reveals a fundamental state of mind for any speaker, "for any *sapiens*." It is not stated where such linguistic superiority – or, if you prefer, such curious coincidence between human nature and the German language – derives from. In *The Open*, the reason for this tacit assumption seems to be a matter of convenience: Heidegger's phenomenological definition of boredom – but not, as we have seen, its structural analysis – is well suited to Agamben's proposed reading of Uexküll's tick. Boredom is thus a temporal state of suspension: *Langeweile* means literally "a long whiling of time" (Dalle Pezze and Salzani 2009, 10). Stripped of its chauvinist and metaphysical assumptions, one could possibly do justice to Heidegger's analysis, by specifically working on an anthropology of boredom that is able to grasp the relation between this state of mind and human nature on broader historical and cultural grounds.

For want of a natural history of boredom (see for instance: Spacks 1995; Svendsen 2004; Toohey 2011), we could be content with, for now at least, distinguishing two traditions. One finds its flag in the Latin word *tedium*. This is the cultural tradition Heidegger and Agamben draw on and that ranges, roughly speaking, from 1st-century CE Latin stoicism to the 17th-century appearance of *to bore* in the English lexicon in the contemporary sense of the word. Second, we could also try and identify a further tradition in the Western world, probably no less important although still to be pieced together. In order to do so, we will have to seriously consider the etymological origin of the Italian word *noia*, of the French *ennui* or of the Spanish *aburrimiento*. They are all expressions of the Latin *inodiare, in odium esse* (Dalle Pezze and Salzani 2009, 9). Such "being in hate" seems to suggest that boredom also has to do with clashing against something that produces difficulty. Heidegger explicitly considers this line nothing but a superficial form of boredom as suspension. Conflict is thus only a form of denial of boredom, a flight from experiencing a state of mind that is on the contrary fundamental. Unlike what Heidegger believes, what today we call *boredom, Langeweile, noia*, and *ennui* suggests that the two elements of suspension and conflict are equally important aspects of this state of mind and should be placed at the same anthropological level. Perhaps the two traditions, *tedium* and *in odium esse*, both focus on particular aspects of this state of mind. Not only there is the conflict of those who delete boredom, but there is also a conflict inherent in boredom itself. The conflict could not only be a medium to get rid of boredom but also one of its fundamental drives, a structural element of its inner logics. Mainstream understanding portrays boredom as contemplative and as something we fight against. Differently, boredom could be seen as the passion of conflict.

While simplifying the matter to an extent, I try to describe these traditions from a different point of view. Well suited to my purpose, Paolo Virno (2010, 71) defines *tedium* with great clarity as "the naked experience of regress" *ad infinitum*. The logical crescendo described by Heidegger would then not be the refinery that enables us to grasp pure boredom, but the structure itself of this state of mind, whose key element – even more central than time – is the tendency toward a "regression ad infinitum," its "immediate sentimental equivalent" (Virno 2010, 70). Boredom is passion for the infinite regressing that is typical of human language. In this regression, any logical plan is overcome by a superior order and, at the same time, as Virno argues, "a certain limit, either cognitive or pragmatic, is confirmed by its very transgression" (Virno 2010, 71). To give a far too banal example, not only can I get bored in boredom, but I can also get bored by the fact that I am bored. This second-degree boredom indeed is no less boring than first-degree boredom. I thus try to overstep a limit through a metalinguistic structure (I am bored with boredom), but the limit arises again (I am bored). One of the many merits of Virno's elaboration is to insist on boredom as logical-emotional crescendo that, indeed, has no culmination. On the contrary, Heidegger takes as the original moment the transgression of such confrontation with the limit. Through this transgression, called "profound boredom" (Heidegger 1995, 138), the human world is revealed as such, thanks to "indifference enveloping beings as a whole." It is no coincidence that profound boredom is the most metaphysical and the least anthropological of the three types of boredom (i.e., "being in limbo," "being left empty," "profound boredom") described by the philosopher because it is precisely what provides access to "something essential" (Heidegger 1995, 160).

And here we come to the second aspect, probably the most controversial, associated with boredom as *inodiare*. My proposal is to see boredom also as a form of clash, a conflict whose limit keeps coming back. Boredom is not absolute difference but, rather, a "frenzied difference," a "turbulent overcoming," as Virno argues (2010, 70, 72), and a "delicate monster" in the words of Dalle Pezze and

Salzani (2009). Why is this difference "frenzied"? Why the "turbulent" overcoming or the delicacy associated to a "monster"? I offer my contribution to the debate with the following hypothesis. Adjectives are as important as the nouns they refer to. Boredom is a frenzied or turbulent condition – or a monster – because when I am bored, the limit is greatly indefinite, and for this, it is something that hinders me because it is not accurately localized or discreetly isolable. Let me suggest an inversion: what if it is precisely the concept that Heidegger despises the most, the *Zeitvertreib*,[5] that plays an agonistic element constitutive of boredom as regression *ad infinitum*? In this case, the clash is not necessarily the superficial antidote for those who do not wish to get bored because they want to go dancing and have fun. Tedious are the dynamics of those who "bump" into a limit that keeps coming back. Precisely because it has no zenith, boredom is always bound to the presence of a difficulty and a conflict that leads to its – always temporary – overcoming. Such interpretation of boredom allows going past Heidegger's account of boredom as *tedium* and, at the same time, not to confine the linguistic-cultural line that connects boredom to hatred or to annoyance to a superficial misunderstanding. The etymology that makes boredom a derivative form of *inodiare* does not signal any difficulty or a generic disgust: it is a sediment of the difficulty of the metalinguistic turn to a superior logical level that characterizes boredom. The point is not to distinguish, following Heidegger, between "simple" and "existential" (Toohey 2011, 130) boredom. We need instead to seriously consider and define in detail something that is recurrent in the descriptions and in the words used, by many a language and in many a historical period, in relation to this state of mind. Such difficulty is not the appropriate expression of who is not yet refined enough to grasp what could be the real essence of a state of mind. If boredom is actually a linguistic twist, the obstacle intrinsic to boredom is bound to the conflicting friction produced by this twist itself.

In this regard, it is possible to find a particularly interesting number of cues in ancient Greece, one of the major linguistic and cultural traditions in the Western world. In this tradition, such logical difficulty, in fact, appears to suggest conflict in the public arena. As is known, "there is no word for boredom in ancient Greek" (Bruss 2012, 313). Nonetheless, this absence is still unclear today (neither Heidegger nor Agamben examines it). How come the ancient Greeks do not have a suitable word to indicate such a crucial state of mind? The first hypothesis is that they were too primitive to do so. The second appears to be less prejudiced as it tends to consider the Greeks, more simply, as "a remarkably foreign people" (Bruss 2012, 314). A more in-depth analysis has led, instead, to identify a wide variety of candidates as close relatives of the current term boredom. Such variety seems to indicate a different focus on the experience of boredom rather than a presumed historical and cultural backwardness. The possibilities include *alus* (agitation), *apatheia* (apathy), *akedia* (indifference, torpor), *koros* (satiety, surfeit), *aniaô* (grieve, distress), and *enochleô* (trouble, annoy; Toohey 1988). The nearest seems to be *alus*, which is close to a "boredom disgusting" (Toohey 1987, 200). However, one should specify that this is a late coinage associated with Plutarch, who lived in the 2nd century CE (Toohey 1987, 1988, 155ff.). Some of the other terms are characterized by a generic quality (*koros, aniaô, apatheia*) or strongly prejudiced by a subsequent reading (see, for instance, *akedia*, the Medieval *acedia*: Agamben 1977; Mazzeo 2012). Most of them, however, share a common attribute that should not be unduly underestimated: they are all characterized by difficulty.

But there is probably more. Some of the Greek terms whose meanings are closer to the contemporary word boredom seem to bring such difficulty, which we have assumed to be logical-linguistic, toward explicitly political and conflicting features,[6] those of an actual *in odium esse*. The word *ochlos* (as well as the whole semantic family of the verb *enochleô*) is probably the most interesting, since it may be the old link of a tradition about boredom that needs still to be investigated. When investigating Antiphon's (5th-century BCE) use of this word, *ochlos* emerges as a relative of boredom which is indeed connected with the terms "crowd, mob, or multitude" (Bruss 2012, 321). Boredom can be the desire of the multitude: of the thoughts that crowd the metalinguistic turns of an ongoing regression *ad infinitum*, of the human beings that trigger it by flocking there. Ancient Greece provides us with a representation of boredom that is the opposite of a desire for isolation and conciliation, and analogous to the lethargic diapause of the tick in Uexküll. It is not the desire of the organism pacified in its suspension, nor "the hatred of the genius for half measures" (Lepenies 1992, 118). It is the desire, as Antiphon argues, of those who are still not able to grasp the conflicting moment, the *kairòs*, the right moment for the rhetorical intervention and the political clash (Lepenies 1992, 323).

Notes

1 See also, for instance, Uexküll (1920, 39), on this polarized distinction which assigns the *Familienumwelt* to women and the *Berufumwelt* to men.
2 For a discussion of Uexküll's political ideas and his relation with National Socialism, see Harrington (1996). For a good overview of Uexküll's ideas on the national State and his political importance: Heredia (2011) and Heredia, Chapter 2, in this volume.
3 Regrettably, Uexküll does not mention his source in detail. In support of this identification, see Le Bot (2016, 199, fn. 8).
4 In one of his courses at the College de France, Merleau-Ponty (1995, 227, 229) correctly refers to *léthargie* in reference to Uexküll's tick, emphasizing Uexküll's aim to focus on the fact that "space and time are relative" (Oliver 2009, 214).
5 The English translation suggests "passing the time," which does not sound effective. The German *vertreiben* means "to drive away," as rightly indicated by the translator in a footnote: Heidegger (1995, 93, note 1).
6 As Walter Benjamin argues, for instance: "Boredom becomes an element in the process of production with its acceleration (by machinery). The *flâneur* protests, with his obstentatious langour [*Gelassenheit*] against the process of production" (Benjamin 1985, 47). For examples of the intertwining of boredom and social conflict: Lepenies (1992, cap. IV).

References

Agamben, Giorgio (1993) [1977] *Stanzas: Word and Phantom in Western Culture*. Translated by Ronald L. Martinez. Minneapolis/London: University of Minnesota Press.

Agamben, Giorgio (2004) [2002] *The Open: Man and Animal*. Translated by Kevin Attell. Stanford: Stanford University Press.

Apanaskevich, Dmitry A. and Oliver, James H., Jr. (2014) 'Life cycles and natural history of ticks'. In: Daniel E. Sonenshine and R. Michael Roepp (eds.) *Biology of Ticks*. Vol. 1. Oxford: Oxford University Press, 59–73.

Attell, Kevin (2015) *Giorgio Agamben: Beyond the Threshold of Deconstruction*. New York: Fordham University Press.

Belozerov, Valentin N. (2008) 'Diapause and quiescence as two main kinds of Dormancy and their significance in life cycles of mites and ticks (Chelicerata: Arachnida: Acari)'. Part 1: 'Acariformes'. *Acarina* 16 (2), 79–130.

Benjamin, Walter (1985) [1961] 'Central Park'. Translated by Loyd Spencer and Mark Harrington. *New German Critique* 34, 32–58.

Brentari, Carlo (2015) *Jakob von Uexküll: The Discovery of the Umwelt between Biosemiotics and Theoretical Biology*. Dordrecht/Heidelberg/New York/London: Springer.

Bruss, Kristine (2012) 'Searching for boredom in ancient Greek rhetoric: Clues in Isocrates'. *Philosophy and Rhetoric* 45 (3), 312–334.

Calarco, Matthew (2008) *Zoographies: The Question of the Animal from Heidegger to Derrida*. New York: Columbia University Press.

Dalle Pezze, Barbara and Salzani, Carlo (2009) 'The delicate monster: Modernity and boredom'. In: Barbara Dalle Pezze and Carlo Salzani (eds.) *Essays on Boredom and Modernity*. Amsterdam/New York: Rodopi, 5–34.

de la Durantaye, Leland (2009) *Giorgio Agamben: A Critical Introduction*. Stanford: Stanford University Press.

Esposito, Roberto (2008) [2007] *Bios: Biopolitics and Philosophy*. Translated by Timothy Campbell. Minneapolis/London: University of Minnesota Press.

Geroux, Robert (2012) 'Umwelt, biology, and bio-politics: A foray into Uexküll's world' (book review). *Humanimalia* 3 (2), 142–147.

Harpaz, Isaac (1984) 'Shimon Fritz Bodenheimer (1897–1959): Idealist, scholar, scientist'. *Annual Review of Entomology* 29, 1–23.

Harrington, Anne (1996) *Reenchanted Science: Holism in German Culture from Wilhelm II to Hitler*. Princeton: Princeton University Press.

Heidegger, Martin (1995) [1929/1930] *The Fundamental Concepts of Metaphysics: World, Finitude, Solitude*. Translated by William McNeill and Nicholas Walker. Bloomington and Indianapolis: Indiana University Press.

Heredia, Juan Manuel (2011) 'Etología animal, ontología y biopolítica en Jakob von Uexküll'. *Filosofia e História da Biologia* 6 (1), 69–86.

Le Bot, Jean-Michel (2016) 'Renouveler le regard sur les mondes animaux. De Jakob von Uexküll à Jean Gagnepain'. *Tétralogiques* 21, 195–218.

Lepenies, Wolf (1992) [1969] *Melancholy and Society*. Translated by Jeremy Gaines and Doris Jones. Cambridge, MA: Harvard University Press.

Mazzeo, Marco (2010) 'Il biologo degli ambienti: Uexküll, il cane guida e la crisi dello Stato'. In: Jakob von Uexküll. *Ambienti animali e ambienti umani. Una passeggiata in mondi sconosciuti e invisibili*. Translated by Marco Mazzeo. Macerata: Quodlibet, 7–33.

Mazzeo, Marco (2012) *Melanconia e rivoluzione. Antropologia di una passione perduta*. Roma: Editori Internazionali Riuniti.

Merleau-Ponty, Maurice (1995) [1957–1958] 'Le Concept de nature. L'animalité, le corps humaine, passage à la culture'. In: Maurice Merleau-Ponty. *La nature. Note. Cours au Collège de France*. Paris: Éditions du Seuil, 167–259.

Oliver, Kelly (2009) *Animal Lessons: How They Teach Us to Be Human*. New York: Columbia University Press.

Prozorov, Sergei (2014) *Agamben and Politics: A Critical Introduction*. Edinburgh: Edinburgh University Press.

Spacks, Patricia Meyer (1995) *Boredom: The Literary History of a State of Mind*. Chicago/London: University of Chicago Press.

Svendsen, Lars Fr. H. (2004) *A Philosophy of Boredom*. Translated by John Irons. London: Reaktion Books.

Toohey, Peter (1987) 'Plutarch, Pyrrh. 13: ἅλυς ναυτιώδης'. *Glotta* 65 (3/4), 199–202.

Toohey, Peter (1988) 'Some ancient notions of boredom'. *Illinois Classical Studies* 13 (1), 151–164.

Toohey, Peter (2011) *Boredom: A Lively History*. New Haven/London: Yale University Press.

Uexküll, Jakob von (1920) *Staatsbiologie: Anatomie-Physiologie-Pathologie des Staates*. Hamburg: Hanseatische Verlagsanstalt.

Uexküll, Jakob von (1973) [1928] *Theoretische Biologie* (2nd. ed). Frankfurt a. M.: Suhrkamp.

Uexküll, Jakob von (2010) [1934, 1940] *A Foray into the Worlds of Animals and Humans with a Theory of Meaning*. Translated by Joseph D. O'Neil. Minneapolis/London: University of Minneapolis Press.

Virno, Paolo (2010) *E così via all'infinito. Logica e antropologia*. Torino: Bollati Boringhieri.

Wadiwel, Dinesh (2004) 'Animal by any other name? Patterson and Agamben discuss animal (and human) life'. *Borderlands* 3 (1) (book review).

Winthrop-Young, Geoffrey (2010) 'Afterword: Bubbles and web: A backdoor stroll through the readings of Uexküll'. In: Jakob von Uexküll (2010) [1934, 1940] *A Foray into the Worlds of Animals and Humans with a Theory of Meaning*. Translated by Joseph D. O'Neil. Minneapolis/London: University of Minneapolis Press, 209–243.

13 Jakob von Uexküll and the study of primary meaning-making[1]

Kalevi Kull

1 Uexküll studies: paving the way to post-Darwinian, semiotic biology

It is remarkable how many scholars have found Jakob von Uexküll's understanding of living beings and his approach to biology valuable and worth learning from. Since the "rediscovery" of Uexküll in the late 1970s, the number of works about him shows a continuous growth. Even after the publication of a large interdisciplinary volume about Uexküll's work in 2001 (*Semiotica* Vol. 134), the interest toward his work has not diminished. Bibliographical records on Uexküll after 2001[2] include

- monographs about Uexküll (Mildenberger 2007; Gens 2014; Brentari 2015),
- books that provide theoretical interpretations of his work (Buchanan 2008; Clausberg 2006; Favareau 2010; Hoffmeyer 2008; Kliková and Kleisner 2006; Maran, Martinelli, and Turovski 2011; Martinelli 2010; Mildenberger and Hermann 2014; Block 2016),
- special issues of journals (*Journal of Comparative Psychology* 116 [2], 2002; *Sign Systems Studies* 32 [1/2], 2004),
- more than a hundred important research articles and essays that interpret Uexküll's work (list available on request), and
- new translations of Uexküll's texts (at least into English, French, Estonian, Spanish, Italian, Slovenian).

The archive of the Jakob von Uexküll Centre in Tartu has also considerably grown,[3] and a series of Jakob-von-Uexküll Lectures is organized at the University of Tartu.

Why is it that the work of a scholar, who, although rather well-known and popular in the first decades of the 20th century, was later not only forgotten but whose views were considered largely inappropriate (if not wrong) in the light of the formulation of the Modern Synthesis and the success of quantitative biology in the 1930s, reappear on the scene a hundred years later? There is no doubt that within the first two decades of the 21st century more has been written about Uexküll than throughout the whole of the 20th century.

Browsing through recent publications on Uexküll, at least two sources of interest can be uncovered. One of them stems from the surge of ecological themes in the humanities and the appearance of new fields of inquiry, such as environmental humanities, human–animal studies, post-humanities, and ecocritical studies. Uexküll's framework fits these approaches rather well. Here we find essays using Uexküll's notions in cinema research, literature studies, or ecophilosophy. This even extends to the referencing of Uexküll's ideas in art projects. Another, no less deep reason, is indirectly connected to the first. I would point here to the profound ongoing change in biological knowledge that can be observed in the last decades. This paradigmatic change is rocking the fundamentals of evolutionary biology, in particular. Instead of taking genetic mutations and natural selection as the major factors causing the observable changes in the world of life, an alternative explanation assigns a key role to the activity of living agents, notably to the choices that the organisms make based on the meanings assigned by them. This new theoretical biology is a semiotic biology, and it finds some formulations of its principles – or hints to these – in the works of Jakob von Uexküll.[4]

The current change in biology is even more profound than just described. It is not limited to modifications to the evolutionary theory, the leading theory in biology if not since Darwin, then at least since the 1930s. Biology is now reintegrating subjectivity into its models. After many earlier attempts to find a place for intentions, feelings, and mind in organic systems, the biosemiotic project (if developed with care) looks very promising. At the core of semiotic biology (or biosemiotics) lies the mechanism of choice-making as actuated by living beings. Far from being entirely discovered, in other words, far from being described in all sufficient details, this mechanism, or rather the bundle of organic choice-mechanisms, can also be described as the mechanism of interpretation, or meaning-making, or semiosis. In this respect, what can be confidently said is that Uexküll's model of a functional circle together with the phenomena the functional circle creates – the meanings, the signs, and the *Umwelt* – is clearly an account on the core of semiotic biology.

Finding and describing the primary mechanism of meaning-making turns out to be a rather difficult task – possibly belonging to the most difficult (or puzzled) problems science has faced. Why so? A synonym for meaning-making is semiosis. The central event of semiosis is interpretation. "Demonstrating that an empirically realistic simple molecular system can exhibit interpretive properties is the critical first step toward a scientific biosemiotic theory," says Terrence Deacon (2015, 310). Interpretation is, by definition, a nondeterministic process – it includes the choice between distinct possibilities, between options. Thus, one needs to identify a molecular (physiological) system that itself can choose. Choice assumes the simultaneous existence of possibilities. Simultaneity that can still embed a process needs to have an extension. Thus, what is to be demonstrated is no less than the origin of the phenomenal present – certainly a "hard problem." This, as it turns out, is at the same time the problem of free choice. To approach this problem, the proper vantage point for its analysis and solution is obviously semiotics, and particularly biosemiotics, for the biosemiotics' field of study explicitly includes the simple – and the simplest – mechanisms of meaning-making or semiosis – the primary emergence of sign relations and subjectivity.

Pioneering studies on mechanisms of meaning-making in simple organisms (if there is such a thing as a "simple organism") were carried out by Jakob von Uexküll. Uexküll is not only seen as a classic reference in biosemiotics (Favareau 2010), but also – since the 1980s – as one of the major classics in general semiotics, together with Charles Sanders Peirce, Ferdinand de Saussure, Charles Morris, Roman Jakobson, Juri Lotman, Umberto Eco, Thomas Sebeok, and few others (Krampen *et al.* 1987; Deely 2004; 2012, 214). This shows to what extent fundamental biosemiotic principles are an important part of general semiotics. Also traditional biology, one should add, has included Uexküll into its list of classics (e.g., Hassenstein 2001). In conclusion, browsing through the writings listed earlier leads to a firm observation: within the last couple of decades, Uexküll has become a classic author. He may even become popular.[5] My remarks here focus on the role of Jakob von Uexküll's ideas in the study and understanding of primary mechanisms of meaning-making. As Uexküll maintained that *Umwelt* and meaningfulness stem from functional circles, this chapter investigates closely the notion of a functional circle. Rather than providing a comprehensive survey of the problem of primary meaning-making, I briefly outline the course of thought that leads to the contemporary hypothesis (or possible conclusion) stating that subjectivity and meaning-making appear together with the phenomenal present, early in the evolution of life, and that Uexküll's concepts provide a good starting point to this reasoning.

2 The problem of the primary mechanism of meaning-making

According to Thomas Sebeok's dictum, semiosis is coextensive with life (Umiker-Sebeok and Sebeok 1980, 1). Sebeok starts his textbook on semiotics with the statement: "The phenomenon that distinguishes life forms from inanimate objects is semiosis" (Sebeok 2001, 1). This statement, however, should be taken as a hypothesis, until the minimal mechanism of semiosis has not been properly described and identified.

The problem of the minimal mechanism of semiosis was introduced into semiotics in explicit and clear terms by Umberto Eco (1976), who formulated the concept of lower semiotic threshold (see Rodríguez Higuera and Kull 2017). According to Eco, what distinguishes the semiosic world from the nonsemiosic is the existence of codes, in other words, those correspondences that are not products of physical or stereochemical forces but are acquired: the correspondences that exist due to historical or social reasons, due to interpretation.

Code-making has been taken as semiosis's key feature also by Marcello Barbieri. According to him, molecular coding is a type of semiosis that exists in cells (Barbieri 2015, 33). Jesper Hoffmeyer and Claus Emmeche (1991) have argued that minimal meaning-making requires code duality. This means that one code is not enough for semiosis; at least two codes, or code plurality, are required. The issue, however, concerns more than just the number of codes. According to the code plurality point of view, the assumption that "the meaning of functional information is grounded in a communication system, which is a set of compatible communicating agents" (Sharov 2013, 347) is insufficient. Differently, meaning-making requires a certain degree of incompatibility – or partial nontranslatability, as Juri Lotman puts it – among codes or communicating agents.

However, the code concept is not indispensable for the description of meaning-making. The concept of code was not used, for instance, by Charles Sanders Peirce, nor is it used by Terrence Deacon (2011) in his explanation of crossing the threshold between morphodynamics and teleodynamics – the latter identified as semiosic – sometimes presented as the origin of mind from matter. The emergence of meaning-making or semiosis has also been understood as the origin of subjectivity or phenomenal experience – or interpretation, in Umberto Eco's terms.

Subjectivity, as well as interpretation, is characterized by the existence of possibilities, of options, of non-automaticity. Provided that the explanation of a phenomenon means finding an algorithm that can produce the phenomenon, the explanation of subjectivity would require finding an algorithm for a non-algorithmic situation, that is to say, for a process that is not automatic in the sense of algorithmicity. The problem of the distinction between code and interpretation corresponds then to the issue of determinacy versus radical indeterminacy, which has been – and still is – cause of many serious debates in linguistics and semiotics (Cobley 2017). For the study of minimal semiosis, one can find helpful insights, at least to some extent, in Charles Sanders Peirce's semiotic model. Peirce describes the minimal sign as qualisign, or quale, or tone, and according to him, there cannot be signs without phenomenal qualities. He also finds that "there is a fair ana-logical inference that all protoplasm feels. It not only feels but exercises all the functions of mind" (Peirce 1965, VI, 167, CP 6.255). Different interpretations of Peirce's texts are possible. Peirce is above all a gradualist or, as he would put it, a synechist; "synechism," according to Peirce, being "the doctrine that all that exists is continuous" (Peirce 1965, I, 70, CP 1.172). The problem of continuity *versus* threshold has been solved – as Jesper Hoffmeyer has duly emphasized – by the introduction of the concept of lower semiotic threshold "zone" (Kull *et al.* 2009). Based on this notion, the boundary between the nonsemiosic and the semi-osic may not be sharp. Within the lower-threshold zone, the characteristics of mind may not appear all in one step. Accordingly, the emergence of symbolicity, which is a big step accounting for the basis of human language, can be understood as developing in substeps in its threshold zone. It is to be concluded, then, that the width or the steepness of the zone may vary.

3 Uexküll's accounts on functional circle and *Umwelt*

Jakob von Uexküll is responsible for introducing a few fundamental concepts into biology, which have made it possible to study the semiotic aspect of life in general: functional circle, *Umwelt*, meaning, perceptual sign, action sign, and so on.[6] The identification of the limits of their applicability – together with specifi-cation of their definitions – would help define the limits of semiosis in general.

The basic as well as the minimal mechanism of meaning-making, according to Uexküll, is the functional circle [*Funktionskreis*]. While the term first appeared presumably only in 1919 (Uexküll 1919, 144f.), the formulation of this concept started with Uexküll's work on neural feedback mechanisms of muscular activity and tonus in the physiological laboratory in Heidelberg and at the Naples Biological Station in the 1890s (see Mildenberger 2007; Brentari 2015; Kull 2001; see also Heredia's and Köchy's contributions, Chapters 1 and 3, in this volume). A further step toward the notion of a functional circle was achieved through the formulation of the rules of the excitation process (Uexküll 1904), one of which became known as Uexküll's law. In its almost-complete form, the concept of functional circle as the fundamental model of behavior and meaning-making was described in the second edition of Uexküll's monograph Umwelt *and Inner-World of Animals* [*Umwelt und Innenwelt der Tiere*, Uexküll 1921; see also the commented publication of this book in Mildenberger and Hermann 2014]; an updated version is instead available in the second edition of his *Theoretical Biology* (Uexküll 1928).

Most of Uexküll's examples describe *Funktionskreise* in multicellular animals. However, he also studied some unicellular organisms such as *Amoeba, Paramecium*, and *Didinium*[7] which are protists and do not belong to the *Animalia* kingdom.[8] In the chapter "Paramaecium" of Umwelt *and Inner-World of Animals* (Uexküll 1909, 39–53, 1921, 32–44), Uexküll provides a detailed description of the behavior of the unicellular *Paramecium caudatum* as a species of *Infusoria*. Speaking of *Paramecium*, he says it is an organism with only one functional circle. This means that, according to Uexküll, even a single cell has enough complexity to have a *Funktionskreis*. Indeed, Uexküll speaks explicitly also about the functional circle and subjectness of cells [*Zellsubjekt*] in multicellular organisms (Uexküll 1928, 191f., 1931, 386; see also Baer 1988, 210).

The book *A Foray into the Worlds of Animals and Humans* (Uexküll and Kriszat 1934) provides an example that establishes a comparison and distinction between the surroundings [*Umgebung*] and the *Umwelt* of a *Paramecium* (also in Uexküll and Kriszat 1956, 49; Uexküll 1992, 342). Uexküll states that the functional circle and *Umwelt* appear together. Only if an organism has receptor units and effector units may it construct and command its *Umwelt* (Uexküll 1982, 33). Thus, a cell as an organism which has its receptor and effector organelles may have a functional circle and an *Umwelt*.[9] In this respect, it is very instructive to read what Uexküll writes about a slime mold [*Schleimpilz*], belonging to the order *Dictyosteliida*:

> We are forced to attribute an *Umwelt*, however limited, to the free-living fungus-cells [*Amöben* K.K.], an *Umwelt* common to each of them, in which the bacteria contrast with their surroundings, as meaning-carriers, as food and, in doing so, are perceived and acted upon. On the other hand, the fungus, composed of many single cells, is a plant that possesses no animal *Umwelt* – it is surrounded only by a dwelling-integument [*Wohnhülle*] consisting of meaning-factors.
>
> (Uexküll and Kriszat 1956, 113; Uexküll 1982, 35f.)

What Uexküll means is that single cells which have receptor and effector units (i.e., organelles) – hence a functional circle – may also have an *Umwelt*, while a collective of cells like the fruit body of a mushroom or a multicellular plant as a whole does not need to have an *Umwelt*. A vegetative living body as a cell colony may lack its colony-level functional circle, while the cells themselves of which the colony consists do have their functional circles and *Umwelten*.

Uexküll did not work with bacteria, but based on his reasoning, it is fair to infer that he would have assigned a functional circle and an *Umwelt* also to the bacteria which have receptors and flagella. Moreover, since Uexküll's time, many more types of cellular receptors and effectors have been discovered, which makes it difficult to find a living cell without a possible functional circle. Studies on plants have shown the existence of receptors and effectors and specific conductance between them in a vascular plant organism (Calvo Garzón and Keijzer 2011; Witzany and Baluška 2012), which means that some simple functional circles (and, accordingly, *Umwelten*) may exist also at the supracellular level of plants. Uexküll's claim that plants do not have *Umwelten* should therefore be dismissed, or if anything, handled with caution.

Another important notion related to the notion of *Umwelt* is that of *Plan*. "Plan versus matter is the watchword of the new science of life," states Uexküll (2001, 123). What is a plan? Uexküll explains that *Umwelten*

do not interact mechanically but are still connected according to a plan as the notes of an oratorio are harmonically connected. It is thus musical and not mechanical laws that we need to study if we want to find out about the laws of Life.

(Uexküll 2001, 117)

He then adds that "the structure of the sensory organs is itself dependent on the total building-plan of the animal, whose *Umwelt* is composed in harmony with other *Umwelten*" (Uexküll 2001, 119). And finally he claims that "all properties of living creatures we find connected to units according to a plan, and these units are contrapuntally matched to the properties of other units" (Uexküll 2001, 122).

Based on the examples Uexküll provides, one can conclude that his concept of plan [*Plan*] corresponds rather closely to what semioticians of the 20th century have termed structure and his concept of rule [*Regel*] to what has been defined in biosemiotics as code. Similarities can be found in particular cases, for instance, with Barbieri's (2015) account of codes. Based on such reading, Uexküll turns out to be a structuralist or even a code biologist. However, while the latter focuses on intraorganismal codes, Uexküll describes both ecological rules (or "codes") that relate organisms to their *Umwelten* and other organisms, and the developmental rules.

As far as Uexküll's usage of the term *meaning* [*Bedeutung*] is concerned, one can trace a slow development from common word to clear, strategical concept. Based on its contexts of usage in Uexküll's texts across the years, one can see that "meaning" becomes a systematically applied concept only in the 1930s, most remarkably in the book *The Theory of Meaning* [*Bedeutungslehre*] (Uexküll 1940).

Differently, the term *sign* [*Zeichen*] appears endowed of conceptual relevance already in the Umwelt *and the Inner-World of Animals* (e.g., Uexküll 1909, 59, 192, 250, etc.). Uexküll also speaks about perception signs [*Merkzeichen*] (Uexküll 1909, 88, 203) and the sign language [*Zeichensprache*] of the brain (Uexküll 1909, 192, 194). The term *action sign* [*Wirkzeichen*] was instead adopted somewhat later (Uexküll 1928, 14, 15, etc.).

4 Cycle or circle

It is worth remarking that Uexküll's technical term *Funktionskreis* has been translated into English in two ways: functional *cycle* (e.g., Krampen 1981, 194;[10] Sebeok 2001, 100, 144) and functional *circle* (e.g., Thure von Uexküll and Wesiack 1997; Hoffmeyer 2008, 173, 215; Brentari 2015). Some English translations of Uexküll's works use *functional cycle* (e.g., Uexküll 1992, 2010), while others use *functional circle* (Uexküll 1926, 1982). Are there arguments for preferring one to the other?

The concept of "cycle" usually stands for a closed "sequence" of processes, a loop of feedback. Sebeok (2001, 100) describes this in the following terms: "Everything in this phenomenal world, or self-world, is labeled with the subject's perceptual cues and effector cues, which operate via a feedback loop that Uexküll called the functional cycle."[11]

The concept of "circle," on the contrary, depicts a loop as a whole, without assuming the sequentiality of the processes in it. Rather than a sequence, a circle is a *Gestalt*, a whole that is perceived as one.

Thus, in order to decide whether the *Funktionskreis* is a cycle or a circle, one should decide whether the *Funktionskreis* as the mechanism of meaning-making is sequential or momentary, diachronic or synchronic. Uexküll himself would probably say that it is both:

> But since the structure is constructed by the impulse-sequence in conformity to a plan, its action in the outer world is both in accordance with plan and also automatic. The question concerning the conformity to a plan is the business of biology; the question concerning the mechanical running belongs to physiology.[12]
>
> (Uexküll 1920, 95, 1926, 125)

Indeed, if the *Funktionskreis* was only a row of processes (sequential, belonging exclusively to physiology), it could not create phenomena and it could not create an *Umwelt*.

A lack of distinction between sequential and instantaneous, cycle and circle, has led, in some cases, to misinterpretations concerning animal behavior. For instance, concerning the famous tick example reported by Uexküll, the sequence of events alone is indeed insufficient for a meaning-making mechanism. At some point, the tick should face choice; it should make decisions between options; only then it can have a subjective world and an *Umwelt*.

In conclusion, what seems to be a merely secondary detail of translation, hides in itself a major difference. *Umwelt* as a subjective world can only be there if the

Funktionskreis is a "circle" – a whole, a *Gestalt*, a simultaneous event. Certainly, the *Funktionskreis* is also a "cycle" – a sequence of processes. However, since a cycle is not enough to create an *Umwelt*, it cannot only be a cycle.

5 *Funktionskreis* and some of its further developments: *Regelkreis, Gestaltkreis, Situationskreis, Lebenskreis*

Uexküll's notion of *Funktionskreis* has generated a remarkable series of similar or derivative concepts. Some of these are listed and investigated in what follows, although a more detailed comparative and integrative analysis of their implicit models would contribute greatly to the elucidation of the core mechanisms of meaning-making. Before Uexküll, already John Dewey had emphasized the element of circularity in the processes that are responsible for the organisms' behaviors, notably in the article in which he presents a critique of the reflex concept (Dewey 1896). In this respect, it is worth remarking that in the second edition (1921) of Umwelt *and Inner-World of Animals*, Uexküll decided to replace the chapter on reflex – see the first edition (1909) of the book – with the chapter on the *Funktionskreis*. A link between Uexküll and Dewey is also suggested by Robert Innis's comments on the topic:

> Dewey had followed up Peirce's crucial hints about the semiotic implications of 'firstness' or 'quality.' His major contribution to semiotic issues was the development of the insight that there was a grasp of significance prior to all forms of explicit sign-reading and that indeed sign-action was truly a form of *action* in which the inquiring organism intervened *into* the environing field, in a continuous circuit of constructive responses. Dewey's great insight was identical in substance, on human level, to Jakob von Uexküll's widely discussed 'functional circle.'
>
> (Innis 2009, 94)

More precisely, one could argue that "John Dewey's work is singled out as offering a rare, early means of linking the work of Uexküll with that of Peirce, illuminating the conceptual framework of each" (Clements 2018, 30). These connections were later developed also by Thomas Sebeok, Thure von Uexküll, and Jesper Hoffmeyer.

Affinities can also be detected between Jakob von Uexküll's model and that of Jean Piaget. This was repeatedly remarked by Thure von Uexküll, who writes that

> adequate behavior [– is] meaning behavior controlled by tested programs. This is presented as a circular process by the model of the functional circle, which is described by Piaget as "*sensorimotor circular reaction.*"
>
> (Uexküll and Wesiack 1997, 29)

Thure von Uexküll lays emphasis on Piaget's idea of circularity, based on which a stimulus presupposes a need, or "a readiness to react." Furthermore, this would mean that "the reflex can only be described as a circular event, in which a neutral phenomenon receives a property which it does not have independently from the reacting organ, and which it loses again after the completion of the reflex, i. e. with the cessation of the readiness to react" (Uexküll 1986, 122).

In general terms, the functional cycle can be seen as an early description of the feedback loop that would later be thoroughly studied in cybernetics (see Lagerspetz 2001; Hassenstein 2001, 354; see also Köchy's contribution, Chapter 3, in this volume). In Germany, Hermann Schmidt has introduced the study of cybernetic control systems, using the term *Regelkreis* since the early 1940s. Arnold Gehlen, who was influenced by Uexküll and was himself a supporter of Schmidt's work, used the term *Handlungskreis*, thus providing a direct link between Uexküll and cybernetics (see Bissell 2011). Seen as a cycle, Uexküll's model of a functional circle reveals a clear algorithmic aspect. However, although a priority in Uexküll's agenda, cybernetic descriptions have not been able to provide a full description of meaning-making or semiosis. In this regard, Thure von Uexküll and Wolfgang Wesiack write,

> What does the model of the functional circle describe [...]? It differs from that of the control system by taking into consideration specific receptors and effectors, as well as their connection through a nerve system in the subject, and the corresponding phenomena on the part of the object. It describes how *impressions on receptors* (*elementary sensations*) trigger *responses in the effectors*, which (as impulses to action) lead to a change in the activities of the receptor (of the elementary sensations). Thus continuous feedback occurs between receptor and effector activities (perceiving and operating) which is projected toward something which takes place outside the organism. Here we encounter the new aspect: J. v. Uexküll says that the impressions on the receptor (elementary sensations) are 'moved outside' as properties (*perceptual cues*) of something which comes into being outside the organism – that is, as properties of the 'objects' of a subjective universe.
>
> (Uexküll and Wesiack 1997, 21)

Further impact of Uexküll's notions can be detected in Viktor von Weizsäcker's work,[13] who introduced the concept of *Gestaltkreis*. Weizsäcker corroborates the function of circles as the mechanism creating the *Gestalt*, describing its temporal and atemporal aspects. Holistic presentation of an object is possible only atemporally, in the subjective present. We see it as developing Jakob von Uexküll's model. The first formulation of *Gestaltkreis* (Weizsäcker 1933) was followed by a book with the same title (Weizsäcker 1940), and soon after by another book on *Gestalt und Zeit* (Weizsäcker 1960). As stated by Weizsäcker (1940, 126), "the life process is not a succession of cause and effect, but a *decision-making*." The closeness of Viktor von Weizsäcker's *Gestaltkreis* to the biosemiotic view has also been noticed by Friedrich Rothschild (1994).[14] A strong mutual influence between the fathers of psychosomatic medicine, Viktor von Weizsäcker, and Jakob von Uexküll's son, Thure von Uexküll, can be clearly held responsible for this (Stoffels 2003, 93).

To Thure von Uexküll, one owes also another extension of the concept of *Funktionskreis*. In 1979, he introduced, in fact, the concept of *Situationskreis* (Stoffels 2003; Uexküll and Wesiack 1997). This was meant as a model serving the description of more complex situations than those that Jakob von Uexküll had in mind: "for the level of the human we need [...] an even more complex model, the *situational circle*" (Uexküll and Wesiack 1997, 29). Compared to the functional circle, the situational circle adds the rehearsal of meaning or phantasy (imagery) into the mechanism (Baer 1988, 28). Here is a brief account of this model:

> The situational circle differs from the functional circle by an obligatory interposition of the imagination. The programs for assigning meaning ("perceiving") and meaning utilisation ("effecting") are at first tested as assumptions of meaning and testing of meaning before the ego releases them for sensorimotoricity. In doing so, the situation (which corresponds to the perceptual cue of the problem situation in the functional circle) is experimentally restructured in the imagination: that means the assigning of meaning is first done as a (hypothetical) assumption of meaning whose consequences can be probed (in the imagination by means of "testing acts").
>
> (Uexküll and Wesiack 1997, 32)

To this list one should also add the concept of *Lebenskreis*, as used by Helmuth Plessner (Schmieg 2017; see also Krüger's contribution, Chapter 5, in this volume). Since Plessner's concept qualifies as an extension to ontogeny, it does not focus on the atemporal aspect of the *Kreis* that one finds emphasized by both Uexküll and Weizsäcker. Already used by Wilhelm Dilthey – together with several other *Leben* terms – the term *Lebenskreis* has been occasionally translated as "sphere of life" (see Bianco 2019).[15] The list of influential models that can be seen as versions of Uexküll's *Funktionskreis* would not be complete without mentioning Pyotr Anokhin's "functional system," which provides a rather detailed neurophysiological description (Anokhin 1974); George Miller, Eugene Galanter, and Karl Pribram's *test–operate–test–exit* or *T-O-T-E* (Miller, Galanter, and Pribram 1960); Robert Rosen's relational model (Rosen 1991); and René Thom's linking of cyclicity to semiosis (Thom 1983). Finally, while relying on Uexküll's basic model, Eugen Baer (1988) specifies its types: symbiotic circle, situational circle, gerontological circle, and individuated circle. More about Baer's work in the following section.

6 Circle in the cycle: meaning-making and subjectivity are coextensive with the phenomenal present

All the previously listed concepts (models) are related to Jakob von Uexküll's attempt to describe something more than just a control loop; although possibly not yet clearly enough, they all make steps toward the description of the conditions of subjectness, of the primary meaning-making, of the assumptions required by a system in order for it to become a subject.

A recent approach in biosemiotics focuses on the notion of "agent" as a helpful starting point when modeling primary meaning-making (Sharov 2013; Tønnessen 2015). In this respect, Aleksei Sharov (2013, 345) defines the agent as follows: "An agent is a system with spontaneous activity that *selects* actions to pursue its goals" (my emphasis – K.K.). Emphasis should be given to whether the agent selects or chooses. Selection is made on the basis of trial and error, which can consist of sequential algorithmic processes. Choice, on the contrary, presupposes a situation of confusion in which no algorithm is provided. Such a situation is possible in the case of a nonsequential presentation of options, as it occurs in the phenomenal present. Together with simultaneity, the operation of "choice" between incompatible options is a necessary consequence of the subjective present that does not require any additional existence of goals (Kull 2018).

Meaning-making requires the simultaneity of options. This is what makes interpretation – as a choice between possibilities – fundamentally different from both deterministic and stochastic interactions. Simultaneity assumes the present. This is an essential element for acquiring relations as well as for the emergence of meaning – or for the irreducibility of sign relations in the sense of Peirce.

As Sharov (2013, 343) remarks, "[t]he hallmark of mind is a holistic perception of objects, which is not reducible to individual features or signals." Indeed, the phenomenal present is the same as the capacity to perceive something simultaneously, that is to say, as a whole.

This idea goes back to the work of John Dewey. As he writes, "[t]he sensory quale gives the value of the act, just as the movement furnishes its mechanism and control, but both sensation and movement lie inside, not outside the act" (Dewey 1896, 359). Dewey also adds that "[t]he circle is a coordination, some of whose members have come into conflict with each other" (Dewey 1896, 370). In conclusion, the circle that includes a conflict inside an act creates the sensory quality.

Here it is appropriate to point to the work of Eugen Baer, who describes additional important links between the functional cycle, synchronicity, and semiosis, based on Uexküll's theory. First, he demonstrates the connection between the models of Uexküll and Ferdinand de Saussure. He writes that in *Theory of Meaning* (Uexküll 1940),

> [m]eaning is [...] presented as a circle (*Kreis*) whose periphery is perceived by the observer as a code, i.e. a marvelous fit or dialog between self and not-self, but whose center remains ungraspable, non-phenomenal, much like Saussure's value which can only appear as difference. We remember that for Saussure the minimal structure of meaning is circular.

(Baer 1988, 208)

Further on, Baer detects a concordance with Martin Heidegger's view "that the circle of meaning performs the 'roundness of a whole,' not the either-or remnant of an excluded and therefore absent middle" (Baer 1988, 209). Baer's reference is here to the following passage:

I take the circle [*Zirkel*] as a sign that here the roundness of a whole [*das Runde eines Ganzen*] is to be thought of, in a kind of thinking, to be sure, for which 'logic,' measured by the freedom of contradiction, can never be the standard.

(Heidegger 1958, 60)[16]

The relation between the temporal moment and the *Funktionskreis* is explicitly spelled out also by Uexküll's himself, as he claims that "[i]f to represent the *Umwelt* of an animal in a certain moment as a circle, then you can add each subsequent moment as a new *Umwelt*-circle" (Uexküll 1928, 70, my transl.). While already marked by Uexküll, more emphasis on the importance of the temporal aspect of meaning-making, in terms of simultaneity or subjective present, is introduced by Weizsäcker. Along similar lines, Julius Fraser, also a follower of Uexküll, has seen in the subjective present a criterial characteristic of life: "The creation and maintenance of the organic present is a necessary and sufficient condition of life" (Fraser 1999, 65, 2001).

While sequential, algorithm-based descriptions are limited to physical continuous time, subjectivity is coextensive with the existence of possibilities, of simultaneous options and semiosis, which require the final present. In this respect, semiosis stops the time in the present and together with this creates meaning. Echoes of this distinctive assumption can be found already in Uexküll's distinction between physiological and biological and the idea that only the latter can see the wholes and meanings. In conclusion, if described as a mechanism of sequential physiological processes, the *Funktionskreis* is a *cycle*, while, in order to describe its capacity to be the condition for the phenomenal world, it should be seen as a *circle*. Among the several models of semiosis available (see, for instance, the review by Krampen 1997), Uexküll's model is unique in its circularity, as represented by the *Funktionskreis*. This allows to approach the identification of phenomenal worlds in the organisms of different species in a more detailed way than maybe any other model of semiosis. As Uexküll says in *The Theory of Meaning*,

> [i]t is through a combination of perception and action that a subject endows things with meaning and thereby makes them signs that are meaningful in the subject's *Umwelt*. Since all action by a subject begins with a perception feature of the sign that is then turned into an action feature by the subject, we can conceive of this process as a circular complementarity of perception and action between the sign and the subject that I call a functional circle.
>
> (Uexküll 1940, 9)

The circularity of the Uexküllian model also allows to establish parallels to the hermeneutic model of meaning-making (Clément, Scheps, and Stewart 1997; Chebanov 1998; Chang 2004). Life is in this respect taken as a self-reading text. Finally, circularity also fosters epistemological potential and may provide a fundamental hint on how to ground meaning.

Conclusion: Uexküll and biosemiotics

Uexküll demonstrates that the functional circle is the primary mechanism of meaning-making, which is responsible for creating an *Umwelt*, in other words a phenomenal world. He describes *Umwelten* in animals and assigns them also to unicellular organisms who have receptor and effector organs. However, there is no agreement among semioticians about the precise criteria allowing the identification of an *Umwelt*. Opinions vary on whether the phenomenal world is a characteristic of animals (as is rather commonly assumed), or of eukaryotes (Aleksei Sharov's hypothesis; Sharov 2017, 12), or of all living organisms (as Jesper Hoffmeyer has suggested; Hoffmeyer 2001, 396 fn. 3).

Based on our brief investigation of Jakob von Uexküll's major concept of a functional circle in the context of contemporary biosemiotics, it can be argued that, since it closely matches the semiotic methodology for qualitative research in biology and psychology, Uexküll's approach turns out to be not only foundational but also productive. Uexküll's work makes it ultimately possible to introduce hermeneutic and phenomenological aspects into biology without losing the connection to physiology.

The main point we aimed to emphasize in this analysis is the importance of the inclusion of the aspect of time in the interpretation of the functional circle as the meaning-making mechanism. It is arguably insufficient to describe a functional circle as a cyclic sequence of processes, and a better description would point to it also as a momentary circle. For a functional circle to be capable of meaning-making, it should, in fact, include the process of interpretation, which entails choosing between options. This has meaningful consequences. First, it means that the functional circle is a key component of agency. Second, options can exist only simultaneously, in the present. The present is itself the feature that characterizes the subjective realm. So understood, the functional circle can be taken as the basic mechanism that creates the subjective time. The present moment turns out to be the fundamental basis of semiosis and of the *Umwelt*. Semiosis (sign process or meaning-making or interpretation) and *Umwelt* emerge together, a functional circle that includes the conflict between codes as its mechanism.

In the current context it is remarkable that, in recent years, several leading semioticians have expressed their dissatisfaction with the existing fundamental models of semiosis. These have been found either too limited in their fields of application, too abstract to be properly applied, or too undifferentiated in order to specify semiosis itself. In order to reach an agreement between different approaches in semiotics, the general concepts of semiotics have to be made both theoretically clearer and empirically more grounded. Based on the same premises, Jakob von Uexküll developed his own methodology, qualifying then undeniably as a pioneer of biosemiotics.

Notes

1 I am grateful to Francesca Michelini and Kristian Köchy for their helpful comments and their enthusiasm, professionalism, and patience in working with this project.

I thank Ene-Reet Soovik and Tessa Marzotto Caotorta for improving the text. The work is related to PRG314.

2 A bibliography of works about Uexküll until 2001 was published in Kull (2001, 39–59).

3 The archive received a large collection of Jakob von Uexküll's correspondence and manuscripts in 2012, formerly kept by Uexküll's family.

4 See a similar conclusion in Winthrop-Young (2010, 242f.), and in Peterson *et al.* (2018).

5 Eagleman (2012, 145) admits, "I think it would be useful if the concept of the *Umwelt* were embedded in the public lexicon. It neatly captures the idea of limited knowledge, of unobtainable information, and of unimagined possibilities."

6 For the concepts functional circle and *Umwelt*, see also Toepfer (2011, articles "Umwelt" and "Verhalten").

7 Spelled Didimium, in Uexküll (1909, 1921).

8 However, Uexküll calls *Paramecium* an animal (*Tier*; e.g., in Uexküll 1921, 42); its German name is *Pantoffeltierchen*.

9 J. Delafield-Butt, who accepts the possibility that *Paramecium* may experience a phenomenological world, writes, "At the heart of the difference between a mechanical and process explanation of cell behavior lies the existence of individual novelty" (Delafield-Butt 2008, 256).

10 Also "function cycle" in Krampen (1981, 190).

11 For Uexküll, when he described "geschlossener Kreislauf, den man den *Funktionskreis des Tieres nennen kann*" (Uexküll 1928, 100), he characterizes the aspect of cyclicity.

12 "Da aber das Gefüge durch die Impulsfolge planmässig gebaut ist, so ist auch seine Wirkung in der Aussenwelt zugleich planmässig und zwangläufig. Die Frage nach der Planmässigkeit beschäftigt die Biologie, die Frage nach der Zwangläufigkeit die Physiologie" (Uexküll 1920, 95).

13 Thure von Uexküll has confirmed this connection (Uexküll 1987, 127).

14 As Chien (2004) remarked, Uexküll's own usage of the term *Gestalt* may not have received much influence from *Gestalt* psychology. However, about a possible link between Jakob von Uexküll and a pioneer of *Gestalt* psychology, Christian von Ehrenfels, see Christians (2016, 202f., 291). On the influence of the Leipzig school of *Gestalt* psychology on Uexküll, see Schmidt (1980, 181–183) and Cheung (2006, 240f.).

15 On Dilthey and biosemiotics, see also Mul (2016).

16 For more on the fundamental aspect of time in medical semiotics, see Freda, De Luca Picione, and Martino (2015).

References

Anokhin, Peter K. (1974) *Biology and Neurophysiology of the Conditioned Reflex and Its Role in Adaptive Behavior*. Oxford: Pergamon Press.

Baer, Eugen (1988) *Medical Semiotics*. Lanham: University Press of America.

Barbieri, Marcello (2015) *Code Biology: A New Science of Life*. Cham: Springer.

Bianco, Giuseppe (2019) 'Philosophies of life'. In: Warren Breckman and Peter E. Gordon (eds.) *The Cambridge History of Modern European Thought*. Cambridge: Cambridge University Press, 153–175.

Bissell, Christopher Charles (2011) 'Hermann Schmidt and German "proto-cybernetics"'. *Information, Communication and Society* 14 (1), 156–171.

Block, Katharina (2016) *Von der Umwelt zur Welt: Der Weltbegriff in der Umweltsoziologie*. Bielefeld: Transcript Verlag.

Brentari, Carlo (2015) *Jakob von Uexküll: The Discovery of the Umwelt between Biosemiotics and Theoretical Biology*. Dordrecht/Heidelberg/New York/London: Springer.

Buchanan, Brett (2008) *Onto-Ethologies: The Animal Environments of Uexküll, Heidegger, Merleau-Ponty, and Deleuze*. Albany: State University of New York Press.

Calvo Garzón, Paco and Keijzer, Fred (2011) 'Plants: Adaptive behavior, root brains and minimal cognition'. *Adaptive Behavior* 19, 155–171.

Chang, Han-liang (2004) 'Semiotician or hermeneutician? Jakob von Uexküll revisited'. *Sign Systems Studies* 32 (1/2), 115–138.

Chebanov, Sergey V. (1998) 'The role of hermeneutics in biology'. In: Peter Koslowski (ed.) *Sociobiology and Bioeconomics: The Theory of Evolution in Biological and Economic Theory*. Berlin: Springer, 141–172.

Cheung, Tobias (2006) 'Cobweb stories: Jakob von Uexküll and the *Stone of Werder*'. In: *Place and Location: Studies in Environmental Aesthetics and Semiotics*. Vol. 5. Tallinn: Estonian Academy of Arts, 231–253.

Chien, Jui-Pi (2004) 'Schema as both the key to and the puzzle of life: Reflections on the Uexküllian crux'. *Sign Systems Studies* 32 (1/2), 187–208.

Christians, Heiko (2016) *Crux Scenica – Eine Kulturgeschichte der Szene von Aischylos bis YouTube*. Bielefeld: Transcript Verlag.

Clausberg, Karl (2006) *Zwischen den Sternen: Lichtbildarchive. Was Einstein und Uexküll, Benjamin und das Kino der Astronomie des 19. Jahrhunderts verdanken*. Berlin: Akademie Verlag.

Clément, Pierre, Scheps, Ruth, and Stewart, John (1997) 'Umwelt et interprétation'. In: Jean-Michel Salanskis, Francois Rastier, and Ruth Scheps (eds.) *Herméneutique: Textes, Sciences*. Paris: Presses universitaires de France, 209–252.

Clements, Matthew (2018) *A World Beside Itself: Jakob von Uexküll, Charles S. Peirce, and the Genesis of a Biosemiotic Hypothesis*. PhD thesis. London: Birkbeck, University of London.

Cobley, Paul (2017) 'Integrationism, anti-humanism and the suprasubjective'. In: Adrian Pablé (ed.) *Critical Humanist Perspectives: The Integrational Turn in Philosophy of Language and Communication*. Abington: Routledge, 267–284.

Deacon, Terrence W. (2011) *Incomplete Nature: How Mind Emerged From Matter*. New York: W.W. Norton.

Deacon, Terrence W. (2015) 'Steps to a science of biosemiotics'. *Green Letters: Studies in Ecocriticism* 19 (3), 293–311.

Deely, John (2004) 'Semiotics and Jakob von Uexküll's concept of umwelt'. *Sign Systems Studies* 32 (1/2), 11–34.

Deely, John (2012) 'The Tartu synthesis in semiotics today viewed from America'. *Chinese Semiotics Studies* 8, 214–226.

Delafield-Butt, Jonathan T. (2008) 'Towards a process ontology of organism: Explaining the behaviour of a cell'. In: Mark Dibben and Thomas Kelly (eds.) *Applied Process Thought I: Initial Explorations in Theory and Research*. Frankfurt a. M.: Ontos, 237–260.

Dewey, John (1896) 'The reflex arc concept in psychology'. *Psychological Review* 3 (4), 357–370.

Eagleman, David M. (2012) 'The umwelt'. In: John Brockman (ed.) *This Will Make You Smarter: New Scientific Concepts to Improve Your Thinking*. New York: Harper Perennial, 143–145.

Eco, Umberto (1976) *A Theory of Semiotics*. Bloomington: Indiana University Press.

Favareau, Donald (ed.) (2010) *Essential Readings in Biosemiotics: Anthology and Commentary*. Berlin: Springer.

Fraser, Julius Thomas (1999) *Time, Conflict, and Human Values*. Urbana: University of Illinois Press.

Fraser, Julius Thomas (2001) 'The extended umwelt principle: Uexküll and the nature of time'. *Semiotica* 134 (1/4), 263–273.

Freda, Maria Francesca, De Luca Picione, Raffaele, and Martino, Maria Luisa (2015) 'Times of illness and illness of time'. In: Lívia Mathias Simão, Danilo Silva Guimães, and Jaan Valsiner (eds.) *Temporality: Culture in the Flow of Human Experience*. Charlotte: Information Age Publishing, 231–256.

Gens, Hadrien (2014) *Jakob von Uexküll, explorateur des milieux vivants: Logique de la signification*. Paris: Hermann.

Hassenstein, Bernhard (2001) 'Jakob von Uexküll'. In: Ilse Jahn and Michael Schmitt (eds.) *Darwin & Co.: Eine Geschichte der Biologie in Portraits*. Vol. 2. München: Beck, 344–364.

Heidegger, Martin (1958) *The Question of Being*. Translated by William Kluback and Jean T. Wilde. New York: Twayne Publishers.

Hoffmeyer, Jesper (2001) 'Seeing virtuality in nature'. *Semiotica* 134 (1/4), 381–398.

Hoffmeyer, Jesper (2008) *Biosemiotics: An Examination into the Signs of Life and the Life of Signs*. Scranton: University of Scranton Press.

Hoffmeyer, Jesper and Emmeche, Claus (1991) 'Code-duality and the semiotics of nature'. In: Myrdene Anderson and Floyd Merrell (eds.) *On Semiotic Modeling*. Berlin: Mouton de Gruyter, 117–166.

Innis, Robert E. (2009) 'My way through signs'. In: Peer Bundgaard and Frederik Stjernfelt (eds.) *Signs and Meaning: 5 Questions*. New York: Automatic Press/VIP, 87–99.

Kliková, Alice and Kleisner, Karel (eds.) (2006) *Umwelt: koncepce žitého světa Jakoba von Uexkülla*. Prague: Pavel Mervart.

Krampen, Martin (1981) 'Phytosemiotics'. *Semiotica* 36 (3/4), 187–209.

Krampen, Martin (1997) 'Models of semiosis'. In: Roland Posner, Klaus Robering, and Thomas A. Sebeok (eds.) *Semiotics: A Handbook on the Sign-Theoretic Foundations of Nature and Culture*. Vol. 1. Berlin: Walter de Gruyter, 247–287.

Krampen, Martin, Oehler, Klaus, Posner, Roland, Sebeok, Thomas A., and Uexküll, Thure von (eds.) (1987) *Classics of Semiotics*. New York: Plenum Press.

Kull, Kalevi (2001) 'Jakob von Uexküll: An introduction'. *Semiotica* 134 (1/4), 1–59.

Kull, Kalevi (2018) 'Choosing and learning: Semiosis means choice'. *Sign Systems Studies* 46 (4), 452–466.

Kull, Kalevi, Deacon, Terrence, Emmeche, Claus, Hoffmeyer, Jesper, and Stjernfelt, Frederik (2009) 'Theses on biosemiotics: Prolegomena to a theoretical biology'. *Biological Theory* 4 (2), 167–173.

Lagerspetz, Kari Y. H. (2001) 'Jakob von Uexküll and the origins of cybernetics'. *Semiotica* 134 (1/4), 643–651.

Maran, Timo, Martinelli, Dario, and Turovski, Aleksei (2011) *Readings in Zoosemiotics*. Berlin: De Gruyter Mouton.

Martinelli, Dario (2010) *A Critical Companion to Zoosemiotics: People, Paths, Ideas*. Berlin: Springer.

Mildenberger, Florian (2007) *Umwelt als Vision: Leben und Werk Jakob von Uexkülls (1864–1944)*. Stuttgart: Franz Steiner.

Mildenberger, Florian and Hermann, Bernd (eds.) (2014) *Jakob von Uexküll: Umwelt und Innenwelt der Tiere*. Berlin: Springer Spektrum. [Includes a faximile publication of Uexküll 1921, at pages 14–242].

Miller, George A., Galanter, Eugene, and Pribram, Karl H. (1960) *Plans and the Structure of Behavior*. New York: Holt, Rinehart, Winston.

Mul, Jos de (2016) 'The syntax, pragmatics and semantics of life: Dilthey's hermeneutics of life in light of contemporary biosemiotics'. In: Christian Damböck and Hans-Ulrich Lessing (eds.) *Dilthey als Wissenschaftsphilosoph*. Freiburg: Verlag Karl Alber, 156–175.

Peirce, Charles Sanders (1965) *Collected Papers of Charles Sanders Peirce*. Vols. 1–2. Cambridge: The Belknap Press of Harvard University Press.

Peterson, Jeffrey V., Thornburg, Ann Marie, Kissel, Marc, Ball, Christopher, and Fuentes, Agustín (2018) 'Semiotic mechanisms underlying niche construction'. *Biosemiotics* 11 (2), 181–198.

Rodríguez Higuera, Claudio Julio and Kull, Kalevi (2017) 'The biosemiotic glossary project: The semiotic threshold'. *Biosemiotics* 10 (1), 109–126.

Rosen, Robert (1991) *Life Itself: A Comprehensive Inquiry into the Nature, Origin, and Fabrication of Life*. New York: Columbia University Press.

Rothschild, Friedrich S. (1994) 'Parallels to biosemiotics in Viktor von Weizsaecker's writings'. In: Friedrich S. Rothschild. *Creation and Evolution: A Biosemiotic Approach*. Jerusalem: J. Ph. Hes, 90–95.

Schmidt, Jutta (1980) *Die Umweltlehre Jakob von Uexexternal's in ihrer Bedeutung für die Entwicklung der vergleichenden Verhaltensforschung*. Marburg: Görich & Weiershäuser.

Schmieg, Gregor (2017) 'Die Systematik der Umwelt: Leben, Reiz und Reaktion bei Uexküll und Plessner'. In: Thomas Ebke and Caterina Zanfi (eds.) *Das Leben im Menschen oder der Mensch im Leben? Deutsch-Französische Genealogien zwischen Anthropologie und Anti-Humanismus*. Potsdam: Universitätsverlag Potsdam, 355–368.

Sebeok, Thomas A. (2001) *Signs: An Introduction to Semiotics* (2nd ed.). Toronto: University of Toronto Press.

Sharov, Alexei (2013) 'Minimal mind'. In: Liz Swan (ed.) *Origins of Mind*. Dordrecht: Springer, 343–360.

Sharov, Alexei (2017) 'Molecular biocommunication'. In: Richard Gordon and Joseph Seckbach (eds.) *Biocommunication: Sign-Mediated Interactions between Cells and Organisms*. New Jersay: World Scientific, 3–35.

Stoffels, Hans (2003) 'Situationskreis und Situationstherapie: Überlegungen zu einem integrativen Konzept von Psychotherapie'. In: Rainer-M. E. Jacobi and Dieter Janz (eds.) *Zur Aktualität Viktor von Weizsäckers*. Würzburg: Königshausen & Neumann, 89–102.

Thom, René (1983) 'Structures cycliques en sémiotiques'. *Actes sémiotiques* 5 (47/48), 38–58.

Toepfer, Georg (2011) *Historisches Wörterbuch der Biologie: Geschichte und Theorie der biologischen Grundbegriffe. Band 3: Parasitismus – Zweckmäßigkeit*. Stuttgart: J. B. Metzler.

Tønnessen, Morten (2015) 'The biosemiotic glossary project: Agent, agency'. *Biosemiotics* 8 (1), 125–143.

Uexküll, Jakob von (1904) 'Die ersten Ursachen des Rhythmus in der Tierreihe'. *Ergebnisse der Physiologie* 3 (2), 1–11.

Uexküll, Jakob von (1909) *Umwelt und Innenwelt der Tiere* (1st ed.). Berlin: Springer.

Uexküll, Jakob von (1919) 'Biologische Briefe an eine Dame'. *Deutsche Rundschau* 178, 309–323, 179, 132–148, 276–292, 451–468.

Uexküll, Jakob von (1920) *Theoretische Biologie* (1st ed.). Berlin: Gebrüder Paetel.

Uexküll, Jakob von (1921) *Umwelt und Innenwelt der Tiere* (2nd ed.). Berlin: Julius Springer.

Uexküll, Jakob von (1926) [1920] *Theoretical Biology*. Translated by Doris L. Mackinnon. London: K. Paul, Trench, Trubner & Co.; New York: Harcourt, Brace & Company.

Uexküll, Jakob von (1928) *Theoretische Biologie* (2nd ed.). Berlin: Julius Springer.

Uexküll, Jakob von (1931) 'Die Rolle des Subjekts in der Biologie'. *Die Naturwissenschaften* 19, 385–391.

Uexküll, Jakob von (1940) *Bedeutungslehre*. Leipzig: J. A. Barth.

Uexküll, Jakob von (1982) [1940] 'The theory of meaning'. *Semiotica* 42 (1), 25–82.

Uexküll, Jakob von (1992) [1934] 'A stroll through the worlds of animals and men: A picture book of invisible worlds'. *Semiotica* 89 (4), 319–391.

Uexküll, Jakob von (2001) [1937] 'The new concept of Umwelt: A link between science and the humanities'. *Semiotica* 134 (1/4), 111–123.

Uexküll, Jakob von (2010) [1934, 1940] *A Foray into the Worlds of Animals and Humans with a Theory of Meaning*. Translated by Joseph D. O'Neil. Minneapolis/London: University of Minneapolis Press.

Uexküll, Jakob von and Kriszat, Georg (1934) *Streifzüge durch die Umwelten von Tieren und Menschen: ein Bilderbuch unsichtbarer Welten*. Berlin: Springer.

Uexküll, Jakob von and Kriszat, Georg (1956) [1934, 1940] *Streifzüge durch die Umwelten von Tieren und Menschen: Ein Bilderbuch unsichtbarer Welten. Bedeutungslehre*. Hamburg: Rowohlt.

Uexküll, Thure von (1986) 'From index to icon: A semiotic attempt at interpreting Piaget's developmental theory'. In: Paul Bouissac, Michael Herzfeld, and Roland Posner (eds.) *Iconicity: Essays on the Nature of Culture: Festschrift for Thomas A. Sebeok on His 65th Birthday*. Tübingen: Stauffenburg, 119–140.

Uexküll, Thure von (1987) 'Gestaltkreis und Situationskreis'. In: Peter Hahn and Wolfgang Jacobs (eds.) *Viktor von Weizsäcker zum 100. Geburtstag*. Berlin: Springer, 126–131.

Uexküll, Thure von and Wesiack, Wolfgang (1997) 'Scientific theory: A bio-psycho-social model'. In: Thure von Uexküll (ed.) *Psychosomatic Medicine*. München: Urban & Schwarzenberg, 11–42.

Umiker-Sebeok, Jean and Sebeok, Thomas A. (1980) 'Introduction: Questioning apes'. In: Thomas A. Sebeok and Jean Umiker-Sebeok (eds.) *Speaking of Apes: A Critical Anthology of Two-Way Communication with Man*. New York: Plenum Press, 1–59.

Weizsäcker, Viktor von (1933) 'Der Gestaltkreis, dargestellt als physiologische Analyse des optischen Drehversuchs'. *Pflüger's Archiv für die gesamte Physiologie des Menschen und der Tiere* 231 (1), 630–661.

Weizsäcker, Viktor von (1940) *Der Gestaltkreis: Theorie der Einheit von Wahrnehmen und Bewegen*. Leipzig: Georg Thieme.

Weizsäcker, Viktor von (1960) *Gestalt und Zeit* (2nd ed.). Göttingen: Vandenhoeck & Ruprecht.

Winthrop-Young, Geoffrey (2010) 'Bubbles and web: A backdoor stroll through the reading of Uexküll'. In: Jakob von Uexküll. *A Foray into the Worlds of Animals and Humans with a Theory of Meaning*. Translated by Joseph D. O'Neil. Minneapolis/London: University of Minneapolis Press, 209–243.

Witzany, Günther and Baluška, František (eds.) (2012) *Biocommunication of Plants*. Berlin: Springer.

14 Jakob von Uexküll's theory of *Umwelt* revisited in the wake of the third culture

Staging reciprocity and cooperation between artistic agents

Jui-Pi Chien

1 Introduction

This chapter explores the conundrum regarding our capacity and willingness to absorb other people's ideas or styles of working that appear strange to us. The topic has been an issue in the interchange between humanities and sciences in view of divided observations concerning our motivation for relating to other people. This chapter proposes to deal with the problem by way of reexamining Uexküll's theory of *Umwelt*, seeking to deepen our appreciation of instincts, emotions, and intuitions in our absorption of alternative views. It is revealed that followers and interpreters of Uexküll's work, certain French philosophers, in particular, have actually embraced his functional cycle as a sort of mental work that induces pleasure, knowledge, and self-governance, regardless of the exact identities or even of the very being of others we are engaging with. From this point of view, we appear as really generous and brilliant agents by virtue of our capacity of ignoring certain traits or disparities that may reduce our chances of relating to others. It is argued that such a viewpoint is compatible with the idea of genuine altruism and reciprocity, much promoted in anthropology, primatology, ecology, semiotics, and neuropsychology today. The updated sense of *Umwelt* serves to explain why we may still work out intriguing ideas and relationships even with the least amount of feedback or payback from others. It also allows us to revise such assumptions as social pressure, social contract, and natural selection in theorizing our motivation for achieving altruistic acts.

While revisiting Uexküll's theory in the contexts of reciprocity and cooperation, this study means to draw on the German term *Umwelt* for a couple of reasons. To begin with, the term was actually coined in the context of poetry writing in the 19th century, and since then, it has been associated with the sentiment of carrying out philosophical contemplation shared by poets and philosophers (Sutrop 2001). Although recognized as part of physiology and zoology, Uexküll's work has been much assessed in terms of a philosophical attempt that theorizes the relationship between animals and humans. Thus, Uexküll is seen not only to have revived a sort of humanistic concern for living beings but also to have equated animals and humans based on the shared ability to shape subjective universes [*Umwelten*]. It is indeed thanks to our sharpened approach to diverse worldviews and behaviors that Uexküll has made his theory of *Umwelt* heuristic and enticing. Moreover, in the reception of Uexküll's ideas across disciplinary borders, scholars have also retained the term *Umwelt* to suggest Uexküll's work as a source of inspiration. When exploring various types of interface in technical terms, for instance, scientists coined terms such as "*Umwelt* overlap" to illustrate changing conditions of interaction and communication between intelligent agents (Warkentin 2009; Ferreira and Caldas 2013). As they aim to discover newly devised approaches and acquired skills occurring specifically within multi-agent systems, studies of this kind have greatly enriched the value of *Umwelt* as a heuristic term. Following such prospects of revitalizing the strength of this German word, this study pushes for perspectives that enable us to appreciate our biological instincts of reaching out to strange universes, namely, other people's ideas and opinions.

2 Reciprocity and cooperation in the functional cycle: between the inner world and the outer world

When it comes to the interchange between humanities and sciences, the third culture can be seen as a movement for scientists, cultural and literary theorists to consider how they can absorb and adapt information found in the sciences. The movement serves to overcome any sort of divide arbitrarily set between humanistic and scientific styles of learning, working, and presenting that are labeled as two separate cultures. While theorists and cultural elites were standing on their bedrocks of seeing and observing, they have also agreed to open up their minds – empathizing with strange ideas and engaging in mutual reciprocal communication – so that they may contribute to the well-being of the public and academia (Brockman 1995; Snow 2012, 70f., 76f.; Shaffer 1998, 3; Žižek 2002, 30, 32). In this light, the third culture appears to have invited us to imagine a certain demanding yet rewarding style of working. Rather than insisting on our own biased or patterned ways of reasoning, we should apply our skill and intelligence to stage cooperation between diverse viewpoints in our own work.

Notions of reciprocity and cooperation encompass a wide range of studies in both humanities and sciences. They have been employed in various fields with the aim of revising our concepts about the links between animals and humans and those between our strong and low-arousal emotions. When used as conceptual tools, these notions serve to emphasize emotive and cognitive capacities shared by animals and humans in coping with diverse novel situations (Vessel, Starr, and Rubin 2013; Starr 2013; Smith 2017). In the fields of primatology and anthropology, in particular, scholars cooperate to engage with the origin of human society, language, and morality in terms of the mammalian sentiments of helping and sharing (Dortier 2015; Tomasello 2016). It is argued that our spontaneous reaction to give aid sheds more light on our innate capacity of empathizing with someone else's situation and well-being than on our own profit and self-serving ends.

Exploring the cooperation between humanities and sciences in terms of similar horizons, this study seeks to reinvent meaningful contexts for our appreciation of Uexküll's theory of *Umwelt* today.

While contemplating the true cause of our altruistic behavior, primatologists hesitate whether they should adopt the ideas of a social contract and social pressure advocated in the 18th and 19th centuries (Darwin 2009b, 393f.; Tomasello 2016, 54f.). Instead of following the strand of thinking that enlarges selfishness and rivalry and that doubts more or less the honesty and reliability of human nature, primatologists turn to alternative discourses that focus on the operation of certain mammalian traits in human societies such as empathy, self-motivation, affection, and creativity (Flack and de Waal 2000; Waal 2000, 2006, 2016; Tomasello 2016). This study notably draws on such a context of theorizing that regards altruistic behavior as the result of self-motivated and self-governed intelligence shared by animals and humans. Such a strand of thinking – much valued by Hume, Kant, Darwin, and Peirce – not only argues for the necessity of integrating communal values and concerns as part of our own motives and behavior but also emphasizes the merits of skillful and imaginative play in attaining genuinely altruistic acts.

The idea of genuine altruism should serve as a powerful means in staging a kind of cooperation between humanities and sciences today. It is argued that our deepest sense of morality is not at all governed by the calculation or accounting of specific benefits but, rather, by the need for emotional support and long-term relationships within and among communities (Axelrod and Hamilton 1981; Waal 2000, 2006; Tomasello 2016, 149). Something close to arbitrarily altruistic instincts – based on which we feel ready to cooperate with both friends and strangers – calls for the kind of attitude we should revitalize in our own work. In addition, the very appealing sense of reciprocity is linked to our capacity to fully adopt someone else's perspective without assuming any reward or payoff (Jeannerod 2005; Kitcher 2006; Carter 2014). Such an approach to working out novel situations or getting along with individuals of different traits is deemed as the attitudinal or emotional reciprocity thanks to which we are supposed to be "blind" to our own benefits. As a matter of fact, with the guidance of our intuitive sympathetic sensation, we not only mimic other individuals' emotions and behaviors but also attribute our evaluation of situations to their needs.

The psychoanalyst Jacques Lacan's reading of Uexküll's *Innenwelt–Umwelt* cycle enables us to appreciate our potential ways of attribution while interacting with others and strange oddities. Uexküll's model appears highly heuristic to Lacan in the way that it serves to bridge madness, instincts, and civic codes of conduct on the same horizons of judging and interpreting (Lacan 1975, 337f.). Similarly, we are thought to carry ourselves around more or less spontaneously whether we are psychotic patients or supposedly balanced people. When falling prey to psychosis, we attribute our imagination and feelings solely to ourselves: we do not even care to accept any feedback from others, since we deem ourselves to be fully in charge of such a situation. When living up to a certain kind of law, justice, and equality, as put forward by Rousseau in his theory of social contract, we simply forget or sacrifice part of ourselves so that we may benefit from the greater well-being emerging in a well-governed society (Rousseau 2004, 44f.). Whether or not we care about our own interests or benefits, we behave naturally and swiftly just like animals thriving in nature. Thus, from the Lacanian perspective, the Uexküllian scheme appears to be a fairly instinctual and automatic model that sheds light on the coexistence of two opposing styles of attribution.

The contexts of biology and zoology help elucidate the fact that we actually deal with asymmetrical communication when observing the conditions in which we would like to reciprocate. We and our potential partners may have completely different interests to fulfill or goals to achieve, but we together still keep the dynamics going out of our instincts. On the one hand, the idea of asymmetry emphasizes our capacity and incentives to cooperate with diverse forms and beings under all circumstances, irrespective of past relationships, our probable profit and benefit. On the other hand, our instincts and emotions – seen as the clue to our own law of self-governance – foster our strength and intelligence of reaching out to others agreeably without at all sacrificing ourselves or doing harm to others. Thus, we should follow primatologists and anthropologists' advice on adopting the noncalculating yet eye-opening model so as to render our theorization highly heuristic. Even though Rousseau seems to suggest that we can expect certain benefits through checking our own power, he actually argued for the necessity of enhancing our own sense of autonomy. We should, by all means, seek to govern ourselves well considering the fact that others may fail to do their duties as citizens, deputies, or lawmakers – they may occasionally become selfish and thus ignore our demands and well-being (Rousseau 2004, 45, 112f.).

We may wonder about the legitimacy of regarding instincts as part of our motivation of interacting with others. It may appear quite daunting to justify the extreme situation that we are so much giving and cooperative that we end up ruining our integrity or losing our precious lives. From certain evolutionary biologists and geneticists' perspectives, such behavior appears like a complete *faux pas*, since we unwisely reduce our chances of survival – our genes to be selected and inherited – while excessively helping and caring for others (Dortier 2015, 205–212). However, in the realms of neuropsychology and philosophical inquiry, such a paradox still appears highly heuristic and meaningful in its own right. According to Paul Ricœur, for example, we actually experience pleasure and delight while instinctively and generously interacting with others. Such experience is seen as "happy forgetting," or, rather, the "festive character" of reciprocity, through which we not only regain certain profound memories but also ignore the disparities that may have emotionally driven others and us apart. We are empowered to draw on large portions of otherness and strangeness while engaging in devising ways of mingling with others that may go beyond our expectation (Ricœur 2004, 2005, 170f., 244f.).

Uexküll's functional cycle, as shown in its sophisticated form, speaks up for the kind of pleasant and healthy model we are advocating in this study (Uexküll 2010, 49). It serves to illustrate the noncalculating and nonjudgmental approach we need while absorbing and adapting information found in various fields of study. This model also revises our assumption concerning the conditions in which we would like to cooperate: we actually have the instincts to endear strangers irrespective of blood ties, social relationships, and physical environments (Axelrod and Hamilton 1981; Uexküll 2010, 323f.; Lenzi *et al.* 2009; Tomasello 2016). Moreover, we are thought to be much more capable and resourceful than we imagine in the situation of empathizing and cooperating with complete strangers or enemies. We are willing to forget the kind of harm or torment that others have inflicted on us to such an extent that we may increase our chances of recognizing and teaming up with potential partners (Ricœur 2004). Equipping ourselves with such an attitude of working, we manage to modify our position and perception while dealing with the difficult conundrum concerning the cooperation between agents and their concerns, actualized in diverse communities or environments. In other words, to really sharpen our perception and observation, we should not only reach out to attractive notions but also reflect on how we can possibly deal with certain strange ideas.

3 *Umwelt*, niche construction, and the ethics and aesthetics of cooperation

When seeking to mediate humanistic and scientific styles of working via the approach of altruistic reciprocal communication, we should first and foremost overcome the conundrum of profit, selfishness, and selection. To begin with, we are supposed not to expect any payoff while interacting with our partners. However, we may not be certain about the direction of working when it occurs that our partners are not very responsive or, rather, stingy and selfish in terms of giving. Such a situation once again emphasizes the phenomenon of asymmetry that may utterly invalidate the principle of equality and justice surmised in the theory of social contract. According to primatologists and anthropologists, we have the right to make our own choices concerning whether we should still embrace selfish or tricky partners. There should be no likelihood – as geneticists and evolutionary biologists once criticized – that we will ruin our integrity or reduce our chances of survival in either the short or the long run (Brockman 1995, 36; Sterelny 2007, 51–65). The kind of selection that is functioning in the process of reciprocal communication should be motivated from deep inside our mind rather than being imposed from the outside such as the idea of a social contract. Judging from the bottom-up and basing on an individual-specific perspective, we should be able to choose those partners who are agreeable and cooperative in our own terms (Brockman 1995, 62, 66; Merleau-Ponty 2003, 194; Tomasello 2016, 18–20).

The idea of making conscientious choices from deep inside enables us to play down the rampant power of selfishness so much propagated in the trend of gene selection. Even though there are tricky liars and free riders in the community, we may just decide not to imitate such behavior that appears detrimental to the well-being of the community. In accordance with the conceded viewpoint reached between geneticists and paleontologists, we gather the clue that choices made by individuals and the community, respectively, should be equally evolvable and valuable. We actually need both types of selection to push for success and well-being attained through cooperation between individuals (Sterelny 2007, 62, 65; Tomasello 2016, 19). Thus, our work is entitled to a big chance of success if we can develop certain traits and behavior – such as grooming and alarm calling in the world of animals – that may greatly benefit the prosperity and well-being of the community. We indeed work out certain forms of profit together with our cooperators that serve to nourish individuals in the community. As we consider the fact that the community seeks to bring them under control time after time, selfish genes cannot always have their own way.

We develop favorable traits and behaviors through imitating others' words and actions, recalling our own memories and experiences, and seeking to integrate others' perspectives with our own. We may honor such a process as a kind of symbolic gift giving, in which we spontaneously and unceasingly make efforts to mingle with others, paying the least attention to whether they are able to respond or to pay back (Ricœur 2005, 244f.). It is also about our duty and reverence for a rigorous engagement with diverse perspectives that enables us to initiate any cooperation. On the one hand, we are more or less affected by the community (such as feeling alarmed or pressured), and on the other hand, we wish to keep our own ingenuity so that we can devise our own approach that differs more or less from all the judgments and achievements already made in the community. On our path toward polishing our skills and intelligence of mingling with others, our mental work such as emulating (comprising imitating and simulating) should serve as the common ground between the three types of concerns (we, you, and me) that constitute the core of our motivated aesthetic selection (Kant 2000, 195f.; Tomasello 2016, 112–115).

Emulation allows us not only to freely acquire certain innovative ideas through observing and contemplating others' judgments and achievements but also to make substantial efforts to act out our own ideas that may win the approval of the community. Rather than submitting ourselves to you- and we-concerns (probably adopted as a quick way out due to social pressure), we willingly and conscientiously manage to absorb and to integrate as many diverse perspectives as possible (Kant 2000, 174f.; Tomasello 2016, 82f.). We value such an interchange of positions (between we, you, and me) as a kind of training and learning without self-constraint that serves to widen our horizons of showing sympathy and empathy, on the one hand, and to improve our way of thinking, imagining, and reasoning, on the other (Kant 2000, 175, 180; Flack and de Waal 2000, 20; Kitcher 2006, 133). We have the duty to create endearing relationships for the community to contemplate and to appreciate. Just like animals that have discerning eyes and the intelligence to make the best choices for the survival of their species, we should also become swift-minded geniuses who possess such skills and spontaneity as to develop perspectives that may appeal to the community (Kant 2000, 186–196; de Waal 2006, 52–58; Hume 2007, 76–78; Darwin 2009b, 394–405; Tomasello 2016, 113–115; Mathôt, Grainer, and Strijkers 2017).

In the light of the current theory of niche construction, the success of our work is closely connected with various forms of feedback we gather from the diverse beings we are interacting with (Odling-Smee and Laland 2011; Laland, Odling-Smee, and Endler 2017). The feedback is thought to be indispensable in the way that it not only serves to modify our own perception, but also enables us to foresee precisely what we are achieving together with other beings (Darwin 2009b; Kendal, Tehrani, and Odling-Smee 2011; Laland, Odling-Smee, and Endler 2017). Thus, from the perspectives of ecology and biology, feedback appears to be the logical and necessary result that we can expect from working with our partners. Rather than assuming prominent traits of individuals to be powerful in the descent of species (as it occurs in certain mistaken judgments of natural selection), the theory of niche construction values the strength of carrying on substantial interaction between agents as the true cause of selection and transmission (Sterelny 2007, 99–100; Darwin 2009a, 2009b; Odling-Smee and Laland 2011, 220–222; Laland, Odling-Smee, and Endler 2017). Considering that we behave like such procreative agents, managing to build up and to mediate communities, we hardly worry about being judged or criticized for the truth or falsity in our own work. Rather, we spontaneously formulate and test different hypotheses that may serve to modify our relationships with you- and we-concerns.

The presumably compact functional cycle in Uexküll appears to suggest the kind of logical and necessary result that we should work out together with our partners. It serves to observe our strength while acting and reacting for the best of communities. We are perceived to be capable of creating and updating networks that intriguingly include both friends and enemies across communities. Nevertheless, the major distinction between the Uexküllian scheme and the theory of niche construction is the fact that, in the former, concrete beings or entities are not required for completing and perfecting the cycle. It appears rather like a mental scheme in which our instincts and memories play a key role in the way we stage and sharpen the cycle independently and impressively (Merleau-Ponty 2003, 191–194). Just like how we empathize with other beings, we may simply work out the cycle without paying attention to the exact identities of those we are helping, yet we are quite likely to gain pleasure and satisfaction from our immediate responses or actions. Such self-motivated behavior – irrespective of the constraint of specific triggers or adaptive pressures thought to influence our choices and selection in certain environments – might enable us to create relationships and scenarios of cooperating that ignore the boundaries between friends and enemies. We may expand our horizons by adopting different attitudes or seeking to invent approaches that enable us to become liaised with seeming enemies in our imagination (Merleau-Ponty 2003; Ricœur 2004; Uexküll 2010, 112f., 167f.).

Compared with the logic of selfish genes that discourages us from giving and sharing, the logic of generous and persistent givers cum players – what we gather from the biological, ecological, and neuropsychological aspects of *Umwelt* and niche construction – appears rather encouraging. We are supposed to take risks to explore any possibility that serves to blur the boundaries between beings, entities, and disciplines. It is argued that our capacity of devising strategies in coping with such a situation claims and recycles the mental reservoir that alternates between profound recalling and forgetting (Ricœur 2004, 414–418; Bergson 2004, 211f.,

232). On the one hand, we may all respond to the same perceptual cues – for example, audial and visual stimuli we spot on partners or in other disciplines – but we may just decode signals and process messages with our own unique approaches. On the other hand, we may all have the same goal or task to carry out, but we are quite likely to sort out our own tactics that draw on diverse beings, entities, and disciplines. Our altruistic instincts enable us to be very choosy yet flexible in modulating our perception of cues and in forming our rationales as to suit various situations. Like animals that are able to stay unaffected by certain things just for the perfection of their functional cycles (Uexküll 2010), we should also train ourselves to let go or to put aside certain strange ideas or details that may impede our absorption and appreciation of other beings or disciplines.

The sort of forgetting referred to here is more about our strength of inventing approaches for getting along with others than our fear of losing precise and precious memories. Rather than forcing ourselves to deal with strange ideas or details immediately, we decide to put them aside for a while so that we still have chances to review them in later contexts or situations. We may find our memories of this sort very useful when we recall the fact that we actually experienced certain joyful sensations in previous reciprocal communication. When this occurs to us at a later time – whether or not we are together with the same beings or entities – we may easily render our deep memories (concerning our own life experiences) accessible and available to the way we argue for our appreciation of oddities without at all bothering to figure out how they have worried us previously (Ricœur 2004, 414–417). Such actual functioning of forgetting (nuisances) and recalling (strategies) serves to justify why we should adopt the noncalculating approach to staging reciprocity and cooperation between artistic agents. Contemplating and imagining strange oddities on a longer timescale actually enables us to widen our horizons and to devise creative approaches for mingling with others. This is also seen as a cognitively and emotionally less demanding approach than that of selfish calculation, in which partners are assumed to reciprocate exactly and immediately (Schweinfurth and Taborsky 2017). We may just lose chances to sharpen our mind and intelligence if we insist on claiming or replicating what we believe others have been indebted to us.

4 Semiotic perspectives on the functional cycle and altruistic reciprocal communication

Attempts of engaging with Uexküll's theory of *Umwelt* around the world elucidate the demand of formulating and testing hypotheses shared by scientific and humanistic styles of inquiry. They also reveal how people have willingly exerted their altruistic instincts in overcoming biased opinions and ideological agendas, on the one hand, and absorbing nourishing ideas Uexküll provided, on the other. Indeed, Uexküll during his lifetime has crossed many national and linguistic boundaries while formulating his hypotheses, but the testing and the reception of his ideas are rather constrained by the needs for reciprocal communication between academic disciplines. Measured in semiotic terms, the convergence of two or three disciplines at a certain time in history may lead to a great change of discourses on the part of the addressees or receivers. Within such a context of vibrant demand, the notion of *Umwelt* appears like a new code that can be penetrated and divided as several subcodes according to the interests of receivers. They drew on the given codes in their respective disciplines (what they already know), and they constantly came up with ideas that serve to embrace the functional cycle (what they think the world should know). Accordingly, there has been incessant interaction between humanities and sciences that cooperate to strengthen the cycle as a favorable scientific or semiotic approach.

The operation of reciprocal communication in the form of functional cycle enables us to justify the making of scientific inquiries in light of our observation of motivated receivers (also known as creative helpers and players). Instead of resorting to causal logic (such as the induction and deduction of certain laws and principles) as the only legitimate approach to sound reasoning, we observe changing intelligent systems made up by the cybernetics between senders and receivers in devising and realizing attractive propositions about certain objects or phenomena (Wright 1971, 158f.; Peirce 1998, 150f., 240f.). To begin with, successful receivers manage to overcome the temporal or geographical contingency of not being able to exactly hear from senders concerning their feedback, yet they work hard from deep inside to modify our perception of senders' messages. They come up with premises or hypotheses that serve, on the one hand, to update potential wishes or desires on the part of senders and to convince us of adopting renovated thoughts or perceptions they have virtually and creatively worked out together with senders, on the other. Finally, our engagement with such dynamics serves to increase a sort of equilibrium between intelligent systems that might be more or less missing before our intervention. We may also draw on the sharpened reasoning and perception to discover a lot more generous helpers who are capable of transforming strange oddities into nourishing ideas that can be shared across disciplines.

Basing on the presumably strong connection between sound reasoning and behaving in our creative output, we consider the functional cycle as a model that values our perception as an efficient starting point while we are seeking to regain truth and meaning from our dealings with strange oddities (Uexküll 2001; Peirce 1998, 155, 250f.; Deleuze 1989, 31f.). Rather than adopting an organized viewpoint to start with, we exert our intuitions, emotions, and memories to gather rough ideas about the kind of oddities we have experienced. This implies that we allow for a dose of chances such as estimating and conjecturing in our work while taking twists and turns in devising and revising hypotheses. Since we are fully in charge of the ways we perceive, we do not even worry if our changing motivations and perceptions are morally justifiable or not. It is not until we start to draw on our knowledge of laws and principles (well accepted in the community) in devising propositions that we may be more or less judged for the consistency in our reasoning. Nevertheless, the validity of our work is not likely to be vitiated – sufficient truth and meaning may well be expected – if we persist to carry out certain perspectives that we spontaneously intuit and apply throughout the course of our inquiry (Peirce 1998, 250f.). Extending such an approach to our discovery of altruistic artistic agents, we avoid criticizing the truth or falsity in their work. Rather, we emphasize how they perceive their relationships with other disciplines and how they illustrate rigorous reasoning that may serve to bridge fields of study of any kind.

The sort of hypothesis much engaged with in the interchange between humanities and sciences is about the traits of empathizing and of aesthetic judgment shared by animals and humans (Hume 2007; Darwin 2009a, 2009b, 2009c; Uexküll 2010; Waal 2006, 2016). Scientists and philosophers have also taken these traits as the grounds for our strength in absorbing strange ideas or details. Let us consider once again how our mammalian instincts, emotions, and memories may boost our strength of reasoning and interpreting in terms of sign functions. According to Saussure and Peirce, the law (*langue, interpretant*) claims the key to our recognition of the legitimacy of sign activities. On the one hand, it strengthens the subtle link between grammatical rules and the community of speakers (Saussure 1959, 77f.), and on the other, it improves our capacity of associating one trait or detail with another as well as our approach to objects. When observing revelations of such compact sign entities in the context of non- or preverbal situations, we may just regard our emotions as an aspect of the law that governs how we form our habit of reasoning and behaving. Theoretically, we expect emotions to overcome the barriers of linguistic specificities that used to divide animals from humans. Presumably, animals and humans may just react more or less in the same way in low-arousal situations. Gradually, the more we practice dealing with oddities, the more we are likely to naturalize our relationships with them in later encounters. We may spontaneously take perspectives to justify the legitimacy of their occurrences in alternative larger contexts.

Emotions seen as something that emanates from changing situations or contexts may enable us to come up with brilliant and convincing interpretations in due course (Eco 1979, 194f.; Smith 2017, 143). The sort of intelligence that we share with animals concerning feeling, perceiving, and reacting allows us to argue for the merits of self-discipline and skillful play while we are pursuing certain courses of observing, imagining, and interpreting (Darwin 2009a, 86; Gombrich 1984b). To begin with, we should be profoundly motivated and affected by our situations, even though at the beginning we may not be clear about what we are achieving (Eco 1979, 183; Darwin 2009b, 108). While actually looking into situations and seeking to sort out our lines of thinking, we keep the kind of distance that allows us to imagine and to devise strategies of coping with strange oddities. By integrating objective findings with our thoughts and imagination, we come to form certain hypotheses that enable us not only to align certain details we have selected but also to invite desirable subjective responses such as pleasure and understanding to the propositions we are putting forward to the community (Gombrich 1984a; Peirce 1998, 251–254). In both the humanities and the sciences, we work to regulate our thinking and imagining as if we were detectives who have the sentiments of weaving clues and evidence together with their selected viewpoints. Apart from a logical building up of our observations, we should also consider sensitive readings of our own emotions as part of the evaluation of valid interpretations (Gombrich 1984a, 17).

It is essential that we appreciate our absorption and integration of laws, communal concerns, and diverse novel situations as various attempts at creating logical and psychological entities of signs (so valued by Saussure and Peirce). Such an attitude enables us to become fully aware of the duty of managing the intelligibility of our propositions (Darwin 2009a, 162). We should be curious about learning and trying out different laws, principles, and perspectives so that we can readily devise strategies to engage with objects. Meanwhile, we make efforts to exert our influence on objects through mapping and inventing contexts that serve to prompt new sensations and understandings of them in the community. The truth and intelligibility of our propositions are therefore measured by the extent to which we are able to modify the perceptions of certain objects advocated in the community (Eco 1979; Peirce 1998, 151–155; Smith 2017, 118–123). Actually, we work hard to help individuals regain emotive and cognitive ties with certain objects that were just disparaged or ignored. We also seek to revise certain biased impressions against seemingly unpleasant phenomena already formed in the community. More precisely, we constantly work to update the law (*langue* and *interpretant*) so that other individuals may learn from our propositions how they can possibly assume their own positions. Drawing on the trait of emotional and attitudinal reciprocity shared by animals and humans, we work not only to overcome fear and anxiety of our own but also to expound refined thoughts that may inspire the community to develop balanced and intriguing viewpoints of certain phenomena.

To really widen our horizons of staging and appreciating altruistic reciprocal communication, we should alternate between paying close attention to certain ideas (or the grouping of certain details) and forgetting those that may not instantly appear relevant in view of the perspectives we would like to carry out (Eco 1979; Peirce 1998; Ricœur 2004; Darwin 2009a, 162). For example, while receiving and absorbing implications of the functional cycle in the context of our mind and well-being, creative agents appear to have cooperated in putting forward the sort of genuine and healthy communication that we are likely to achieve in our life and work. In the first instance, the cycle was simply judged as the operation of an automatic machine that explains well aphasic and psychotic patients' verbal and nonverbal traits (as argued by Cassirer and Lacan). So it appears that these patients always follow and replicate a certain invariable code or principle to express themselves. They were thus more or less regarded as animals that are short of the real human intelligence of adopting multiple perspectives and principles. However, at second thought, creative agents modified our perception of the heuristic function of the cycle so as to enlarge on how we may attain sophisticated ideas and behavior through trials and errors (as shown in Goldstein and Canguilhem). Then the cycle was seen to provide us with a suitable basis to bridge animals and humans on many aspects. Finally, the vicious reasoning that associates patients with wrongly assumed deprived animals vanishes from our horizons, and we start to imagine how we can turn noncommunication into engaging interaction even if we are coping with extremely unfavorable and unpleasant situations.

Actually, we cannot dispense with either invariable or variable codes in conceiving our intellectual relationships with others: they both enable us to polish our way of integrating others' reasoning and behavior as part of our own rigorous thinking and imagining. On the one hand, we have the gift to closely mimic others' words and actions, that is, to remember exactly those who have inspired us in one way or another. On the other hand, we need time to really deepen our appreciation of others' words and actions so that we can forget oddities or nuisances and reciprocate with our own kind and intelligent remarks that may benefit the community as well (Jakobson 1971; Schweinfurth and Taborsky 2017). Just imagine the two types of mental work dancing together – imitation is woven with empathetic understanding, the former is modified or revised by the latter, and so on and so forth – through the evolving functional cycle. This should enable us to overcome the conundrum of profit, selfishness, and selection said to have plagued the merits of theorizing and practicing altruistic behavior. By way of recalling and applying the law – communal laws continually updated by our own law of self-governance – we work hard to carry out the sort of perspectives that enable the community to take in strange and disconcerting ideas. Just like nature has been cycling and recycling ugly and monstrous forms while creating adorable new species (Beer 1998, 23, 26f.), the functional cycle may empower us to a certain extent to achieve something similar in the wake of the third culture. Time and again, we absorb ideas and viewpoints found in other disciplines spontaneously and independently even though the feedback or the payback can be quite minimal or deprived in most encounters.

References

Axelrod, Robert and Hamilton, William D. (1981) 'The evolution of cooperation'. *Science* 211 (4489), 1390–1396.

Beer, Gillian (1998) 'Has nature a future?'. In: Elinor S. Shaffer (ed.) *The Third Culture: Literature and Science.* Berlin/New York: Walter de Gruyter, 15–27.

Bergson, Henri (2004) [1912] *Matter and Memory.* New York: Dover.

Brockman, John (1995) *The Third Culture.* New York: Simon and Schuster.

Carter, Gerald (2014) 'The reciprocity controversy'. *Animal Behavior and Cognition* 1 (3), 363–380.

Darwin, Charles (2009a) [1871] *The Descent of Man and Selection in Relation to Sex.* Vol. 1. Cambridge: Cambridge University Press.

Darwin, Charles (2009b) [1871] *The Descent of Man and Selection in Relation to Sex.* Vol. 2. Cambridge: Cambridge University Press.

Darwin, Charles (2009c) [1890] *The Expression of the Emotions in Man and Animals.* Cambridge: Cambridge University Press.

Deleuze, Gilles (1989) [1985] 'Recapitulation of images and signs'. Translated by Hugh Tomlinson and Robert Galeta. In: Gilles Deleuze. *Cinema 2: The Time-Image.* Minneapolis: University of Minnesota Press, 25–43.

Dortier, Jean-François (ed.) (2015) *Révolution dans nos origines* (Sciences Humaines Éditions). Paris: Seuil.

Eco, Umberto (1979) 'Peirce and the semiotic foundations of openness: Signs as texts and texts as signs'. In: Umberto Eco. *The Role of the Reader: Explorations in the Semiotics of Texts*. Bloomington: Indiana University Press, 175–199.

Ferreira, Maria I. A. and Caldas, Miguel G. (2013) 'The concept of Umwelt overlap and its application to cooperative action in multi-agent systems'. *Biosemiotics* 6 (3), 497–514.

Flack, Jessica C. and de Waal, Frans (2000) 'Any animal whatever: Darwinian building blocks of morality in monkeys and apes'. *Journal of Consciousness Studies* 7 (1/2), 1–29.

Gombrich, Ernst (1984a) 'Focus on the arts and humanities'. In: Ernst Gombrich *Tributes: Interpreters of Our Cultural Tradition*. Ithaca, NY: Cornell University Press, 11–27.

Gombrich, Ernst (1984b) 'The high seriousness of play: Reflections on. *Homo ludens* by J. Huizinga (1872–1945)'. In: Ernst Gombrich. *Tributes: Interpreters of Our Cultural Tradition*. Ithaca, NY: Cornell University Press, 139–163.

Hume, David (2007) [1748] *An Enquiry Concerning Human Understanding*. Oxford: Oxford University Press.

Jakobson, Roman (1971) [1954] 'Two aspects of language and two types of aphasic disturbances'. In: Roman Jakobson. *Studies on Child Language and Aphasia*. The Hague: Mouton, 49–73.

Jeannerod, Marc (2005) 'How do we decipher others' minds?'. In: Michael A. Arbib and Jean-Marc Fellous (eds.) *Who Needs Emotions? The Brain Meets the Robot*. Oxford/ New York: Oxford University Press, 147–169.

Kant, Immanuel (2000) [1790, 1793] *Critique of the Power of Judgment*. Translated by Paul Guyer and Eric Matthews. Cambridge: Cambridge University Press.

Kendal, Jeremy, Tehrani, Jamshid J., and Odling-Smee, John (2011) 'Human niche construction in interdisciplinary focus'. *Philosophical Transactions of the Royal Society B Biological Sciences* 366, 785–792.

Kitcher, Philip (2006) 'Ethics and evolution: How to get here from there'. In: Stephen Macedo and Josiah Ober (eds.) *Primates and Philosophers: How Morality Evolved*. Princeton, NJ: Princeton University Press, 120–139.

Lacan, Jacques (1975) [1932] *De la psychose paranoïaque dans ses rapports avec la personnalité*. Paris: Éditions du Seuil.

Laland, Kevin, Odling-Smee, John, and Endler, John (2017) 'Niche construction, sources of selection and trait coevolution'. *Interface Focus* 7, 1–9.

Lenzi, Delia, Trentini, Cristina, Pantano, Patrizia. Macaluso, Emiliano, *et al.* (2009) 'Neural basis of maternal communication and emotional expression processing during infant preverbal stage'. *Cerebral Cortex* 19, 1124–1133.

Mathôt, Sebastiaan, Grainer, Jonathan, and Strijkers, Kristof (2017) 'Pupillary responses to words that convey a sense of brightness or darkness'. *Psychological Science* 28 (8), 1116–1124.

Merleau-Ponty, Maurice (2003) [1995] 'Animality: The study of animal behavior'. In: Maurice Merleau-Ponty. *Nature: Course Notes from the Collège de France*. Translated by Robert Vallier. Evanston, IL: Northwestern University Press, 167–199.

Odling-Smee, John and Laland, Kevin N. (2011) 'Ecological inheritance and cultural inheritance: What are they and how do they differ?'. *Biological Theory* 6, 220–230.

Peirce, Charles S. (1998) 'On phenomenology: Pragmatism as the logic of abduction: What makes a reasoning sound?'. In: Nathan Houser and Christian J. W. Kloesel (eds.) *The Essential Peirce: Selected Philosophical Writings (1893–1913)*. Vol. 2. Bloomington: Indiana University Press, 145–159, 226–241, 242–257.

Ricœur, Paul (2004) 'Ch. 3. Forgetting'. In: Paul Ricœur. *Memory, History, Forgetting*. Translated by Kathleen Blamey and David Pellauer. Chicago/London: University of Chicago Press, 412–456.

Ricœur, Paul (2005) [2004] *The Course of Recognition*. Translated by David Pellauer. Cambridge, MA/London/England: Harvard University Press.

Rousseau, Jean-Jacques (2004) [1762] *The Social Contract*. Translated by Maurice Cranston. London: Penguin Books.

Saussure, Ferdinand de (1959) [1971] *Course in General Linguistics*. Translated by Wade Baskin. New York: McGraw-Hill Book Co.

Schweinfurth, Manon K. and Taborsky, Michael (2017) 'The transfer of alternative tasks in reciprocal cooperation'. *Animal Behaviour* 131, 35–41.

Shaffer, Elinor (1998) 'Introduction: The third culture-Negotiating the two cultures'. In: Elinor S. Shaffer (ed.) *The Third Culture: Literature and Science*. Berlin/New York: Walter de Gruyter, 1–12.

Smith, Murray (2017) *Film, Art and the Third Culture: A Naturalized Aesthetics of Film*. Oxford: Oxford University Press.

Snow, Charles Percy (2012) [1998] 'The two cultures: A second look'. In: Charles Percy Snow. *The Two Cultures*. Cambridge: Cambridge University Press, 53–107.

Starr, G. Gabrielle (2013) *Feeling Beauty: The Neuroscience of Aesthetic Experience*. London/Cambridge, MA: MIT Press.

Sterelny, Kim (2007) *Dawkins vs. Gould: Survival of the Fittest*. Cambridge: Icon Books.

Sutrop, Urmas (2001) 'Umwelt-Word and concept: Two hundred years of semantic change'. *Semiotica* 134 (1/4), 447–462.

Tomasello, Michael (2016) *A Natural History of Human Morality*. Cambridge, MA/London: Harvard University Press.

Uexküll, Jakob von (2001) [1937] 'The new concept of Umwelt: A link between science and the humanities'. *Semiotica* 134 (1/4), 111–123.

Uexküll, Jakob von (2010) [1934, 1940] *A Foray into the Worlds of Animals and Humans, with A Theory of Meaning*. Minnesota: University of Minnesota Press.

Vessel, Edward A., Starr, G. Gabrielle, and Rubin, Nava (2013) 'Art reaches within: Aesthetic experience, the self and the default mode network'. *Frontiers in Neuroscience* 7, Article 258, 1–9.

Waal, Frans de (2000) 'Attitudinal reciprocity in food sharing among brown capuchin monkeys'. *Animal Behaviour* 60 (2), 253–261.

Waal, Frans de (2006) *Primates and Philosophers: How Morality Evolved*. Princeton, NJ: Princeton University Press.

Waal, Frans de (2016) *Are We Smart Enough to Know How Smart Animals Are?* New York/London: W. W. Norton & Company.

Warkentin, Traci (2009) 'Whale agency: Affordances and acts of resistance in captive environments'. In: Sarah E. McFarland and Ryan Hedger (eds.) *Animals and Agency: An Interdisciplinary Exploration*. Leiden/Boston: Brill, 23–43.

Wright, Georg Henrik von (1971) *Explanation and Understanding*. Ithaca/New York: Cornell University Press.

Žižek, Slavoj (2002) 'Cultural studies versus the third culture'. *The South Atlantic Quarterly* 101 (1), 19–32.

Afterword

A future for Jakob von Uexküll

Ezequiel A. Di Paolo

It may not seem like big news to readers of this book, but in a time when machine metaphors overwhelmingly dominate scientific and popular discourse, it comes as nothing short of a revelation to say that meaning is a relation between multiple processes involving emerging agencies and norms. To say that meaning is not a thing, a state, a series of algorithms, a content moved about in vehicles. That meaning is not, as functionalism would have it, a sort of well-packaged information that is processed by cognitive mechanisms in the brain. And, at the same time, to say that meaning is not a sort of magical *a priori* either, coming out of nowhere, casting a veil of significance on a meaningless world. These claims still sound radical within Western thinking. Researchers continue to pin meaning down to things, locations, mechanisms, models, and boxes as if a relation of significance between dynamic material processes could ever be assigned a whereabouts, a boundary, or be compartmentalized within the brain, realized somehow in neural events. "But meaning is not in the head!" many will protest today, and others have said this in different ways in the past, but Jakob von Uexküll said it in a particularly useful manner.

The task today is similar to the task in Uexküll's times, but also rather different. It still demands a critical questioning of dualistic ontologies and epistemologies that continue to nourish research in biology, psychology, neuroscience, AI, and robotics – the mechanistic worldview that schizophrenically wishes to explain the mind by remaining skeptical about its key components: meaning, autonomy, agency, and so on. Critical schools of thought have emerged that oppose these widespread views, and over the past two decades, they have moved beyond criticism and set themselves the task of building concrete positive alternative ways of thinking and doing research. I am talking about embodied perspectives that move scientific inquiry from an exclusive focus on brain mechanisms and their algorithmic counterparts, into situated living bodies, into their ecological surroundings, their enacted activity, history, and social world. These perspectives owe much to Uexküllian thinking and often acknowledge this debt explicitly.

Perhaps the question that remains open today – and this book contributes to its examination – is about Uexküll's place in the present context. It is certainly the case that his ideas are useful and profound and can continue to shape research in enactive, ecological, and embodied approaches to life and mind. But, as we would expect, they do so not simply as canonical transpositions of his œuvre into the 21st century with the occasional terminological update. Things have changed not only within science (new discoveries, new methods, new technologies, and new theories) but outside science too (different social and political climate, different communication and information technologies, an advanced economic and ecological crisis of global magnitude). We are faced therefore with the task of assessing and adapting not necessarily all aspects of Uexküll's ideas but perhaps the ones that speak most directly to our urgent needs and current concerns, maybe even discarding some original elements, or changing the accent of meaning into something new, something contemporaneous. All of this is to be done in the spirit that the best recognition of genius is when someone's ideas keep evolving by their own momentum.

The contemporary accents we need to emphasize to continue to transform the scientific understanding of life and mind are at the radical end of the Uexküllian corpus, not at the conservative end, the one Uexküll himself was sometimes more outspoken about. The movement of the ideas themselves, rather than their letter, will give us the richest Uexküllian heritage, the one we need in our time.

Consider the claim that it is in the nature of all life to be "surrounded." That such is, in fact, part of its essence. Nonliving objects, by contrast, are located in place, encircled by relations around them. But objects are relatively resilient to being moved from one environment to another. For organisms, things are not so easy. Living organisms take place "within surroundings," that is, through relations that orient their activity and existence; take place as in "happen," also as in "claim a stake" in the here and now. Organisms are ongoing happenings; organisms claim their place.

From a *non*relational perspective, this sounds odd and unintuitive. We are used, since Aristotle, to finding essences in things themselves once we divest them from contingent particulars. We think of essence as immanent. Even artifacts that are built for a purpose, things that have a wherefore as their reason for being, can be said to be what they are by virtue of how they are put together, to have a mode of existence all by themselves and not relative to their surroundings. Assuming this tacit epistemology, we claim to be able to ascertain essential and inessential properties by mere observation, following a classical logic of necessity and sufficiency, and by linguistic inquiries into the grammar of the words we use to talk about things.

Yet, organisms are networks of relations between "themselves" and their surroundings. Moreover, they are time-extended, self-individuating, and autonomous, thus projecting relations of significance onto their world by enacting the norms they themselves live by. A living organism is grasped inadequately by our perceptions as being this or that anatomical body. Bodies are concrete assemblages of self-sustaining material flows and therefore what counts as essential cannot be an abstract idea we formulate simply by projecting our own views into their existence and their worlds. This entails that perceiving living organisms differs from perceiving nonliving objects, and yet, historically, we have found difficulties in articulating this difference despite many voices remarking on it.

The difference is flatly ignored and living bodies are treated just as any other machine.

We must acknowledge the fact that organisms build worlds different from ours, and we must devise epistemologies that take us closer to understanding these worlds, since the epistemology by which we approach nonliving objects is systematically misleading. This is a Uexküllian heritage we must cherish today.

Uexküll's theory of meaning and his concept of the *Umwelt* help a lot in furthering relational perspectives, new ontologies, and new scientific thinking, that give due justice to living (co)existence in fragile surroundings. What we should question in this legacy are the vestiges of idealism and conservative appeals to the apparent harmony of the living world.

In contraposing harmony to "mindless materiality," as he understands it, Uexküll does us an important service, provided we are careful not to take him too much at face value. We cannot deny the complex web of coherent relations both within and between organisms. Nor can we explain this complexity only as a result of blind mechanisms that fixate accidental changes. But we must also see that "harmony," if and when it is the case, is precarious, conflictive, and always changing. Partial harmonies are at best temporary achievements, modes of existence unwarranted by the very forces that bring them into being, challenged "internally" by these forces and not only by the external "mindless" forces of entropy and decay. If the spider is fly-shaped, this harmony should not hide the conflict that brings it into being. The untrapped fly teases the spider's hunger; the trapped fly is killed by it. In a dialectical view, an *Umwelt* does not entirely coincide with itself. It is self-contradictory as well as unified; its partial harmonies revealed precisely only against the background of inherent contradiction. *Umwelten* have open horizons. This is why things change and evolve. This is why life is, at all scales, both metastable form and perpetual transformation.

This means that, while we must avoid the flattening out of the biological and psychological worlds into a series of mechanisms, we must also be cautious with the theme of the harmony of the world. The harmony metaphor is in its own way a flattening out of biological and psychological phenomena if we understand harmony as a primordial state of mutually counterpunctual relations of meaning ("the spider is fly-like"). Here, what is excluded, to repeat, are the precarious conditions and the ongoing, effortful processes by which meaning is achieved whatever the timescale, whether evolutionary, developmental, or behavioral. This is not a mere addendum but a fundamental condition that warrants the introduction of relations of meaning in a materialist ontology. For it is the ongoing risk and precarious conditions that tend to *dis*harmony and the dissipation of metastable relations that drive the ongoing struggle for sense-making. Otherwise, meaning would be superfluous; it would all boil down to letting self-organizing systems relax into their ultimate attractor states. Nothing would even need to be achieved by living beings – a game with no stakes.

The ongoing individuation, the "constitutive unfinishedness" of the living condition is what makes an *Umwelt* meaningful for organisms in ways that a network of relations is not meaningful "for" nonliving objects whose ongoing existence is not at stake. Lacks and surpluses make relational processes meaningful, but for needs and excesses to exist objectively, it is necessary for material self-individuation to be in place and for vital norms to emerge in processes of organic life, sensorimotor agency, interpersonal relations, and collective history.

The ghost of the harmonious world is the idea that tempts us with guidance for our actions and beliefs in the face of ongoing degradation, be this manifested as the current environmental and political crises or at the "simple" level of the needy creature in search of food and shelter where none is to be found. Also, in the ever-present risk of illness and death. "If only we could steer our activity toward this presumed Ur-harmony of the natural world ...". In spite of beautiful musical metaphors, this haunting harmony is only a normative abstraction, a conservative idea. It makes us think of degrading and conflictive conditions as anomalous simply because they challenge the apparent norms of harmony. It blinds us to the way these "negative" trends are as much the means as the test of life.

It is true that the nonrelational epistemologies of meaninglessness – scientism, with its mechanicisms, functionalisms, representationalisms, and so forth – justify themselves in adopting a "no-nonsense" metaphysics of the spiritless void as the starting point of all scientific inquiry. They do so precisely by looking at the evidence of conflictive and disharmonious nature all too closely. They see in this evidence the indifference of the world. But this view is also abstract. Its error lies in missing the whole by taking only one of its contradictory moments simply because it serves to make a moral point. Scientism is gleefully stoical; it attaches a moral superiority to the "realistic" attitude of confronting meaninglessness the way "rational adults" (read: white male adults) must while ignoring the evidence of living experience that stares them in the face. Scientism does not really pay heed to concrete materiality and its self-renewing, active, and vibrant nature. It is "materialist" in name only. Uexküll is right to point to dynamic *Gestalt* forms of meaning as the evidence that scientism is keen to ignore (or downplay to the status of accidents or illusions).

However, rejecting scientism, we insist, does not necessarily demand a return to a conservative all-encompassing harmony. On the contrary, it is by pushing Uexküllian thought to new frontiers, looking at the agencies entailed in the perspectivism of *Umwelten*, at the precarious, not only time-extended but also time-limited processes of self-organization, at the internal and external conflicts inherent in the living condition, at the struggle and ongoing transformations taking place at all scales, that we can conceive of meaning not as unattained perfection, but as thriving processes of fragile and vulnerable life. It is by these operations that meaning can be finally naturalized, not as a harmonious exception to the assumed random patterns in nature but as the struggling activities of living beings. These activities become meaningful by the very enacting of the organic, sensorimotor, and social normativities they bring forth as the condition for sustaining their precarious and challenged existence.

The future of Uexküll is open and exciting, although probably also riddled with conflicts and contradictions. One of the most active and innovative strands of embodied cognition in the 21st century is the enactive approach, which predicates the relation between agents and worlds in terms of participation and enactments. We bring forth the world together with other creatures. Uexküll anticipates this idea when he discusses the intertwined relation between perception and action, which he nevertheless still sees as conceptually distinct. But is not this joint bringing forth of a world at the same time a dissolution of the static, bubblelike concept of the *Umwelt*? By giving this idea an inherent temporality, a sense of praxis, achievement, and risk, in other words, by grounding the concept of sense-making in the temporal and material tensions of life, enaction tells us that the *Umwelt* is always in the making, that it may not yet entirely surround us, that its potentialities are always in the process of being collectively realized. This anticipation drives our actions. In this sense, to act is always simultaneously to act both within an *Umwelt* and outside it, insofar as material actions galvanize the forces that consciously or unconsciously can and often do change our worlds. Perception/action/emotion, sense-making in general, are therefore liminal concepts; they reaffirm a world by the very fact that they risk changing it. They occur both within a world and at its limits. Living creatures and *Umwelten* are a codefined conceptual pair, very much like the notion of a boundary and the notion of crossing it.

Perhaps the idea of harmonious *Umwelten* is less able to offer us comfort in the 21st century. In the face of the irrecoverable damage we inflict on the planet that hosts us and the limit situations we drive ourselves into through social weathering and outright violence, our epoch looms more dangerous even than the decades in which Uexküll worked. The patterns that recur from that violent era are so much more amplified and so much more destructive and powerful today. We must urgently recover a progressive idea of the *Umwelt* but not as a refuge, not as a conservative move, but as a tool for action. For this, we must always understand it dynamically and dialectically, perhaps in ways Uexküll himself might have disagreed with in the 1930s but maybe, who knows, might have found it acceptable were he alive today witnessing the world with which we have surrounded ourselves.

Index

Printed in the United States
by Baker & Taylor Publisher Services